Handbook of Cell and Gene Therapy

This handbook provides an in-depth review of information across the developmental spectrum of gene and cell therapy products. From introductory information to state-of-the-art technologies and concepts, the book provides insights into upstream processes such as vector design and construction, purification, formulation and fill/finish, as well as delivery options. Planning steps for compliance with current good manufacturing practice (cGMP) to readiness for chemistry, manufacturing and controls (CMC) are also discussed. This book wraps up with examples of successes and pitfalls addressed by experts who have navigated the multiple challenges that are part of any innovative endeavor.

Features

- It is intended as a one-stop resource for the availability of state-of-the-art information related to cell and gene therapy products for researchers, scientists, management and other academic and research institutions.
- Provides the most up-to-date information on the development of gene therapy, from the technology involved to gene correction and genome editing.
- Discusses siRNA, mRNA, and plasmid manufacturing.
- Describes the importance of supplier-sponsor synergies on the path to commercialization.
- Written for a diverse audience with a large number of individuals in the core technologies and supportive practices.

Handbook of Cell and Gene Therapy
From Proof-of-Concept through Manufacturing to Commercialization

Edited by

Hazel Aranha
GAEA Resources Inc., Northport, USA

Humberto Vega-Mercado
Bristol Myers Squibb, Cell Therapy Operations, Global Manufacturing Sciences and Technology, Warren, USA

CRC Press is an imprint of the
Taylor & Francis Group, an **informa** business

First edition published 2023
by CRC Press
6000 Broken Sound Parkway NW, Suite 300, Boca Raton, FL 33487-2742

and by CRC Press
4 Park Square, Milton Park, Abingdon, Oxon, OX14 4RN

© 2023 selection and editorial matter, Hazel Aranha and Humberto Vega-Mercado; individual chapters, the contributors

CRC Press is an imprint of Taylor & Francis Group, LLC

Reasonable efforts have been made to publish reliable data and information, but the author and publisher cannot assume responsibility for the validity of all materials or the consequences of their use. The authors and publishers have attempted to trace the copyright holders of all material reproduced in this publication and apologize to copyright holders if permission to publish in this form has not been obtained. If any copyright material has not been acknowledged please write and let us know so we may rectify in any future reprint.

Except as permitted under U.S. Copyright Law, no part of this book may be reprinted, reproduced, transmitted, or utilized in any form by any electronic, mechanical, or other means, now known or hereafter invented, including photocopying, microfilming, and recording, or in any information storage or retrieval system, without written permission from the publishers.

For permission to photocopy or use material electronically from this work, access www.copyright.com or contact the Copyright Clearance Center, Inc. (CCC), 222 Rosewood Drive, Danvers, MA 01923, 978-750-8400. For works that are not available on CCC please contact mpkbookspermissions@tandf.co.uk

Trademark notice: Product or corporate names may be trademarks or registered trademarks and are used only for identification and explanation without intent to infringe.

ISBN: 978-1-032-25797-6 (hbk)
ISBN: 978-1-032-25798-3 (pbk)
ISBN: 978-1-003-28506-9 (ebk)

DOI: 10.1201/9781003285069

Typeset in Times
by SPi Technologies India Pvt Ltd (Straive)

Dedication

We would like to dedicate our handbook to the people who motivate and support us in our continued effort to improve health care and treatments across the world:

Amy Davis: She provided us with the initial inspiration to pursue this project. Rest in peace Amy!

To my late parents, Gregory and Bibiana Aranha, who were educators and recognized the universality of education.

Hazel Aranha

Money spent on the brain is never spent in vain.

– English proverb

My wife Donna and our children and grandchildren who are the center of my universe.

Humberto Vega

Our contributors from across the industry for providing their knowledge, feedback and guidance.

The team at Taylor & Francis for making this a reality.

Contents

Keywords xi
Preface xiii
Author Biography xv
Contributors xvii
Abbreviations xix

Chapter 1 History of Gene Therapy Products 1
Bridget Heelan

Chapter 2 Gene Editing in Humans: How Bioethics Discussions Can Guide Responsible Research and Product Development 13
Hazel Aranha

Chapter 3 Applying a Risk-Based Approach in the Development and Manufacture of Advanced Therapy Medicinal Products 27
Bridget Heelan

Chapter 4 Gene Therapy Products: Basics and Manufacturing Considerations 35
Humberto Vega-Mercado

Chapter 5 Cell Therapy Products: Basics and Manufacturing Considerations 49
Humberto Vega-Mercado

Chapter 6 Facility and Equipment Considerations 65
Humberto Vega-Mercado

Chapter 7 Analytical Methods for In-Process Testing and Product Release 81
Neil A. Haig, Jamie E. Valeich, Laura DeMaster, Christopher Do, Razvan Marculescu, Jianbin Wang, Bridget S. Fisher, Xiaoping Wu, Akshata Ijantkar, Cayla N. Rodia, Sean D. Madsen, Mohammed Ibrahim, Mel Davis-Pickett, Jennifer F. Hu, Jennifer J. Robbins, James Collins Powell, Divya Shrivastava, and Satyakam Singh

Chapter 8	Validation, Verification and Qualification Considerations.... 135	
	Humberto Vega-Mercado	
Chapter 9	Adventitious Agent Contamination Considerations during the Manufacture of Cell and Gene Therapy Products 153	
	Hazel Aranha	
Chapter 10	Training Approaches to Build Cell and Gene Therapy Workforce Capacity 175	
	Orin Chisholm	
Chapter 11	How to Distribute Cell and Gene Therapies 183	
	Andrea Zobel	
Chapter 12	Regulatory Compliance and Approval................................. 197	
	Siegfried Schmitt	
Chapter 13a	Regulatory Landscape in US, EU and Canada 209	
	Kirsten Messmer	
Chapter 13b	Regulatory Landscape in South America 225	
	Heloisa Mizrahy	
Chapter 13c	Regulatory Landscape in Australia and New Zealand.......... 233	
	Orin Chisholm	
Chapter 13d1	Regulatory Landscape in Singapore 243	
	Stefanie Fasshauer	
Chapter 13d2	Regulatory Landscape in Malaysia....................................... 255	
	Stefanie Fasshauer	
Chapter 13e	Regulatory Landscape in China.. 269	
	Kai Zhang	

Contents

Chapter 13f Regulatory Landscape in Japan 285
Hazel Aranha

Chapter 13g Regulatory Landscape in India 295
Arun Bhatt and Rajesh Jain

Chapter 14 Avoiding Pitfalls during Advanced Therapy Development 311
Kirsten Messmer

Chapter 15 Going Forward – Existing and Evolving Technologies (CRISPR, mRNA, siRNA) 321
Humberto Vega-Mercado and Hazel Aranha

Glossary 333
Index 337

Keywords

adeno-associated viral (AAV) vector
adventitious agent contamination
Allogeneic cell therapy
animal derived materials (ADMs)
Aseptic processing
Autologous cell therapy
Biologic License Application
biosafety cabinets
disposable systems
chimeric antigen receptor (CAR)-T cells
CRISPR/Cas9
Cytokine Release Syndrome (CRS)
Downstream processing
environmental monitoring
FMEA (Failure Mode and Effects Analysis)
hematopoietic stem and progenitor cells (HSPC)
identity
Ishikawa (Fishbone) diagram
isolator systems
Massive parallel sequencing
p53 gene
potency
purity
Quality risk management (QRM)
Recombinant Adeno-Associated virus (rAAV) vectors
restrictive access barrier systems (RABS)
risk priority number RPN
Single use systems
sterility by design'
suitability
transcription activator-like effector nuclease TALEN
viability
Viral clearance
Vendor qualification
zinc-finger nuclease (ZFN)

Preface

Cell and gene therapy products represent a seismic shift in medical intervention. Unlike conventional biotherapeutics, the technology seeks to modify or manipulate the expression of a gene or to alter the biological properties of living cells for therapeutic use. By replacing, manipulating or engineering cells and/or genetic material, they offer an opportunity for enhanced patient outcomes through the promise of, potentially, a lifelong cure for at least some diseases and disorders – all by infusion of genetically modified cells.

Conventional biotherapeutics (monoclonal antibodies, recombinant products) have reached a stage of maturation where platform manufacturing technologies can be applied and there is clarity related to regulatory and compliance issues. With cell and gene therapy products, there is no established manufacturing playbook that can be referenced, and the regulations are continuing to evolve. Personalized therapeutics have transitioned from mainly clinical stage in hospital settings to a full commercial enterprise with the routine manufacture of these products guided by established regulatory and industry best practices.

The biggest blind spot in the development life cycle is being able to transition and adopt a realistic understanding of the time, resources and cost it takes to develop such a complex biologic. Decisions made in the early phase are critical to the future viability of the product to an unprecedented degree. There are a multitude of moving parts in terms of product development and cGMP production. Additionally, specifics related to clinical study planning, logistics and execution and, also, meeting specific regulatory expectations must be addressed. Early planning forms an important strategic grounding point for cell and gene therapies. Having a commercial vision must be an objective from the outset and can help form a critical path of understanding.

Our goal in this handbook is to provide an overview of the development, manufacture, testing, and regulatory expectations related to cell and gene therapies. The state-of-the-art information presented here is intended for researchers, scientists, management and other academic and research institutions working in the field. The reader should be able to follow key principles and concepts, including, but not limited to, process design, facility design and construction; technology transfers, analytical methods development and transfers; validation life cycle; and regulatory frameworks around the world. If this handbook saves a modicum of your time, we will have fulfilled our objective.

Technologies and knowledge will continue to evolve, but the basic principles described in our handbook should serve as a baseline for our colleagues across the industry. Progress builds on the history and lessons from the past; the future is shaped by those lessons and history.

To quote American author and playwright Alfred Sheinwold, *"Learn all you can from the mistakes of others. You won't have time to make them all yourself."*

Author Biography

Dr. Hazel Aranha is a biopharma professional with over 35 years' experience in industry, academia and consulting. She is a subject matter expert in the area of adventitious agent contamination in biopharmaceuticals and cell and gene therapy products. Her company, Gaea Resources Inc, has provided consulting and auditing services in the US, Europe and Asia.

Hazel brings deep domain expertise to assist clients in achieving strategic and operational objectives. Her projects have included opportunity mapping and competitive market analysis, due diligence for potential acquisitions and gap analysis to identify key risks and propose mitigation strategies. She has conducted customized training to address unmet needs in the biopharma sector.

Hazel has a master's degree in virology, PhD in environmental microbiology and holds Regulatory Affairs Certification (RAC) for both the US and European Union. She has to her credit 2 books, more than 45 publications and 5 book chapters. She is on the review board of multiple biotechnology journals. Her past assignments have included positions at Sartorius Stedim N.A., Catalent Pharma Solutions, Wyeth Vaccines (Pfizer) and Pall Corporation. She holds professional memberships in the Parenteral Drug Association (PDA) and Regulatory Affairs Professional Society (RAPS).

Dr. Humberto Vega-Mercado is Executive Director of Bristol Myers Squibb (BMS), Cell Therapy Operations, Global Manufacturing Sciences and Technology (GMS&T), Head of External Manufacturing Drug Product, Critical Materials, Packaging Technology, Labeling, Cell Therapy Capability Center, Apheresis and Non-Cell Products Technical Groups based in Warren, New Jersey. He has been in the pharmaceutical and food industries for over 34 years in multiple roles, including Research Assistant-Food Processing; Process and Manufacturing Head of Sterile and Non-sterile Pharmaceutical Operations; Sr. Scientist and Associate Director of Validation, Technology

Head – Vaccine Technology and Engineering; Associate Director of MS&T; and Sr. Director of MS&T. He holds BS and MS degrees in chemical engineering from the University of Puerto Rico and a PhD in engineering science from Washington State University. He holds professional memberships in the PDA, Institute of Food Technologists (IFT) and International Society for Pharmaceutical Engineering (ISPE).

Contributors

Hazel Aranha
GAEA Resources Inc.
Northport, USA

Arun Bhatt
Clinical Research & Drug
 Development – Consultant
Mumbai, India

Orin Chisholm
Principal Consultant, Pharma Med
UNSW Sydney, Australia

Laura DeMaster
Bristol Myers Squibb Cell Therapy-
 Analytical Development
Seattle, USA

Christopher Do
Bristol Myers Squibb Cell Therapy-
 Analytical Development
Seattle, USA

Stefanie Fasshauer
PharmaLex
Peng Chau, Hong Kong

Bridget S. Fisher
Bristol Myers Squibb Cell
 Therapy-Analytical
 Development
Seattle, USA

Neil A. Haig
Bristol Myers Squibb Cell Therapy-
 Analytical Development
Seattle, USA

Bridget Heelan
Technical, Independent Consultant
London, UK

Jennifer F. Hu
Gentibio
Seattle, USA

Mohammed Ibrahim
Bristol Myers Squibb Cell Therapy-
 Analytical Development
Seattle, USA

Akshata Ijantkar
Bristol Myers Squibb Cell Therapy-
 Analytical Development
Seattle, USA

Rajesh Jain
PharmaLex India Pvt. Ltd
Mumbai, India

Sean D. Madsen
Bristol Myers Squibb Cell Therapy-
 Analytical Development
Seattle, USA

Razvan Marculescu
Bristol Myers Squibb Cell Therapy-
 Analytical Development
Seattle, USA

Kirsten Messmer
Agency IQ
Garner, USA

Heloisa Mizrahy
HMC Consultoria Ltda. –
 Independent Consultant
Rio de Janeiro, Brazil

Mel Davis-Pickett
Bristol Myers Squibb Cell Therapy-
 Analytical Development
Seattle, USA

James Collins Powell
Bristol Myers Squibb Cell Therapy-Analytical Development
Warren, USA

Jennifer J. Robbins
Bristol Myers Squibb Cell Therapy-Analytical Development
Warren, USA

Cayla N. Rodia
Bristol Myers Squibb Cell Therapy-Analytical Development
Seattle, USA

Siegfried Schmitt
Parexel
Braintree, UK

Divya Shrivastava
Bristol Myers Squibb Cell Therapy-Analytical Development
Warren, USA

Satyakam Singh
Simnova Biotherapeutics
Cambridge, USA

Jamie E. Valeich
Bristol Myers Squibb Cell Therapy-Analytical Development
Seattle, USA

Humberto Vega-Mercado
Bristol Myers Squibb, Cell Therapy Operations, Global Manufacturing Sciences and Technology
Warren, USA

Jianbin Wang
Bristol Myers Squibb Cell Therapy-Analytical Development
Seattle, USA

Xiaoping Wu
Bristol Myers Squibb Cell Therapy-Analytical Development
Seattle, USA

Kai Zhang
Fresenius Kabi
Beijing, China

Andrea Zobel
World Courier GmbH
Berlin, Germany

Abbreviations

AABB	Association for the Advance of Blood & Biotherapies
AAC	assay acceptance criteria
AAV	adeno-associated virus vectors
ACMG	American College of Medical Ethics and Genomics
ADA	adenosine deaminase
ADM	animal-derived material
ALL	Acute Lymphomblastic Leukemia
AM	ancillary materials
ANVIZA	Agência Nacional de Vigilância Sanitária
API	active pharmaceutical ingredient
APS	aseptic process simulations
ASFA	American Society for Apheresis
BAR	B-cell-targeting antibody receptor
BCMA	B-cell maturation antigen
BLA	Biological License Application
BSC	Biological Safety Cabinet
CAPA	corrective actions and preventive actions
CAR	chimeric antigen receptor
CGD	chronic granulomatous disease
CDMO	Contract Development and Manufacturing Organizations
CEA	carcinoembryonic antigen
CFR	Code Federal Register
cGMP	Current Good Manufacturing Practices
CMC	Chemistry, Manufacturing and Control
CMO	Contract Manufacturing Organizations
CMV	cytomegalovirus
COC	Chain of Custody
COI	Chain of Identity
CNC	controlled-non-classified
CTL	Contract Testing Laboratory
CRISPR	clustered regularly interspaced short palindromic repeats
CRS	Cytokine Release Syndrome
DLBCL	diffuse large B-cell lymphoma
DNA	deoxyribonucleic acid
DOE	Design of Experiment
DP	Drug Product
EAE	experimental autoimmune encephalomyelitis
EBMT	European Group for Blood and Marrow Transplantation
EBV	Epstein-Barr virus
EM	Environmental Monitoring
EMA	European Medicine Agency

EMPQ	Environmental Monitoring Performance Qualification
EPAR	European Public Assessment Report
FACS	fluorescence-activated cell sorting
FACT	Foundation for the Accreditation of Cellular Therapy
FMEA	Failure Mode and Effect Analysis
FOXP3	forkhead box protein P3
FRS	Functional Requirements Specifications
FVIII	factor VIII
GvHD	graft-versus-host disease
HBV	hepatitis B virus
HCT/Ps	human cells, tissues, or cellular or tissue-based products
HCV	hepatitis C virus
HEK	Human cell line (e.g., 293T)
HEPA	High Efficiency Particulate Air
HIV	human immunodeficiency virus
HLA	human leukocyte antigen
HPV	human papilloma virus
HSPC	hematopoietic stem and progenitor cells
HVAC	Heating, Ventilation and Air Conditioning
IA	murine major histocompatibility complex class II molecule I-A
ICCBBA	International Council for Commonality in Blood Banking Automation
ICH	International Council for Harmonization
Ig	immunoglobulin
IL	interleukin
IND	Investigational New Drug
IQ	Installation Qualification
ISBT	International Society of Blood Transfusion
ISCT	International Society for Cellular Therapy
ISO	International Organization for Standardization
ISPE	International Society of Pharmaceutical Engineering
JACIE	Joint Accreditation Committee of ISCT and EBMT
LLOQ	lower limit of quantitation
LOD	limit of detection
LOQ	limit of quantification
LVV	lentiviral vectors
MAL	Material Airlock
MACS	magnetic-activated cell sorting
MBP	myelin basic protein
MCL	mantle cell lymphoma
MERV	Minimum Efficiency Reporting Value
MFI	mean fluorescent intensity (MFI)
MHRA	Medicine and Healthcare Product Regulatory Agency
MOA	Mechanism of Action, Mode of Action
NOD	non-obese diabetic

NOR	Normal Operation Range
OBD	opinion group of the Bioethics and Law Observatory
OOS	out of specification
OQ	Operation Qualification
OTC	ornithine transcarbamylase
PAI	Pre-approval Inspection
PAL	Personnel Airlock
PCV	porcine circovirus
PD	Product Development
PDA	Parenteral Drug Association
PIC/S	Pharmaceutical Inspection Co-operation Scheme
PID	primary immune deficiency diseases
PLI	Pre-licensing Inspection
PPQ	Process Performance Qualification
PQ	Performance Qualification
PV	Process Validation
QA	Quality Assurance
QbD	Quality by Design
QC	Quality Control
QRM	Quality Risk Management
RABS	Restricted Access Barrier Systems
RCR	replication-competent retrovirus
SAC	sample acceptance criteria
SCID	severe combined immunodeficiency
SIN	self-inactivating
SME	Subject Matter Expert
SmPCs	summary of product characteristics
SOP	standard operating procedure
SSM	spillover spreading matrix (SSM)
SV40	Simian Virus 40
T1D	type 1 diabetes
TALEN	transcription activator-like effector nucleases
TCR	T cell receptor
Teff cell	effector T cell
TNP	2,4,6-trinitrophenyl
T_{reg} cell	regulatory T cell
TT	Technology Transfer
ULOQ	upper limit of quantitation
URS	User Requirement Specifications
US-FDA	United States Food and Drug Administration
USP-NF	United States Pharmacopeia–National Formulary
VMP	Validation Master Plan
ZFN	zinc-finger nucleases

1 History of Gene Therapy Products

Bridget Heelan
Technical, Independent Consultant, London, UK

CONTENTS

Historical Timeline and Key Technological Advances 1
Key Historical Trials and Lessons Learned 2
Key Successes and Regulatory Approvals 4
    Definition of Gene Therapy in the EU 4
    Definition of GTMP in the US 5
References 10

HISTORICAL TIMELINE AND KEY TECHNOLOGICAL ADVANCES

The discovery that other than conjugation, it was possible to transform the characteristics of organisms, and their progeny was first described by Griffith (1928) and later it was shown that the transforming factor was deoxyribonucleic acid DNA (Avery et al. 1944), a process termed transformation (reviewed in Johnston et al. 2014). It was later discovered that bacteriophages were capable of carrying DNA between bacteria via a process termed transduction (Zinder & Lederberg 1952). The understanding from these experiments in prokaryotic organisms was extended into studies using mammalian cells and viruses (Szybalska & Szybalski 1962; Rogers et al. 1973).

The first trial of a therapeutic gene was performed in 1990 in patients with severe combined immunodeficiency (SCID) due to adenosine deaminase (ADA) deficiency. The main features of ADA-SCID are profound immunodeficiency with recurrent infections often involving opportunistic pathogens and, in the absence of treatment, most infants with this diagnosis die in the first year or two of life. Treatment at that time was human lymphocyte antigen-matched sibling bone marrow transplantation or, in the absence of a suitable donor, regular replacement of ADA enzyme through infusions of pegylated ADA (PEG-ADA). Anderson et al. (1990) attempted immune reconstitution in two patients with ADA-SCID who were maintained on PEG-ADA. The trial involved administering infusions of autologous T lymphocytes, which were genetically corrected by insertion of a normal ADA gene using retroviral-mediated gene transfer. The retroviral vector LASN was the first therapeutic vector used in a gene therapy clinical trial and contained adenosine deaminase gene complementary DNA (cDNA). Both patients showed some evidence of immune reconstitution with a range of immunological tests, with Patient 1 achieving normalization of her T-cell count out to four years follow-up, but the response in Patient 2 was minimal. Nonetheless, despite the very low final proportion of vector containing T cells in Patient 2, both CD4+ and CD8+ T-cell populations from both subjects remained consistently positive for integrated vector sequences out to day 1,480 for Patient 1 and day 1,198 for Patient 2. Clinically, Patient 1 normalized in growth and had a reduction in infections, whereas Patient 2 had no discernable long-term clinical improvement. Although both patients remained on PEG-ADA enzyme replacement, the dose was decreased by more than half up to four years post-treatment. Although the possibility of insertional mutagenesis remained a concern, no indication of malignancy in either patient was seen (Blaese et al. 1995).

Following this study, vectors capable of higher levels of gene transfer and transgene expression were developed. Removal of selectable markers and adaptations of the promoters and use of different viral vectors were constructed with the goal of improved performance over the PA317/LASN vector initially used in ADA-SCID (Onodera et al. 1998). At the same time, adenoviral vectors were modified to enable safer expression of the transgene (Yeh & Perricaudet 1997). Additional work resulted in the improvement of adeno-associated virus (AAV)-based vectors, plasmid formulations and retroviral vectors (reviewed in Cooney et al. 2018).

KEY HISTORICAL TRIALS AND LESSONS LEARNED

Every realm of medicine has its defining moment, and for gene therapy, this was the case involving the death of 18-year-old Jesse Gelsinger (Stolberg 1999). The trial was a phase 1 single ascending dose trial in patients with ornithine transcarbamylase (OTC) deficiency, a condition of variable severity but life-threatening in those severely affected early in life. The patients enrolled in this phase 1 trial were late-onset patients whose symptoms could largely be managed by diet and ammonia scavengers. Ethically it was considered more suitable to have these adults consent to the trial than to enroll severely affected infants. Jesse was in the high-dose group such that other subjects in the lower-dose groups had already been treated. There were some severe adverse events in patients in the dose group below the dose group that Jesse was in. Within four days of receipt of the gene therapy, Jesse died of multiorgan failure, secondary to an overwhelming immune reaction to the adenoviral vector. This tragic case highlighted the potential danger of gene therapy, the need for strict risk minimization of anticipated risks, the need for strict protocol adherence and rapid communication with regulators in the event of severe adverse events as well as searching for vectors with a lower immunogenicity risk.

Using a gammaretroviral gene therapy in X-linked SCID, a different and delayed adverse safety profile emerged. X-linked SCID is caused by mutations in the IL2RG gene, which leads to defective expression of the common cytokine receptor gamma chain, which is a subunit shared by multiple cytokine receptors, including the interleukin (IL)-2, IL-4, IL-7, IL-9, IL- 15 and IL-21. Patients are profoundly immunosuppressed and if undiagnosed present with recurrent infections, persistent infections and infections with opportunistic organisms in infancy. Following an early trial where retrovirally mediated transfer of the IL2RG gene into CD34+ cells resulted in sustained correction of X-linked severe combined immunodeficiency disease in four of five patients, one of the four patients presented with a lymphocytosis consisting of a monoclonal population of g/d T cells at a routine 30-month checkup. Before this finding, an additional four patients had received this treatment. On analysis of the monoclonal T-cell population, one proviral integration site was found, located on the short arm of chromosome 11 within the LIM domain-only 2 (LM02) proto-oncogene locus. This proviral integration within the LM02 locus was associated with aberrant expression of the LMO-2 transcript in the monoclonal T-cell population, similar to the aberrant expression of LMO-2 described in acute lymphoblastic leukemia arising from T cells with a/b receptors, usually with the chromosomal translocation t(11;14). The patient did well on chemotherapy. The investigators considered this finding as the consequence of the insertional mutagenesis event, a risk that is potentially associated with retrovirally mediated gene transfer but that had previously been considered to be very low in humans (Hacein-Bey-Abina et al. 2003). A later update on these patients treated found that four out of nine developed T-cell leukemia 31–68 months after treatment. Chemotherapy led to remission in three patients but the fourth died. Examination of the leukemic cells found insertional mutagenesis was near the LM02 locus in two cases, near the proto-oncogene BM11 in the third case and in the fourth patient near a third proto-oncogene CCND2 (Hacein-Bey-Abina et al. 2008). An additional case of T-cell leukemia was described in an X-SCID patient 24 months after successful gene therapy with a gammaretroviral vector, again with integration upstream of the proto-oncogene LM02 (Howe et al. 2008). In this report, the authors conclude that leukemogenesis was likely precipitated by the acquisition of other genetic abnormalities unrelated to vector insertion, including a gain-of-function mutation in NOTCH1, deletion of the tumor suppressor gene locus cyclin-dependent kinase 2A (CDKN2A), and translocation of the TCR-β region to the STIL-TAL1 locus.

Stein et al. (2010) reported on two young adults who received gene therapy for their underlying immunodeficiency of X-linked chronic granulomatous disease (X-CGD), a disease characterized by recurrent severe bacterial and fungal infections due to a functional defect in the microbe-killing activity of phagocytic neutrophils due to defects in a gene called gp91-phox. After the initial resolution of bacterial and fungal infections, both subjects showed silencing of transgene expression due to methylation of the viral promoter, myelodysplasia with monosomy seven as a result of insertional activation of ecotropic viral integration site 1 (EVI1) and oncogenic transcription factor.

These insertional mutagenesis events increased the need for more tailored risk minimization. To minimize these risks, the vector design can be modified to prevent cis activation of genes that flank the integration sites, and new assays have been optimized to better assess these risks. Alternatively, vectors can be used that do not integrate or achieve targeted genomic integration into specific chromosomal loci (Schneider et al. 2010). Potential strategies could include removing the long terminal repeat (LTR) enhancer element or using different vectors such as lentiviruses (Modlich & Baum 2009). Use of nonintegrating vectors or targeted integration at the desired loci is also a potential strategy that could be considered.

Use of self-inactivating (SIN) viral vectors, devoid of long terminal repeats promoter/enhancer function, in recent gene therapy protocols has reduced the risk for insertional mutagenesis and clonal dominance in primary immune deficiency diseases (PID) (Kumar et al. 2016).

Gene editing approaches have gathered significant pace in recent years, including transcription activator-like effector nucleases (TALENs), and CRISPR/Cas9. These techniques are expected to allow precise editing of mutations, targeting specific gene loci or genomic safe harbors (GSHs) in the genome able to accommodate the integration of new genetic material in a manner that ensures that the newly inserted genetic elements function predictably and do not cause alterations of the host genome and so would be without risk for mutagenesis (Papapetrou & Schambach 2016). While GSHs could represent a universal platform for gene targeting and thus expedite clinical development, so far, no site of the human genome has been fully validated. Clinical translation of putative loci will require extensive validation (Pavani & Amendola 2021).

KEY SUCCESSES AND REGULATORY APPROVALS

To understand how regulators view gene therapy trials, the definitions used by the European Union (EU) and the United States (US) are provided in the following section.

Definition of Gene Therapy in the EU

In the EU, the definition of a Gene Therapy Medicinal Product (GTMP) is provided in Directive 2009/120/EC amending Directive 2001/83/EC, part IV of Annex I.

A GTMP means a biological medicinal product that has the following characteristics:

(a) it contains an active substance that contains or consists of a recombinant nucleic acid used in or administered to human beings with a view to regulating, repairing, replacing, adding or deleting a genetic sequence;
(b) its therapeutic, prophylactic or diagnostic effect relates directly to the recombinant nucleic acid sequence it contains, or to the product of the genetic expression of this sequence.

GTMPs do not include vaccines against infectious diseases.

This definition includes GTMPs containing recombinant nucleic acid sequence(s) or genetically modified microorganism(s) or virus(es). It also includes genetically modified cells, although in this case the quality requirements for

History of Gene Therapy Products

somatic-cell therapy medicinal products and tissue-engineered products also apply. Of note, Directive 2001/20/EC states that no gene therapy trials may be carried out which result in modifications to the subject's germ line genetic identity, and the EU clinical trials regulations maintain that position.

DEFINITION OF GTMP IN THE US

The United States Food and Drug Administration (US FDA) generally considers human gene therapy products to include all products that mediate their effects by transcription or translation of transferred genetic material, or by specifically altering host (human) genetic sequences. Gene therapy is a medical intervention based on modification of the genetic material of living cells. Some examples of gene therapy products include nucleic acids (e.g., plasmids, in vitro transcribed ribonucleic acid (RNA)), genetically modified microorganisms (e.g., viruses, bacteria, fungi), engineered site-specific nucleases used for human genome editing and ex vivo genetically modified human cells. Gene therapy products meet the definition of "biological product" in section 351(i) of the Public Health Service (PHS) Act (42 U.S.C. 262(i)) when such products are applicable to the prevention, treatment, or cure of a disease or condition of human beings (Federal Register Notice: Application of Current Statutory Authorities to Human Somatic Cell Therapy Products and Gene Therapy Products (58 FR 53248, October 14, 1993). Cells may be modified ex vivo for subsequent administration or may be altered in vivo by gene therapy products given directly to the subject. The administration of cells that have undergone ex vivo genetic manipulation is considered a combination of somatic-cell therapy and gene therapy The genetic manipulation may be intended to prevent, treat, cure, diagnose or mitigate disease or injuries in humans. Other gene therapy products such as chemically synthesized products meet the drug definition but not the biological product definition, similar to the EU definition.

The first country to approve a gene therapy was China in 2003. Gendicine™ is an adenoviral gene vector carrying the tumor suppressor p53 gene and is licensed for head and neck squamous cell carcinoma (Patil et al. 2005).

The first European Medicine Agency (EMA) approval of a gene therapy was in 2012 for Glybera®. The indication was for adults with familial lipoprotein lipase deficiency (LPLD) suffering from severe or multiple pancreatitis attacks despite dietary fat restrictions. This product was an AAV vector containing the gene encoding the lipoprotein lipase (LPL). Unlike the majority of gene therapies described earlier, this was an in vivo gene therapy with administration directly to the patients with a one-time series of multiple (up to 60) intramuscular injections in the legs. The marketing authorization expired in 2017, as the marketing authorization holder decided not to apply for a renewal of the marketing authorization due to a lack of demand for the product, and as a result, Glybera® is no longer on the EU market.

Since the approval of Glybera®, 12 more gene therapy products have been approved in the EU. The products, which received a positive opinion from EMA, are shown in Table 1.1.

TABLE 1.1
List of Gene Therapy Products Approved in the EU

Name	Description	Indication	Date of Approval	Link to EPAR[1]
[2] Glybera® (Alipogene tiparvovec)	Contains the human LPL gene variant LPLS447X in a vector. The vector comprises a protein shell derived from adeno-associated virus serotype 1 (AAV1), the cytomegalovirus (CMV) promoter, a woodchuck hepatitis virus posttranscriptional regulatory element and AAV2 derived inverted terminal repeats.	For adult patients diagnosed with familial LPLD and suffering from severe or multiple pancreatitis attacks despite dietary fat restrictions.	25.10.2012	EPAR: Glybera
Imlygic® (Talimogene laherparepvec)	An attenuated herpes simplex virus type-1 (HSV-1) derived by functional deletion of two genes (ICP34.5 and ICP47) and insertion of coding sequence for human granulocyte macrophage colony-stimulating factor (GM-CSF).	For the treatment of adults with unresectable melanoma that is regionally or distantly metastatic (Stage IIIB, IIIC and IVM1a) with no bone, brain, lung or other visceral disease.	16.12.2015	EPAR: Imlygic
Strimvelis®	An autologous CD34+ enriched cell fraction that contains CD34+ cells transduced with retroviral vector that encodes for the human adenosine deaminase (ADA) cDNA sequence from human hematopoietic stem/progenitor (CD34+) cells.	For the treatment of patients with severe combined immunodeficiency due to adenosine deaminase deficiency (ADA-SCID), for whom no suitable human leukocyte antigen (HLA)–matched related stem cell donor is available.	26.05.2016	EPAR: Strimvelis
Kymriah® (tisagenlecleucel)	Immunocellular therapy containing autologous T cells genetically modified ex vivo using a lentiviral vector encoding an anti-CD19 CAR.	For the treatment of: • Pediatric and young adult patients up to and including 25 years of age with B-cell acute lymphoblastic leukemia (ALL) that is refractory, in relapse post-transplant or in second or later relapse. • Adult patients with relapsed or refractory diffuse large B-cell lymphoma (DLBCL) after two or more lines of systemic therapy	23.08.2018	EPAR: Kymriah

History of Gene Therapy Products

Yescarta® (axicabtagene ciloleucel)	CD19-directed genetically modified autologous T-cell immunotherapy. To prepare Yescarta, patient's own T cells are genetically modified ex vivo by retroviral transduction to express a CAR comprising a murine anti-CD19 single chain variable fragment linked to CD28 co-stimulatory domain and CD3-zeta signaling domain. The anti-CD19 CAR-positive viable T cells are expanded and infused back into the patient, where they can recognize and eliminate CD19-expressing target cells.	For the treatment of adult patients with relapsed or refractory DLBCL and primary mediastinal large B-cell lymphoma (PMBCL), after two or more lines of systemic therapy.	23.08.2018	EPAR: Yescarta
Luxturna® (Voretigene neparvovec)	Gene transfer vector that employs an adeno-associated viral vector serotype 2 (AAV2) capsid as a delivery vehicle for the human retinal pigment epithelium 65 kDa protein (hRPE65) cDNA to the retina. Voretigene neparvovec is derived from naturally occurring AAV using recombinant DNA techniques.	For the treatment of adult and pediatric patients with vision loss due to inherited retinal dystrophy caused by confirmed biallelic RPE65 mutations and who have sufficient viable retinal cells.	22.11.2018	EPAR: Luxturna
Zynteglo™ (betibeglogene autotemcel)	Genetically modified autologous CD34+ cell-enriched population that contains hematopoietic stem cells (HSC) transduced with lentiviral vector (LVV) encoding the β^{A-T87Q}-globin gene.	For the treatment of patients 12 years and older with transfusion-dependent β-thalassemia (TDT) who do not have a β^0/β^0 genotype, for whom HSC transplantation is appropriate but an HLA-matched related HSC donor is not available.	29.05.2019	EPAR: Zynteglo
Zolgensma® (Onasemnogene abeparvovec)	A GTMP that expresses the human survival motor neuron (SMN) protein. It is a non-replicating recombinant adeno-associated virus serotype 9 (AAV9) based vector containing the cDNA of the human SMN gene under the control of the cytomegalovirus enhancer/chicken-β-actin-hybrid promoter.	For the treatment of: • patients with 5q spinal muscular atrophy (SMA) with a biallelic mutation in the SMN1 gene and a clinical diagnosis of SMA Type 1, or • patients with 5q SMA with a biallelic mutation in the SMN1 gene and up to three copies of the SMN2 gene.	18.05.2020	EPAR: Zolgensma

(Continued)

TABLE 1.1 (CONTINUED)
List of Gene Therapy Products Approved in the EU

Name	Description	Indication	Date of Approval	Link to EPAR[1]
Tecartus® (Autologous anti-CD19-transduced CD3+ cells)	GTMP containing autologous T cells genetically modified ex vivo using a retroviral vector encoding an anti-CD19 CAR comprising a murine anti-CD19 single chain variable fragment (scFv) linked to CD28 co-stimulatory domain and CD3-zeta signaling domain.	For the treatment of adult patients with relapsed or refractory mantle cell lymphoma (MCL) after two or more lines of systemic therapy including a Bruton's tyrosine kinase (BTK) inhibitor.	14.12.2020	EPAR: Tecartus
Libmeldy™ (atidarsagene autotemcel)	Gene therapy containing an autologous CD34+ cell-enriched population that contains HSPC transduced ex vivo using a lentiviral vector encoding the human arylsulfatase A (ARSA) gene.	For the treatment of metachromatic leukodystrophy (MLD) characterized by biallelic mutations in the arylsulfatase A (ARSA) gene leading to a reduction of the ARSA enzymatic activity: in children with late infantile or early juvenile forms, without clinical manifestations of the disease, and in children with the early juvenile form, with early clinical manifestations of the disease, who still have the ability to walk independently and before the onset of cognitive decline.	17.12.2020	EPAR: Libmeldy
Abecma® (idecabtagene vicleucel)	Genetically modified autologous immunotherapy consisting of human T cells transduced with LVV encoding a CAR that recognizes B-cell maturation antigen.	For the treatment of adult patients with relapsed and refractory multiple myeloma who have received at least three prior therapies, including an immunomodulatory agent, a proteasome inhibitor and an anti-CD38 antibody and have demonstrated disease progression on the last therapy.	18.08.2021	EPAR: Abecma

Breyanzi® (lisocabtagene maraleucel)	A CD19-directed genetically modified autologous cell-based product consisting of purified CD8+ and CD4+ T cells in a defined composition that has been separately transduced ex vivo using a replication-incompetent lentiviral vector expressing an antiCD19 CAR comprising an scFv binding domain derived from a murine CD19-specific monoclonal antibody (mAb; FMC63) and a portion of the 4-1BB co-stimulatory endodomain and CD3 zeta (ζ) chain signaling domains and a nonfunctional truncated epidermal growth factor receptor (EGFRt).	For the treatment of adult patients with relapsed or refractory DLBCL, PMBCL and follicular lymphoma grade 3B (FL3B) after two or more lines of systemic therapy.	04.04.2022	EPAR: Breyanzi
³ Carvykti™ (ciltacabtagene autoleucel)	A genetically modified autologous T-cell immunotherapy consisting of modified T cells bearing a CAR targeting B-cell maturation antigen (BCMA). BCMA is primarily expressed on the surface of malignant multiple myeloma B-lineage cells, as well as late-stage B cells and plasma cells. Upon binding to BCMA-expressing cells, the CAR promotes T-cell activation, expansion, and elimination of target cells.	For the treatment of adult patients with relapsed and refractory multiple myeloma, who have received at least three prior therapies, including an immunomodulatory agent, a proteasome inhibitor and an anti-CD38 antibody, and have demonstrated disease progression on the last therapy.	24.03.2022*	Not yet published

Notes:

1 EPAR: European Public Assessment Report.
2 The marketing authorization for Glybera® was not renewed and is no longer on the market since 2017.
3 On March 24, 2022, the Committee for Medicinal Products for Human Use (CHMP) adopted a positive opinion, recommending the granting of conditional marketing authorization for the medicinal product Carvykti™. The European Commission's final decision is awaited.

The highest number of approved gene therapies are the ex vivo transduced autologous cells; chimeric antigen receptor (CAR)-T cells which are Kymriah®, Yescarta®, Tecartus®, Abecma® Breyanzi® and Carvykti™. These therapies have had a major impact on the outcome for patients with relapsed or refractory leukemia and lymphoma and more recently for patients with relapsed and refractory multiple myeloma. The known serious side effect from CAR-T cells includes Cytokine Release Syndrome (CRS). The monitoring and management of CRS is an integral part of the management of patients receiving these therapies and is detailed in their respective summary of product characteristics (SmPCs).

The other group of ex vivo transduced cells comprise the therapies using ex vivo transduction of autologous hematopoietic stem and progenitor cells (HSPC) where the gene missing in the patient is replaced – namely, Libmeldy™, Zynteglo™ and Strimvelis® for the treatment of metachromatic leukodystrophy, transfusion-dependent β-thalassaemia and ADA-SCID, respectively.

The remaining three therapies are in vivo gene therapy products, which, like Glybera®, are administered directly to the patient. These are Imlygic®, Luxturna® and Zolgensma® for the treatment of unresectable melanoma, vision loss due to inherited retinal dystrophy caused by RPE65 mutations and patients with 5q spinal muscular atrophy, respectively.

Most of the gene therapies approved in the EU are also approved in the US (Abecma®, Breyanzi®, Carvykti™, Imlyglic®, Kymriah®, Luxturna®, Tecartus®, Yescarta® and Zolgensma® [fda.gov.approved-cellular-and-gene-therapy-products]).

Although the history of gene therapy is relatively recent, with the first suggestion that gene therapy for humans could be a possibility in the 1970s (Friedmann & Roblin 1972), this field has come a long way since then. In the recent past since the authorization of Glybera®, there have been a very high number of gene therapy treatments approved by both the EMA and the US FDA. These new therapies have provided highly effective treatments for a range of diseases, each with a high unmet medical need. These include gene replacement, either directly or indirectly (ex vivo via gene therapy modified cells), or utilizing gene therapy to increase the ability of T cells to directly attack cancer cells (CAR-T cells).

In view of the increase in scientific advances, both in terms of the implications for individuals and society for possessing the detailed genetic information made possible by the Human Genome Project, and advances in molecular biology and gene targeting and editing, it is expected that this recent increase in availability of gene therapies will continue to improve the options for the treatment of serious disease.

REFERENCES

Anderson, W. F., Blaese, R. M., & Culver, K. (1990). Points to consider response with clinical protocol, July 6, 1990. *Human Gene Therapy*, *1*(3), 331–362.

Avery, O. T., MacLeod, C. M., & McCarty, M. (1944). Studies on the chemical nature of the substance inducing transformation of pneumococcal types: induction of transformation by a desoxyribonucleic acid fraction isolated from pneumococcus type III. *The Journal of Experimental Medicine*, *79*(2), 137–158.

Blaese, R. M., Culver, K. W., Miller, A. D., Carter, C. S., Fleisher, T., Clerici, M., ... & Anderson, W. F. (1995). T lymphocyte-directed gene therapy for ADA– SCID: initial trial results after 4 years. *Science*, *270*(5235), 475–480.

Commission Directive 2009/120/EC of 14 September 2009 amending Directive 2001/83/EC of the European Parliament and of the Council on the Community code relating to medicinal products for human use as regards advanced therapy medicinal products. *Official Journal of the European Union*, L242-3. http://data.europa.eu/eli/dir/2009/120/oj

Cooney, A. L., McCray, P. B., & Sinn, P. L. (2018). Cystic fibrosis gene therapy: Looking back, looking forward. *Genes*, *9*(11), 538.

European Parliament and the Council of the European Union. (2014). Regulation (EU) No 536/2014 of the European Parliament and of the Council of 16 April 2014 on clinical trials on medicinal products for human use, and repealing Directive 2001/20/EC. *Official Journal of the European Union*, *57*, L58-1. https://ec.europa.eu/health/system/files/2016-11/reg_2014_536_en_0.pdf

Friedmann, T., & Roblin, R. (1972). Gene therapy for human genetic disease? Proposals for genetic manipulation in humans raise difficult scientific and ethical problems. *Science*, *175*(4025), 949–955.

Griffith, F. (1928). The significance of pneumococcal types. *Epidemiology & Infection*, *27*(2), 113–159.

Hacein-Bey-Abina, S., Garrigue, A., Wang, G. P., Soulier, J., Lim, A., Morillon, E., ... & Cavazzana-Calvo, M. (2008). Insertional oncogenesis in 4 patients after retrovirus-mediated gene therapy of SCID-X1. *The Journal of Clinical Investigation*, *118*(9), 3132–3142.

Hacein-Bey-Abina, S., Von Kalle, C., Schmidt, M., Le Deist, F., Wulffraat, N., McIntyre, E., ... & Fischer, A. (2003). A serious adverse event after successful gene therapy for X-linked severe combined immunodeficiency. *New England Journal of Medicine*, *348*(3), 255–256.

Howe, S. J., Mansour, M. R., Schwarzwaelder, K., Bartholomae, C., Hubank, M., Kempski, H., ... & Thrasher, A. J. (2008). Insertional mutagenesis combined with acquired somatic mutations causes leukemogenesis following gene therapy of SCID-X1 patients. *The Journal of Clinical Investigation*, *118*(9), 3143–3150.

Johnston, C., Martin, B., Fichant, G., Polard, P., & Claverys, J. P. (2014). Bacterial transformation: Distribution, shared mechanisms and divergent control. *Nature Reviews Microbiology*, *12*(3), 181–196.

Kumar, S. R., Markusic, D. M., Biswas, M., High, K. A., & Herzog, R. W. (2016). Clinical development of gene therapy: Results and lessons from recent successes. *Molecular Therapy-Methods & Clinical Development*, *3*, 16034.

Modlich, U., & Baum, C. (2009). Preventing and exploiting the oncogenic potential of integrating gene vectors. *The Journal of Clinical Investigation*, *119*(4), 755–758.

Onodera, M., Nelson, D. M., Yachie, A., Jagadeesh, G. J., Bunnell, B. A., Morgan, R. A., & Blaese, R. M. (1998). Development of improved adenosine deaminase retroviral vectors. *Journal of Virology*, *72*(3), 1769–1774.

Papapetrou, E. P., & Schambach, A. (2016). Gene insertion into genomic safe harbors for human gene therapy. *Molecular Therapy*, *24*(4), 678–684.

Patil, S. D., Rhodes, D. G., & Burgess, D. J. (2005). DNA-based therapeutics and DNA delivery systems: A comprehensive review. *The AAPS Journal*, *7*(1), E61–E77.

Pavani, G., & Amendola, M. (2021). Targeted gene delivery: Where to land. *Frontiers in Genome Editing*, *2*, 36.

Rogers, S., Lowenthal, A., Terheggen, H. G., & Columbo, J. P. (1973). Induction of arginase activity with the Shope papilloma virus in tissue culture cells from an argininemic patient. *The Journal of Experimental Medicine*, *137*(4), 1091.

Schneider, C., Salmikangas, P., Jilma, B., Flamion, B., Todorova, L., ... & Celis, P. (2010). Challenges with advanced therapy medicinal products and how to meet them. *Nature Reviews Drug Discovery*, *9*(March), 195–201.

Stein, S., Ott, M. G., Schultze-Strasser, S., Jauch, A., Burwinkel, B., Kinner, A., ... & Grez, M. (2010). Genomic instability and myelodysplasia with monosomy 7 consequent to EVI1 activation after gene therapy for chronic granulomatous disease. *Nature Medicine*, *16*(2), 198–204.

Stolberg, S. G. (1999). The biotech death of Jesse Gelsinger. *New York Times Magazine*, *28*, 136–140.

Szybalska, E. H., & Szybalski, W. (1962). Genetics of human cell lines, IV. DNA-mediated heritable transformation of a biochemical trait. *Proceedings of the National Academy of Sciences of the United States of America*, *48*(12), 2026.

Yeh, P., & Perricaudet, M. (1997). Advances in adenoviral vectors: From genetic engineering to their biology. *The FASEB Journal*, *11*(8), 615–623.

Zinder, N. D., & Lederberg, J. (1952). Genetic exchange in Salmonella. *Journal of Bacteriology*, *64*(5), 679–699.

2 Gene Editing in Humans
How Bioethics Discussions Can Guide Responsible Research and Product Development

Hazel Aranha
GAEA Resources Inc., Northport, USA

CONTENTS

Introduction 13
Risk-Benefit Considerations 14
Bioethics Considerations in Genome Editing 18
Regulatory and Nonregulatory Discussions and Declarations 21
Concluding Comments 23
References 24

[I]n nature there are no rewards or punishment; there are consequences. Robert Ingersoll, "Some Reasons Why," 1881.

INTRODUCTION

In 2018, Chinese biophysicist He Jiankui and colleagues edited the genome of human embryos with the CRISPR-Cas9 editing tool using human embryos obtained from assisted reproductive technology (ART). Their research was approved by the Medical Ethics Committee of the First Affiliated Hospital, Sun Yat-sen University, Guangzhou, China; the study was compliant with the ethical standards of the Declaration of Helsinki and with existing Chinese laws. These scientists intended to edit out the gene responsible for HIV. Nevertheless, the incident caused a furor in the scientific community (Wee, 2019). While He and his colleagues did not violate any existing Chinese laws, they were convicted of 'illegal practice of medicine.' This incident brought to the forefront the absolute imperative for guidance and guidelines that proactively take into consideration ethical and legal safeguards so that technology must be tempered with safeguards and guardrails.

The discovery that clustered regulatory interspaced short palindromic repeats (CRISPR), present in prokaryotes, provided a bacterial adaptive defense mechanism, combined with the discovery of the *Cas* gene, allowed the unveiling of the basic function of the CRISPR system (Jinek et al., 2012), and multiple significant advances have resulted thereafter. The strength of these technologies stems from the ability to create fractures in the desired region of a specific target sequence, allowing one to modify the genome, in practice, in any region (Memi et al., 2018). This provides the power to change the genome by correcting a mutation or introducing a new function.

The CRISPR-Cas9 technology has revolutionized the field of genome editing since its discovery in 2012 (Doudna & Charpentier, 2014). Its advantages include high accuracy, easy handling and relatively low cost compared with previous technologies, such as zinc-finger nuclease (ZFN) and transcription activator-like effector nuclease (TALEN). Less than ten years after discovery, the Nobel Prize in Chemistry was awarded to Emmanuelle Charpentier and Jennifer Doudna (Doudna & Charpentier, 2014). The social impact of CRISPR/Cas9 gene editing technologies has gained significant attention in part due to the remarkable speed of the development of these technologies for use in humans (Doudna, 2020; Salsman & Dellaire, 2017).

Genome editing technologies are often referenced as key developments in future therapeutic and nontherapeutic applications. Central to any conversation relating to the application of such technologies are ethical, legal and social difficulties around their application. This chapter addresses the potential and aspirational applications of gene editing and addresses the current bioethical considerations, along with the opinions of some nonregulatory and regulatory groups.

RISK-BENEFIT CONSIDERATIONS

Every process and product carries with it an associated risk that can range from totally unacceptable to acceptance in the context in which it is applied. The clinical acceptability of biologicals and other pharmaceutical interventions is concomitant with risk assessment and must be guided by risk-benefit analyses (Aranha, 2005). Public perception leans toward acceptance of risk associated with voluntary activities (e.g., skiing) even when they pose risk several orders of magnitude greater than medical interventions that provide the same (or greater) level of benefit with a lower risk profile. Case-in-point: vaccination from a public health standpoint (COVID-19 vaccine) is seen by some as more invasive than a procedure for aesthetic or cosmetic reasons.

There are few, if any, therapy options available for many hereditary diseases. Current research on gene therapy treatment has focused on targeting body (somatic) cells such as bone marrow or blood cells. This type of genetic alteration cannot be passed to one's offspring. Germline gene therapy is targeted to egg and sperm cells (germ cells), allowing the genetic changes to be passed to future generations; these modifications, understandably, are of significant concern.

A plethora of literature reports document the application of CRISPR-Cas9 to multiple inherited and other diseases including Duchenne muscular dystrophy, cystic fibrosis, sickle cell anemia, and cancer (Amoasii et al., 2018; Firth et al., 2015; Koo et al. 2018; Wang & Chang, 2018; Wong & Cohn, 2017). Other studies include prevention/treatment of AIDS by inhibiting the entry of HIV into the cell or by removing the HIV genome integrated into the host genome and use of CRISPR-Cas9 to control malaria by altering the genome of the mosquito (Gantz et al., 2015; Saayman et al., 2015).

Editing the genome of human embryos opens vistas for multiple avenues of human experimentation. While research on human embryos and other applications of genome editing has been debated from a bioethical standpoint for the last quarter of a century (Robertson, 2001), the addition of the CRISPR-Cas9 system to our genome editing toolbox has escalated these discussions because of its high specificity, versatility, accessibility and efficiency (de Lecuona et al., 2017).

CRISPR-Cas9 for human germline editing brings up two key issues: (i) what happens if it fails, i.e., issues arising due to failure of germline genome editing, and (ii) what happens if it succeeds, i.e., issues related to the successful application of germline genome editing.

Although CRISPR–Cas9 allows genome editing with relative ease and has numerous benefits, there are important limitations. Some of the technical limitations include (i) possibilities of limited on-target editing efficiency, (ii) incomplete editing, (iii) inaccurate on- or off-target editing. These limitations have been reported in CRISPR experiments involving animals and human cell lines. Off-target effects may produce loss-of-function mutations at the genetic level. If, for example, occurring in a tumor suppressor gene, the consequences can possibly be amplified on the cellular and organismal level. Off-target events may also lead to gain-of-function mutations, such as activation of oncogenes by chromosomal translocations, resulting in malignant cell transformation, which likely will not be evident in the short term but only be observable as long-term serious adverse events.

The studies by He and his group in the human embryos that were engineered to provide resistance to HIV also provided concrete evidence of these limitations (Jordan, 2019; Raposo, 2019). As the technology continues to rapidly develop, other Cas systems using isolated proteins Cas12 and Cas13 from other bacterial strains have been identified. Advantages of these variants over Cas9 include the ability to perform multiplex genome editing (Cas12), increasing specificity and thus lowering the possibility of off-target effects and the ability to cleave RNA instead of DNA, thus allowing for more specific modulation of protein-production levels (Molina et al., 2020; Murugan et al., 2017).

Effective and safe delivery of genome editing agents to the intended target tissue is essential for the clinical application of CRISPR-Cas9-based therapeutics.

While carrier-free delivery of CRISPR-Cas9 agents has been demonstrated to be fairly effective in certain applications, there are still obvious potential limitations, including unwanted immunostimulation, difficulties in renal clearance, insufficient transport into the cytoplasm and nucleus and unwanted degradation by cellular enzymes.

The applications of CRISPR technologies are limited only by our imagination. Some are in late-stage clinical trials and a few are marketed products. In other cases, research and preclinical studies have demonstrated proof of concept. Table 2.1 provides a brief summary. Some applications are (i) mapping chromosomal primary information to identify regulatory connections underlying gene function; identifying loci associated with myocardial infarction susceptibility and development of bioprobes for specific biomarkers of disease, e.g., prostate-specific antigen (PSA; Taghdisi et al., 2022); (ii) treatment of genetic and nongenetic diseases, e.g., Duchenne muscular dystrophy, cystic fibrosis, cataracts, Parkinson's disease (Amoasii et al., 2018; Firth et al., 2015; Koo et al., 2018; Yang et al. 2016); (iii) cancer therapy to develop targeted and more efficient cancer cell elimination strategies; (iv) approaches to address antibiotic resistance in bacteria; CRISPR-Cas9 may be used to specifically remove resistance genes in gut bacteria, selectively impacting the pathogenic species rather than the beneficial commensal ones in the microbiota of the gastrointestinal tract; (v) detoxification of toxins – diphtheria toxin, ochratoxin; (vi) military applications; while there are not many publications in this area (for obvious reasons), studies have been done to increase tolerance of soldiers to biological and chemical weapons, human performance optimization and identifying new genes that can be associated with treatment of post-traumatic stress disorder (Cornelis et al., 2010; Greene & Master, 2018); (vii) CRISPR-Cas systems have also been used in industrial and food applications; industrial applications of gene modification have been ongoing for a few decades; application of CRISPR technology would facilitate increased yield in the livestock industry and control of insects and weeds (Esvelt et al., 2014); and (viii) vaccine applications for diseases that are ongoing, e.g., Hepatitis B genome targeted and cut by CRISPR-Cas9 to prevent viral gene expression and replication, development of smallpox virus vector (VACV) in the eradication of smallpox, efficiency of marker-free VACV vectors has been increased (Ramanan et al., 2015; Yuan et al., 2015).

The implications of germline alterations are known and recognized; however, when is its application an acceptable alternative to existing pharmaceutical interventions? Using gene editing to spare future generations in a family from having a particular genetic disorder could potentially affect the development of a fetus in unexpected ways or have long-term side effects that are yet unknown and could be manifest in the long term. Because people who would be affected by germline gene therapy are not yet born, how does one obtain informed consent, as they have no choice as to whether to have the treatment or not?

Table 2.2 briefly summarizes some of the risk-benefit issues.

TABLE 2.1
Applications of CRISPR and Other Gene Editing Technologies

Application	Description	Comments
Functional genome screening	Identification of genes impacting biological processes to explore molecular mechanisms' cellular functions for better understanding of disease to facilitate drug target identification. Mapping chromosomal primary information to identify regulatory concerns underlying gene function. Detection of disease markers, e.g., PSA.	• Studies have been conducted in experimental models to evaluate therapeutic efficacy, e.g., cell viability in cancer, detoxification of toxins, such as diphtheria, ochratoxin. • Ability to conduct rapid drug identification and correlate with therapeutic efficacy will allow using the potential of personalized medicine by combining genomics, disease phenotypes and therapeutic targets. This is still aspirational.
Genetic and nongenetic diseases	Addressing mutation responsible for disease, e.g., Duchenne muscular dystrophy, cystic fibrosis. Prevention and treatment of AIDS by inhibiting the entry of HIV into the cell or by removing the HIV genome integrated into the host genome. Treatments for neurological diseases, e.g., Parkinson's disease.	• CRISPR-Cas9 offers hope in treatment of diseases; however, limitations associated with CRISPR-Cas9 in treatment must be addressed before it can be offered as a safe treatment option.
Viral infections	Viruses that induce latent infections or cause persistent infections could be targeted with CRISPR-Cas9.	• A new therapeutic application may be eliminating HIV-1 DNA from CD4+ T cells, and CRISPR–Cas9 may be used as a novel and efficient platform for the cure of AIDS.
Antibiotic resistance in bacteria	CRISPR–Cas9 may be used to specifically remove resistance genes and could be a part of the solution to address antimicrobial antibiotics resistance concerns.	• Can be a powerful tool to counteract antibiotic resistance. CRISPR systems could potentially target only antibiotic-resistant bacteria while preserving commensal ones in the microbiota.
Cancer therapy	Enormous opportunities for increasing knowledge of cancer biology, for developing new preclinical models and for progress in more efficient and targeted cancer cell elimination strategies.	• Tumor heterogenicity can be a problem in treatment because tumors usually consist of different subclones • The lack of a safe and efficient delivery method that can be applied in clinical trials is one of the major problems related to direct targeting in cancer.

(Continued)

TABLE 2.1 (CONTINUED)
Applications of CRISPR and Other Gene Editing Technologies

Application	Description	Comments
Detoxification	CRISPR application to develop resistance to toxins.	• Examples here include mycotoxins (ochratoxin), diphtheria toxin, and anthrax.
Military	Studies have focused on increasing tolerance against biological or chemical warfare. Potential to influence human performance optimization, enhance muscle mass, decrease sensitivity to anthrax toxin.	• How is informed consent obtained in a military culture where adherence to strict norms and chains of command is required? • How does one control rogue nations who may attempt to use these enhancements as a competitive advantage?
Industrial applications	Increase yield in the livestock industry. Control invasive pest species by reverse pesticide and herbicide resistance in insects and weeds or to prevent disease spread. Prevent the spread of genes protecting mosquitoes from harmful malaria parasites and making female mosquitoes infertile in the laboratory.	• Concerns about ecological imbalance.
Vaccine development	Hepatitis B genome targeted and cut by CRISPR-Cas9 to prevent viral gene expression and replication. Smallpox VACV to increase efficiency of marker-free VACV vectors.	• Research ongoing.

BIOETHICS CONSIDERATIONS IN GENOME EDITING

Any approach to effective and legitimate governance of genome editing will require a combination of national and supranational legislative regulation or 'hard' law, in combination with 'soft' ethics (Boldt & Muller, 2008). Regulations, in and of themselves, will not suffice. Ethics and ethical guidelines firmly anchored in and underpinned by human rights values are of value; however, they should be complementary to, rather than a substitute for, the law (Vassy et al., 2017).

The gains garnered from the use of CRISPR-Cas9 and similar systems in the treatment of disease and creation of human disease models to facilitate understanding of the development and molecular mechanisms of diseases and identification of regulatory connections underlying gene function cannot be underestimated. However, there is universal consensus that it should be prohibited for the purposes of eugenics or enhancement of what is perceived as a 'desirable trait.' But, who

TABLE 2.2
Risk-Benefit Considerations in Genome Editing

	Benefits	Risks/Harm
Basic and preclinical research	• Facilitate better understanding of disease development through development of animal models • Ability to modify genomes of specific tissues such as liver and brain tissues • Identification of novel regulatory and developmental pathways that could be leveraged for therapeutic use • Development of novel gene-editing approaches (base editing, RNA targeting) • High throughput screens	• Experimentation involving human embryos is highly controversial; illegal in some countries • Potential for privacy/confidentiality breaches • Development of animal models/chimeric animals imposes on animal dignity and welfare
Translational clinical medicine/therapeutic and nontherapeutic applications	• Multiple benefits (see Table 2.1)	• Serious adverse events (injury, disability, death) to research participants and implications for offspring • Blurry distinction between therapeutic and enhancement applications • Eugenics concerns • Patenting of technology limiting universal access (theoretically) • Potential for inequitable access and exacerbation of inequalities • 'Rogue' players exploiting the technology for personal gain
Access to CRISPR technology	Should be universal access considering that a lot of the research is funded by government and semi-government organizations	• Price gouging • Expensive intervention, making it of limited access to the financially disadvantaged • Potential for misuse, ethics shopping, medical tourism

(Continued)

TABLE 2.2 (CONTINUED)
Risk-Benefit Considerations in Genome Editing

	Benefits	Risks/Harm
Regulations related to clinical research in human subjects	Oversight through both international treaties, national and local laws	• Current lack of appropriate infrastructure, regulatory framework in many countries
National and international laws, regulations and policy	Need for harmonization	• Incomplete under-/overlegislation • Potential to encroach on societal autonomy could limit discovery and progress • Difficult to enforce • 'Rogue' officials who may look the other way

decides what is a desirable trait? In some cultures, a light skin color is favored. Mouse models that have been altered to display a white fur coat have been developed. Researchers have developed dog embryos with higher muscle mass using CRISPR-Cas9 (Zou et al., 2015).

An area of significant concern in germline editing is the future of the modified entity: will they be affected indefinitely, and will the edited genes be transferred to future generations, potentially affecting them in yet unknown and unexpected ways? Who controls the use of germline editing in ART? This, combined with the technical limitations of our editing tools and the complexities of biological systems, makes predictions about the future of an edited organism and gauging potential risks and benefits difficult, if not impossible. This uncertainty hinders accurate risk/benefit analysis and complicates moral decision-making.

Genome editing is not only about social and bioethical issues related to people. There are concerns about the impact on the ecosystem that may occur if the genetically modified organisms (GMOs) produced with CRISPR-Cas9 are released into the ecosystem in a controlled or uncontrolled manner. CRISPR-Cas9 technologies have been used to protect mosquitoes from malaria parasites or make female mosquitoes infertile (Gantz et al., 2015; Hammond et al., 2016). While this would be useful in malaria control, the effect of genetically modified mosquitoes on other organisms with which they are associated in their ecosystems cannot be predicted. The 'gene drive' approach with CRISPR technology would allow us to speed up genetic recombination such that a gene of interest is disseminated through a population much faster than in a Mendelian inheritance rate, providing a parasite-resistance phenotype in mosquitoes that could spread through the population in a non-Mendelian fashion. Gene drives could provide a tool to fight diseases such as malaria, dengue fever, and Lyme disease among others. It would provide an opportunity to control and/or alter reservoirs of infections such as bats and rodents.

It is clear that small-scale research in the laboratory does not fully reflect possible changes in the natural ecosystem (Carroll, 2017). In agriculture, public perception is another concern, and there has been pushback on GMO products in general. Additionally, the fact that GMOs produced with CRISPR-Cas9 are difficult to identify outside the laboratory raises safety concerns (Shinwari et al., 2018). Risk communication to the public is an essential component of our risk evaluation efforts.

Bioethics discussions are related to beneficence, autonomy of patients and justice for all; this implies universal availability and not restrictions based on treatment affordability. Patenting is another area that elicits concern. There is disagreement in the scientific community regarding the patenting or non-patenting of GMOs to be used specifically for therapeutic purposes (Sherkow, 2018; Shinwari et al., 2018). In recent years, deals between the scientific community and the pharmaceutical and biotechnology sectors for the therapeutic use of CRISPR-Cas9 have raised public safety concerns (Carroll, 2017; Shinwari et al., 2018). The guidelines and legislation that will regulate the content and application of these deals should be prepared as quickly as possible and shared with the public. Due to the challenges and bioethical issues of CRISPR-Cas9, the scientific community and other interested bioethical, social, legal and government parties should be provided with a detailed guide for future processing and use of this technology (Shinwari et al., 2018). This allows for a long-term policy to be developed that will support the scientific development of CRISPR-Cas9 technology together with the discussion of the possible problems in advance and preparation of solution plans.

REGULATORY AND NONREGULATORY DISCUSSIONS AND DECLARATIONS

There have been multiple consultations, conferences and declarations advanced. These have been summarized in the literature (Brokowski, 2018; Brokowski & Adli, 2019, 2020; de Lecuona et al., 2017) and published and posted on the websites of various organizations. The paper by deLecuona summarizes the main points and ethical concerns of documents released by various organizations: European Academies' Science Advisory Council, ACMG (American College of Medical Ethics and Genomics) and the National Academies of Sciences, Engineering and Medicine, and OBD (Opinion group of the Bioethics and law observatory). While there are several common points and a few divergent ones, there is a global consensus that suggests that a better understanding of CRISPR is needed before its clinical application. Moreover, the National Academies of Science, Engineering and Medicine document highlights the effects on the human gene pool. Some genes that cause serious genetic diseases have been subject to positive selection to maintain the disease-causing allele in the population because it produces some protection against infectious disease when present in only one copy. If this gene is modified, this protection will be lost.

Multiple issues need to be considered when developing and proposing a normative framework. The guidance should be (i) sensitive to societal and cultural differences in what is considered appropriate and realistic; (ii) responsive, proactive and flexible to the technological transformation in the biotechnology industry; (iii) foster and build public confidence through risk communication and providing adequate protection measures within this context; and (iv) align with global rights-based best practices, principles and standards (Townsend, 2020).

A global instrument to assist in establishing a stable list of nonarbitrary standards and principles that justifies and endorses certain values, ethics and rights should be developed. In the absence of this, patients seeking treatment may resort to a less regulated country, providing an equivalent of medical tourism. Informal mechanisms of international cooperation provide general agreement between countries in the form of non-binding guidelines; they do not impose legal requirements on the individual countries or states to implement specific provisions. Although they have normative value, their enforcement level is often inconsistent in that state practice determines what eventually becomes settled as norms, values and rules.

Most conventions are not self-executing instruments. International instruments or conventions would offer a malleable template or guide for advancing legislation, which can be modified by member states to fit their national requirements (Townsend, 2020). Similar to climate change, countries that accede to and ratify the treaty would commit to establishing a legal regime based on its provisions and the obligations created therein may provide a useful catalyst for the evolution of legal frameworks around the world. Although harmonization is not entirely achievable, an approach of coordination and collaboration between regulators in different countries where feasible offers a solution. It would, also, provide opportunities for identifying common ground on specific substantive or technical aspects and creating a valuable framework for adoption.

In attempting to find a solution, we should not lose sight of the tremendous potential benefit genome editing affords – particularly, somatic cell editing. As the promise of germline editing is still under question, how useful it might be and its safety and efficacy are still to be established. We have an entire toolbox of instruments, guidelines, measures and platforms at our disposal from which to proactively develop ethical and legal safeguards so that technology can be allowed to progress for the benefit of humankind.

A proactive rather than reactive decision-making approach can provide for an ethical and regulatory toolbox of formal and informal rules, guidelines, protocols and practices for appropriate genome editing governance. Additionally, cooperative efforts help determine the acceptable ethical boundaries of genome editing and should be a priority so that the diversity of societal, cultural, and political values becomes an asset rather than a hindrance in reaching consensus.

This perspective is supportive of an 'ecosystem' approach to the regulation of novel biotechnologies. It serves to take advantage of the ecosystem of regulatory actors and develop a road map for responsible translational research that includes 'stringent criteria for use of germline editing and standards for determining whether these criteria have been met, embedded within larger political structures offering vehicles for public input.' This being said, while many approaches may provide partial or incomplete answers, an integrated, holistic solution to global enforcement and remedy is not easily done, but the aforementioned recommendations may go some way in providing a path forward.

The chorus of 'something must be done' sung by scientists, regulators and other stakeholders conveniently does not address exactly what must be done or how we go about doing it. We, therefore, need some worldwide regulatory frameworks endorsing values and universally accepted human rights: the right to life, the rights of the future child, the right to health care, the right to dignity, the right to equality and nondiscrimination and the freedom of scientific progress. These positions have already been discussed (Townsend, 2020) in several declarations, such as the Universal Declaration on the Human Genome and Human Rights, the International Declaration on Human Genetic Data and the Universal Declaration on Bioethics and Human Rights, so we have to start to define normative framework within which scientists may (or may not) lawfully and ethically operate.

CONCLUDING COMMENTS

Mutations are an essential part of evolution. Mutations, which seem deleterious today, may have inclusive fitness tomorrow; others when they exist as a homozygous recessive allele, e.g., the gene for sickle cell anemia, confer selective resistance to disease.

Today, with the availability of advanced genome editing tools and the insights provided by epigenomics we are in a position to drive evolution in a non-Mendelian fashion. While somatic cell modifications in humans provide solutions to many ailments, as of now, with germline modifications, there are significant gaps in information, and, consequently, a defined risk profile has yet to be conclusively established. Researchers, therefore, need to be doubly cautious, and basic stringent regulations should be framed regarding the various aspects of germline gene modifications and any potential conflict with nature for future outcomes.

The pressing question, asked both of regulators and the scientific community alike, is how best to stop 'rogue' actors in what could be viewed as 'dangerous pursuits'? While new regulations and ethical rules to govern research and therapies that involve genome editing, gene transfers or attempts to regulate gene expression and other so-called high-risk biotechnologies in humans (or embryos) have been called for, does the behavior of these actors fall under the umbrella of 'unethical' and/or 'illegal'? Some activities may not be illegal but would still be considered unethical. What we have consensus about is that we cannot allow rogue actors to make these irreversible decisions for us all.

Similar to our approach with Agenda 2030 for sustainability, we need a general agreement about the applicative fields of the editing techniques (e.g., prohibition of modification of embryos) with specific national laws precisely regulating them in each country.

The misuse of genome editing technology, its impact on future generations, the unclear definition considered therapeutic versus for enhancement purposes and the possible accentuation of social inequalities that CRISPR could produce are our main ethical concerns.

In a very competitive scientific world with heterogeneous regulations and cultural environments, a moratorium would be counterproductive. We need a proactive approach to start thinking both nationally and internationally about how to regulate from a bioethical and legal perspective. Inaction and reluctance to voice our opinions constitute an action. Are we going to let the loudest mouths in the room prevail? Collectively, we vacillate between truth and sensationalism, making it less like a balancing of the scales of justice and more like a windmill caught in a tropical storm. As scientists, our understanding of the risk assessment, mitigation and management of new genome editing technologies should be a vocal asset in our risk communication efforts both to the public and the national and international governance bodies.

REFERENCES

Amoasii, L., Hildyard, J. C. W., Li, H., Sanchez-Ortiz, E., Mireault, A., Caballero, D., & Olson, E. N. (2018). Gene editing restores dystrophin expression in a canine model of Duchenne muscular dystrophy. *Science, 362*(6410), 86–91. doi:10.1126/science.aau1549.

Aranha, H. (2005). Virological safety of biopharmaceuticals: A risk-based approach. *Bioprocess Int*, (Supplement November), 17–20.

Boldt, J., & Muller, O. (2008). Newtons of the leaves of grass. *Nat Biotechnol, 26*(4), 387–389. doi:10.1038/nbt0408-387.

Brokowski, C. (2018). Do CRISPR germline ethics statements cut it? *CRISPR J, 1*, 115–125. doi:10.1089/crispr.2017.0024.

Brokowski, C., & Adli, M. (2019). RISPR ethics: Moral considerations for applications of a powerful tool. *J Mol Biol, 431*(1), 88–101. doi:10.1016/j.jmb.2018.05.044.

Brokowski, C., & Adli, M. (2020). Ethical considerations in therapeutic clinical trials involving novel human germline-editing technology. *CRISPR J, 3*(1), 18–26. doi:10.1089/crispr.2019.0051.

Carroll, D. (2017). Genome editing: Past, present, and future. *Yale J Biol Med, 90*(4), 653–659.

Cornelis, M. C., Nugent, N. R., Amstadter, A. B., & Koenen, K. C. (2010). Genetics of posttraumatic stress disorder: Review and recommendations for genome-wide association studies. *Curr Psychiatry Rep, 12*(4), 313–326. doi:10.1007/s11920-010-0126-6.

de Lecuona, I., Casado, M., Marfany, G., Lopez Baroni, M., & Escarrabill, M. (2017). Gene editing in humans: Towards a global and inclusive debate for responsible research. *Yale J Biol Med, 90*(4), 673–681.

Doudna, J. A. (2020). The promise and challenge of therapeutic genome editing. *Nature, 578*(7794), 229–236. doi:10.1038/s41586-020-1978-5.

Doudna, J. A., & Charpentier, E. (2014). Genome editing. The new frontier of genome engineering with CRISPR-Cas9. *Science, 346*(6213), 1258096. doi:10.1126/science.1258096.

Esvelt, K. M., Smidler, A. L., Catteruccia, F., & Church, G. M. (2014). Concerning RNA-guided gene drives for the alteration of wild populations. *Elife, 3*. doi:10.7554/eLife.03401.

Firth, A. L., Menon, T., Parker, G. S., Qualls, S. J., Lewis, B. M., Ke, E., ... Verma, I. M. (2015). Functional gene correction for cystic fibrosis in lung epithelial cells generated from patient iPSCs. *Cell Rep, 12*(9), 1385–1390. doi:10.1016/j.celrep.2015.07.062.

Gantz, V. M., Jasinskiene, N., Tatarenkova, O., Fazekas, A., Macias, V. M., Bier, E., & James, A. A. (2015). Highly efficient Cas9-mediated gene drive for population modification of the malaria vector mosquito Anopheles stephensi. *Proc Natl Acad Sci USA, 112*(49), E6736–E6743. doi:10.1073/pnas.1521077112.

Greene, M., & Master, Z. (2018). Ethical issues of using CRISPR technologies for research on military enhancement. *J Bioeth Inq, 15*(3), 327–335. doi:10.1007/s11673-018-9865-6.

Hammond, A., Galizi, R., Kyrou, K., Simoni, A., Siniscalchi, C., Katsanos, D., ... Nolan, T. (2016). A CRISPR-Cas9 gene drive system targeting female reproduction in the malaria mosquito vector Anopheles gambiae. *Nat Biotechnol, 34*(1), 78–83. doi:10.1038/nbt.3439.

Jinek, M., Chylinski, K., Fonfara, I., Hauer, M., Doudna, J. A., & Charpentier, E. (2012). A programmable dual-RNA-guided DNA endonuclease in adaptive bacterial immunity. *Science, 337*(6096), 816–821. doi:10.1126/science.1225829.

Jordan, B. (2019). CRISPR babies: Technology and transgression. *Med Sci (Paris), 35*(3), 266–270. doi:10.1051/medsci/2019034.

Koo, T., Lu-Nguyen, N. B., Malerba, A., Kim, E., Kim, D., Cappellari, O., ... Kim, J. S. (2018). Functional rescue of dystrophin deficiency in mice caused by frameshift mutations using campylobacter jejuni Cas9. *Mol Ther, 26*(6), 1529–1538. doi:10.1016/j.ymthe.2018.03.018.

Memi, F., Ntokou, A., & Papangeli, I. (2018). CRISPR/Cas9 gene-editing: Research technologies, clinical applications and ethical considerations. *Semin Perinatol, 42*(8), 487–500. doi:10.1053/j.semperi.2018.09.003.

Molina, R., Sofos, N., & Montoya, G. (2020). Structural basis of CRISPR-Cas Type III prokaryotic defence systems. *Curr Opin Struct Biol, 65*, 119–129. doi:10.1016/j.sbi.2020.06.010.

Murugan, K., Babu, K., Sundaresan, R., Rajan, R., & Sashital, D. G. (2017). The revolution continues: Newly discovered systems expand the CRISPR-Cas Toolkit. *Mol Cell, 68*(1), 15–25. doi:10.1016/j.molcel.2017.09.007.

Ramanan, V., Shlomai, A., Cox, D. B., Schwartz, R. E., Michailidis, E., Bhatta, A., ... Bhatia, S. N. (2015). CRISPR/Cas9 cleavage of viral DNA efficiently suppresses hepatitis B virus. *Sci Rep, 5*, 10833. doi:10.1038/srep10833.

Raposo, V. L. (2019). The first Chinese edited Babies: A leap of faith in science. *JBRA Assist Reprod, 23*(3), 197–199. doi:10.5935/1518-0557.20190042.

Robertson, J. A. (2001). Human embryonic stem cell research: Ethical and legal issues. *Nat Rev Genet, 2*(1), 74–78. doi:10.1038/35047594.

Saayman, S., Ali, S. A., Morris, K. V., & Weinberg, M. S. (2015). The therapeutic application of CRISPR/Cas9 technologies for HIV. *Expert Opin Biol Ther, 15*(6), 819–830. doi:10.1517/14712598.2015.1036736.

Salsman, J., & Dellaire, G. (2017). Precision genome editing in the CRISPR era. *Biochem Cell Biol, 95*(2), 187–201. doi:10.1139/bcb-2016-0137.

Sherkow, J. S. (2018). The CRISPR patent landscape: Past, present, and future. *CRISPR J, 1*, 5–9.

Shinwari, Z. K., Tanveer, F., & Khalil, A. T. (2018). Ethical issues regarding CRISPR mediated genome editing. *Curr Issues Mol Biol, 26*, 103–110. doi:10.21775/cimb.026.103.

Taghdisi, S. M., Ramezani, M., Alibolandi, M., Khademi, Z., Hajihasani, M. M., Alinezhad Nameghi, M., & Danesh, N. M. (2022). A highly sensitive fluorescent aptasensor for detection of prostate specific antigen based on the integration of a DNA structure and CRISPR-Cas12a. *Anal Chim Acta, 1219*, 340031. doi:10.1016/j.aca.2022.340031.

Townsend, B. A. (2020). Human genome editing: How to prevent rogue actors. *BMC Med Ethics, 21*(1), 95. doi:10.1186/s12910-020-00527-w.

Vassy, J. L., Christensen, K. D., Schonman, E. F., Blout, C. L., Robinson, J. O., Krier, J. B., ... MedSeq, P. (2017). The impact of whole-genome sequencing on the primary care and outcomes of healthy adult patients: A pilot randomized trial. *Ann Intern Med, 167*(3), 159–169. doi:10.7326/M17-0188.

Wang, K. C., & Chang, H. Y. (2018). Epigenomics: Technologies and applications. *Circ Res, 122*(9), 1191–1199. doi:10.1161/CIRCRESAHA.118.310998.

Wee, S. L. (2019). Chinese scientist who genetically edited babies gets 3 years in prison. *New York: Times*.

Wong, T. W. Y., & Cohn, R. D. (2017). Therapeutic applications of CRISPR/Cas for duchenne muscular dystrophy. *Curr Gene Ther, 17*(4), 301–308. doi:10.2174/1566523217666171121165046.

Yang, W., Tu, Z., Sun, Q., & Li, X. J. (2016). CRISPR/Cas9: Implications for modeling and therapy of neurodegenerative diseases. *Front Mol Neurosci, 9*, 30. doi:10.3389/fnmol.2016.00030.

Yuan, X., Lin, H., & Fan, H. (2015). Efficacy and immunogenicity of recombinant swinepox virus expressing the A epitope of the TGEV S protein. *Vaccine, 33*(32), 3900–3906. doi:10.1016/j.vaccine.2015.06.057.

Zou, Q., Wang, X., Liu, Y., Ouyang, Z., Long, H., Wei, S., ... Gao, X. (2015). Generation of gene-target dogs using CRISPR/Cas9 system. *J Mol Cell Biol, 7*(6), 580–583. doi:10.1093/jmcb/mjv061.

3 Applying a Risk-Based Approach in the Development and Manufacture of Advanced Therapy Medicinal Products

Bridget Heelan
Technical, Independent Consultant, London, UK

CONTENTS

Introduction 27
References 32

INTRODUCTION

The risk-based approach (RBA) for the presentation of data in the Marketing Authorization Application (MAA) dossier to the European Medicines Agency (EMA) is specific to the EMA for Advanced Therapy Medicinal Products (ATMP). The RBA provides a guide as to where in the MAA dossier each risk factor and risk are discussed within the dossier and assists the assessors in reviewing the data and justifications provided. While the guidance is for an MAA submission and whether developers use it or not is optional, it provides a useful framework throughout the development of an ATMP. In the United States, the US Food and Drug Administration (US FDA) does not use the term ATMP but is very active in this field with a range of guidances for cell and gene therapies (https://www.fda.gov/vaccines-blood-biologics/biologics-guidances/cellular-gene-therapy-guidances). For all regulators, the technical aspects of addressing the risk factors of the products and risks to the patients are integral to clinical development, but in this chapter, we discuss the RBA as defined and expected by the EMA.

In Europe, an ATMP is defined as a biological medicinal product [Directive 2001/83/EC, amended by 2003/63/EC Annex 1, Part I] that can be classified as either a Gene Therapy Medicinal Product (GTMP), a Somatic Cell Therapy Medicinal Product (CTMP), a Tissue Engineered Product (TEP) or any combination of the three.

ATMPs present additional complexities during development compared with chemical products and non-ATMP biologics such as monoclonal antibodies. These additional uncertainties about the product itself and the limited translatability from nonclinical studies mean that all data available needs to be reassessed throughout development in order to ensure that every effort is made to mitigate risk to human subjects. To accommodate these challenges and to create flexibility, the European Union introduced the RBA (Kooijman et al., 2013). This approach provides the possibility of omitting guideline-based studies based on risk analyses, as the standards required for a non-ATMP product may not apply fully to ATMPs, and therefore justification for deviations from a standard MAA dossier is often required.

The nature of the risks to patients from ATMPs can be different from those typically associated with other types of pharmaceuticals. The first notable example of this was the death in 1999 of an 18-year-old subject with ornithine transcarbamylase (OTC) deficiency from an overwhelming immune response following gene therapy based on human adenovirus type 5 containing the human OTC cDNA (reviewed by Raper et al., 2003). This finding highlighted the limitations of animal studies with ATMPS, the steep toxicity curve for adenovirus vectors, intersubjects variability in immune responses, and the need to address underlying immunogenicity to vectors

Additional examples of severe adverse events with gene therapy were late-onset T-cell leukemia in subjects who received gene therapy (GT) products for severe combined immunodeficiency (SCID) (reviewed by Fischer & Hacein-Bey-Abina, 2020).

Similarly, serious risks from cell therapy were also described. Following treatment with intracerebellar and intrathecal injection of human fetal neural stem cells, a boy with ataxia telangiectasia was diagnosed with a multifocal brain tumor. Molecular and cytogenetic studies showed that the tumor was of nonhost origin suggesting it was derived from the transplanted neural stem cells (Amariglio et al., 2009). More recently Kuriyan et al. (2017) reported on three patients who developed severe bilateral visual loss after they received bilateral intravitreal injections of autologous adipose tissue–derived stem cells at a clinic in the United States, without FDA oversight. The patients' severe visual loss highlights the need for oversight of such clinics, and such an outcome would not be anticipated in a well-conducted regulatory approved trial, but it nonetheless shows how strict regulation of ATMP studies is needed to ensure patient safety.

In view of risks with cell and GT and to assist developers in optimizing their approach and data submission package at the time of MAA, in 2013 the EMA released the guideline on the RBA according to Annex I, part IV of Directive 2001/83/EC applied to ATMPs.

It is important to note that the application of the RBA in the preparation of an MAA dossier is optional.

Applying a Risk-Based Approach

The RBA is defined as

a strategy aiming to determine the extent of quality, nonclinical and clinical data to be included in the MAA, in accordance with the scientific guidelines relating to the quality, safety and efficacy of medicinal products and to justify any deviation from the technical requirements as defined in Annex I, part IV of Directive 2001/83/EC.

The methodological application of the RBA provided in this guideline is based on the identification of risks and associated risk factors of an ATMP and the establishment of a specific profile for each risk. The RBA should start at the beginning of product development and mature as the knowledge of the product and its characteristics increases.

A **risk** is defined as a potential unfavorable effect that can be attributed to the clinical use of the ATMP and is of concern to the patient and/or to other populations (e.g., caregivers and offspring).

A **risk factor** is defined as a qualitative or quantitative characteristic that contributes to a specific risk following the handling and/or administration of an ATMP.

Risk profiling is defined as a methodological approach to systematically integrate all available information on risks and risk factors in order to obtain a profile of each individual risk associated with a specific ATMP.

The methodological application of the RBA consists of four steps.

1st step: To identify risks associated with the clinical use of the ATMP
2nd step: To identify product-specific risk factors contributing to each identified risk
3rd step: To map the relevant data for each identified risk factor against each of the identified risks
4th step: To conclude on the risk factor – risk relationships

In order to evaluate the contribution of each risk factor to an identified risk (Step 3), the relevant data and justifications (and location in the MAA dossier) regarding each risk factor should be mapped with the help of a two-dimensional table. Examples (nonexhaustive fictitious examples) of two-dimensional tables are provided in the annex of the guideline.

To conclude on the risk factor – risk relationships (Step 4) a narrative text addressing risk-risk factor combinations identified to be of relevance should be composed. Risk factor-risk combinations for which a reasonable relationship has been identified need to be further detailed in respect to their causative scientific relationship – an overview of studies that have been performed to determine the impact of the identified risk factors on the particular risk. Where such studies have been omitted, a scientifically sound justification is needed as to why the quality, nonclinical and/or clinical data are not needed – a conclusion whether the provided scientific data (quality, nonclinical and clinical) and/or published information addressing the individual risk factor-risk combinations are considered adequate and sufficient to support an MAA. The guideline expects that on completion of the profiling of the identified risk-risk factor combinations a specific profile for each risk can be concluded.

Reviewing the example table in the guidelines for the risk of immunogenicity for GTMPs, the risk factors include, poor predictive nature of nonclinical data, vector types (OTC example; Raper et al., 2003), level of impurities, immunogenicity of the transgene itself (higher potential risk in patients with no endogenous component, very high risk in patients with an endogenous component that is nonredundant such that cross immunogenicity could lead to severe consequences both for the patients and the ability for further treatment; mode of administration – more likely to be immunogenic if associated with local inflammation or high doses intramuscularly; biodistribution to nontarget sites, which may be more likely to result in an immune response; dysfunctional transgene proteins may stimulate an immune response).

Each of the risk factors that are considered relevant to the GTMP should be assessed and addressed in the relevant section of the MAA dossier with appropriate justifications and data as per Annex 1 (see Table 3.1 adapted from the Guideline Annex 1 tables for GT, where only the details of the risk of immunogenicity are included).

Considering the risk of tumorigenicity with CTMP, again the risk factors need to be addressed (example in Annex 1) such as the potential for growth and de-differentiation, purity of cell population, genetic stability, presence of growth factors, immune competence of patients and need for immunosuppression.

Each of the risk factors that are considered relevant to the CTMP should be assessed and addressed in the relevant section of the MAA dossier with appropriate justifications and data to demonstrate that the possible risk factors associated with the product have been addressed by taking each of the risk factors in the product and the data available from nonclinical to apply to the safe administration to the patient population of interest as per the Annex 2 (see Table 3.2, adapted from the Guideline Annex 1 tables for human embryonic stem cell-derived cells secreting bioactive substances injected into the Central Nervous System (CNS), where only the details of the risk of tumor formation are included).

While the application of this RBA is considered optional by the EMA in terms of presenting the RBA at the time of MAA, in the author's opinion, this approach to assess risks and risk factors is very useful for ATMP developers throughout the product cycle from initial development, in particular for those new to the field such as small medium enterprises, hospitals and academic researchers. The technical considerations of RBA (which will be specific to the product and patients for each ATMP), which will likely include other risk factors and different risks that are not included in this guideline (as it is not a technical guideline), should be instituted early in development and updated throughout the life cycle as more information and knowledge are gained. It is advised also to seek scientific advice during development to ensure that the data and mitigations for risk factor-risk combinations for which a relationship exists are deemed adequate by regulators for progression and submission of the final application for approval.

TABLE 3.1
Example: Adeno-associated Virus (AAV) Vector Expressing the Human Fictionase Enzyme (FE) Administered i.m. for the Treatment of FE Deficiency Disease

Risk Factor \ Risk	Unwanted Immunogenicity
Recombination/ mobilization	Recombination/mobilization may lead to increased immunogenicity due to higher number of vector/replication competent virus (RCV) particles. Addressed in Common Technical Document (CTD) 3.2.P.5 – Control of Final Product (FP) and CTD 4.2.3 –Toxicology.
Type of transgene and transgene expression levels	The therapeutic gene is of human origin, and the respective endogenous gene product in patients is present but defective. This might cause unwanted immunogenicity. Expression of therapeutic protein addressed and justified in CTD 5.3.5 – Reports of efficacy and safety studies.
Vector type	AAV is known to be immunogenic. Addressed in immunogenicity and toxicity studies (CTD 4.2.3) and clinical safety studies (CTD 5.3.5 – Reports of efficacy and safety studies).
Impurities	AAV can be difficult to purify. Amount and type of impurities may lead to immunogenic reactions. Addressed in CTD 3.2.S.2 (Manufacture), 3.2.S.4 (Control of Drug Substance (DS)), 4.2.3 (Toxicology) and 5.3.5 – Reports of efficacy and safety studies.
Biodistribution	Biodistribution of the vector to nontarget, immunogenic sites. Addressed in biodistribution/immunogenicity studies – CTD 4.2.2 – pharmacokinetics (biodistribution), CTD 4.2.3 – toxicology (immunogenicity), CTD 5.3.5 – Reports of efficacy and safety studies (clinical safety).
Relevance of animal model	Animal model is not predictive for immunogenicity in patients due to differences in immune responses. An additional animal model to address immunogenicity was used. Addressed in CTD 4.2.3 – toxicology (immunogenicity) and in clinical studies CTD 5.3.5 – Reports of efficacy and safety studies.
Patient-related	Immune reaction might be triggered dependent on immune status of the patient. Addressed in nonclinical studies using vector-pretreated animals (CTD 4.2.3 – Toxicology) and in CTD 5.3.5 – Reports of efficacy and safety studies (clinical safety).
Disease-related	Variable levels of dysfunctional protein may be expressed in the patients resulting in immune reactions to the therapeutic protein. Addressed in CTD 5.3.5 – Reports of efficacy and safety studies.
Medical procedure–related	A high local dose administered i.m. might cause local inflammatory response due to immunreaction to a vector component or the expressed therapeutic protein. Addressed in CTD 4.2.3 – Toxicology and 5.3.5 – Reports of efficacy and safety studies.

Adapted from Guideline Annex 1 to address only one risk of immunogenicity.

TABLE 3.2
Example: Human Embryonic Stem Cell-Derived Cells Secreting Bioactive substances Injected into the CNS

Risk Factor	Tumor Formation
Cell starting material	Human embryonic stem cells (hESC) have inherent capability for teratoma formation. Risk addressed in other sections of this table and in CTD 3.2.S.2.3 – Control of Materials.
Culture/feeder cells and growth factors	Culture with Growth Factor (GFs) or hormones to enhance proliferation/trigger differentiation may induce tumor formation. Process-related impurities controlled – CTD 3.2.S.2.3 – Control of materials; 3.2.S.2.5 – Process validation and/or evaluation; 3.2.S.4 – Control of AS.
Cell population, heterogeneity and differentiation potential	Undifferentiated and undesirable lineage commitment cells resulting from nonsynchronized differentiation. Product-related impurities controlled CTD 3.2.S.2.5 – Process validation and/or evaluation; 3.2.S.4 – Control of AS.
Genetic stability	Genetic instability is associated with tumorigenicity. Genetic stability tested. CTD 4.2.2.3 – Pharmacokinetics – Distribution (in vivo tumorigenicity study).
Biodistribution	Tumor formation in different organs – Biodistribution study CTD 4.2.2.3 – Pharmacokinetics – Distribution.
Relevance of animal model	Age, dosing, immunocompetence and duration of the available animal study may not be appropriate for detection of tumor formation – In vivo tumorigenicity study CTD 4.2.3.4 – Toxicology – Carcinogenicity.
Patient-related	Malignancy in patients may result from the patient's medical/treatment history, including age and immunosuppressive status. Reports of efficacy and safety studies and post-marketing studies. CTD 5.3. – Clinical study reports.
Medical procedure–related – concomitant treatment	Risk for tumor formation due to previous use of immune suppressants. In vivo tumorigenicity study. Safety Adverse Events (AEs) reported in CTD 2.5 – Clinical overview, CTD 2.7 – Clinical summary, CTD 5.3 – Clinical study reports.
Medical procedure–related – mode of administration (injection into the brain)	Potential tumor formation at local and/or distant sites resulting from administration procedure. In vivo tumorigenicity study CTD 4.2.3.4 – Toxicology – Carcinogenicity, CTD 5.3 – Clinical study reports and CTD 5.4.

Adapted from Guideline Annex 2 to address only one risk of immunogenicity.

REFERENCES

Amariglio, N., Hirshberg, A., Scheithauer, B. W., Cohen, Y., Loewenthal, R., Trakhtenbrot, L., ... & Rechavi, G. (2009). Donor-derived brain tumor following neural stem cell transplantation in an ataxia telangiectasia patient. *PLoS Medicine*, 6(2), e1000029.

Commission Directive 2009/120/EC of 14 September 2009 amending Directive 2001/83/EC of the European Parliament and of the Council on the Community code relating to medicinal products for human use as regards advanced therapy medicinal products. (2009). *Official Journal of the European Union*, L242, 3–12.

Fischer, A., & Hacein-Bey-Abina, S. (2020). Gene therapy for severe combined immunodeficiencies and beyond. *Journal of Experimental Medicine*, 217(2), e20190607.

Guideline on the risk-based approach according to annex I, part IV of Directive 2001/83/EC applied to Advanced therapy medicinal products. EMA/CAT/CPWP/686637/2011 (Agency EM, London, 2013). https://www.ema.europa.eu/en/documents/scientific-guideline/guideline-risk-based-approach-according-annex-i-part-iv-directive-2001/83/ec-applied-advanced-therapy-medicinal-products_en.pdf

Kooijman, M., Van Meer, P. J. K., Gispen-de Wied, C. C., Moors, E. H. M., Hekkert, M. P., & Schellekens, H. (2013). The risk-based approach to ATMP development–generally accepted by regulators but infrequently used by companies. *Regulatory Toxicology and Pharmacology*, 67(2), 221–225.

Kuriyan, A. E., Albini, T. A., Townsend, J. H., Rodriguez, M., Pandya, H. K., Leonard, R. E., … & Goldberg, J. L. (2017). Vision loss after intravitreal injection of autologous "stem cells" for AMD. *New England Journal of Medicine*, 376(11), 1047–1053.

Raper, S. E., Chirmule, N., Lee, F. S., Wivel, N. A., Bagg, A., Gao, G. P., … & Batshaw, M. L. (2003). Fatal systemic inflammatory response syndrome in a ornithine transcarbamylase deficient patient following adenoviral gene transfer. *Molecular Genetics and Metabolism*, 80(1–2), 148–158.

4 Gene Therapy Products
Basics and Manufacturing Considerations

Humberto Vega-Mercado
Bristol Myers Squibb, Cell Therapy Operations, Global Manufacturing Sciences and Technology, Warren, USA

CONTENTS

Introduction 35
Manufacturing Process 40
Quality Control 43
Final Remarks 45
References 45

INTRODUCTION

Gene therapies are used for transferring genes to express proteins that are not present in cells or the insertion of therapeutic genes at specific targets in the human genome. The work in this field dates to the 1880s when Robert Koch confirmed the relationship between anthrax and the bacterium *Baciilus anthracis* (Koch, 1877), which later was linked to two plasmids in the organism required for its virulence (Little and Ivins, 1999).

The term "plasmid," initially proposed by J. Lederberg (1952), refers to extranuclear structures that can autonomously reproduce. The work by Lederberg and Tatum (1946) paved the way for additional work resulting in the identification of an agent, described as sex factor F, in an *Escherichia coli* strain, as well as the discovery of extrachromosomal antibiotic resistance (R) factors and colicinogenic (Col) plasmids (Fredcricq, 1969; Watanabe, 1963). These factors are responsible for the transmissibility of multiple antibiotic resistance among the enterobacterial. The study of these plasmids contributed to today's understanding of the mechanics of chromosomal gene transfer from donor to recipient microorganisms. Further, Jacob, Brenner, and Cuzin (1963) analyzed the replication and conjugal transfer properties of the F factor, which led to the development of the concept of plasmid replicons (e.g., the entire region of DNA that is independently replicated from a single origin of replication).

Meanwhile, Jacob and Wollman (1958) introduced the term "episome" when referring to extrachromosomal genetic material that may replicate autonomously

DOI: 10.1201/9781003285069-4

or become integrated into the chromosome. However, its uses have evolved and been replaced by the term "plasmid"; specifically, the term episome is used in the context of prokaryotes to refer to a plasmid that is capable of integrating into the chromosome (Hayes, 1969; Van Craenenbroeck et al., 2000). These integrative plasmids may be replicated, and they may exist as independent plasmid molecules. Conversely, Colosimo et al. (2000) use the term episome when referring to nonintegrated extrachromosomal closed circular DNA molecules that may be replicated in the nucleus of eukaryotes cells. Viruses are the most common examples of this, such as herpesviruses, adenoviruses and polyomaviruses. Other examples include aberrant chromosomal fragments, such as double-minute (DM) chromosomes, that can arise during artificial gene amplifications or in pathologic processes (e.g., cancer cell transformation) as discussed by Caroll et al. (1988), Van Roy et al. (2006), Stephen et al. (2011), L'Abbate et al. (2014) and Shimizu (2021).

A vector (e.g., plasmid, cosmid or Lambda phages) is used as a vehicle to carry a foreign nucleic sequence (e.g., usually DNA) into another cell where it can be replicated and/or expressed (NHGRI, 2022). A vector containing foreign DNA is termed "recombinant DNA." The four major types of vectors are plasmids, viral vectors, cosmids, and artificial chromosomes. The most commonly used vectors are plasmids.

Knowledge of plasmids, episomes and vectors has resulted in the development of artificial plasmids that are used as vectors in genetic engineering (e.g., transduction of cells or gene editing). These plasmids are now used to clone and amplify or express genes (Russell and Sambrook, 2001), and they offer several advantages for synthetic biology applications due to their modular structure, manipulation and large gene dosage capacity, as discussed by Oliveira (2009) and Oliveira and Mairhofer (2013).

On the subject of artificial plasmids, viral and nonviral vectors (i.e., lentiviral vectors (LVV) and adeno-associated virus (AAV) vectors) are frequently used for gene and cell therapies. These have been developed and evaluated for their efficiency of transduction, sustained expression of the transgene, and safety (Miyoshi et al., 1998). Retroviruses are efficient tools for gene transfer due to several reasons (Elsner & Bohne, 2017; Munis, 2020):

- Integrate their genetic information stably into the host-cell chromosomes by the conversion of the RNA genome into double-stranded DNA by reverse transcription.
- Target cell specificity is changed by the incorporation of heterologous envelope proteins during vector production.
- Feasibility of generating hybrid vectors by incorporating elements from various viruses.
- All retrovirus genera can result in potential gene therapy vectors. Existing research has focused on three viruses: (a) gammaretroviruses (murine leukemia virus, MLV), (b) lentiviruses (human immunodeficiency virus, HIV) and (c) spumaviruses (human foamy virus, HFV). Human immunodeficiency virus type 1 (HIV-1)–based vectors are widely used due to their ability to infect both dividing and nondividing cells.

However, among potential problems with retrovirus vectors is the requirement for the proliferation of the target cells for integration, limiting their use for gene transfer into nondividing cells such as hepatocytes, myoblasts, neurons and hematopoietic stem cells as summarized by Miyoshi et al. (1998). In contrast, lentiviruses such as HIV-1 can infect nondividing cells as discussed by Weinberg et al. (1991), Lewis et al. (1992) and Bukrinsky et al. (1992). Significant progress has been reported in the production of viruses with enhanced biosafety by means of a third-generation lentiviral packaging system where HIV accessory genes are eliminated and packaging elements are separated into different plasmids (Lee & Cobrinik, 2020). Specifically, Lee and Cobrinik (2020) reported that the viral yield and transducing efficiency of the third-generation packaging system were significantly improved by replacing the Rous Sarcoma Virus (RSV) promoter of the vector pLKO.1 with the Cytomegalovirus (CMV) promoter and optimizing the ratio of viral packaging components. They suggest in their report the method to improve transduction efficiency while retaining the significant biosafety features of the third-generation lentiviral packaging system. pLKO.1 is one of the most widely used lentiviral vectors for short hairpin RNA (shRNA) expression due to its transduction efficacy, low levels of recombination, high-level shRNA expression and use in shRNA libraries (Moffat et al., 2006)

AAV is a nonenveloped virus that can be engineered to deliver DNA to target cells (Naso et al., 2017). AAV is a protein shell surrounding and protecting a small, single-stranded DNA genome. AAV belongs to the parvovirus family and is dependent on co-infection with other viruses, mainly adenoviruses, in order to replicate. It was initially discovered as a contaminant of adenovirus preparations (Rose et al., 1966; Hastie & Samulski, 2015). The use of AAV is proven to be safe for gene therapies due to the ability to generate recombinant AAV particles lacking any viral genes and containing DNA sequences of interest for therapeutic applications. The design of an rAAV vector should consider the packaging size of the expression cassette that will be placed between the two inverted terminal repeats (ITRs) required for genome replication and packaging. Dong et al. (2010) suggest as generally accepted that anything under 5kb (including the viral ITRs) is sufficient in size. The identification of various capsid sequences has resulted from work on cloning AAV variants from human and primate tissues. These capsid sequences are intended to deliver genes to selective cell and tissue types (Limberis et al. 2009). Kotterman and Schaffer (2014) and Buning et al. (2015) discuss the utilization of recombinant techniques involving capsid shuffling, directed evolution and random peptide library insertions for the development of AAVs with unique attributes. In addition, these techniques let to novel AAV variants with reduced sensitivities to neutralizing antibodies (Naso et al., 2017)

Table 4.1 provides a list of additional references covering the development work of plasmids, episomes and vectors since 1943.

The goal of this chapter is to provide a high-level summary of the basic considerations for the manufacture of plasmids and vectors used in gene and cell therapies by means of a historical review of developmental work for the manufacture of plasmids and vectors and general process considerations, equipment and steps required for them.

TABLE 4.1
Summary of Scientific Work Involving Plasmids and Vectors (Little & Ivins, 1999; Lederberg & Tatum, 1946; Watanabe, 1963; Fredcricq, 1969; Jacob et al., 1963; Caroll et al., 1988; Van Roy et al., 2006; Stephen et al., 2011; L'Abbate et al., 2014; Elsner & Bohne, 2017; Naso et al., 2017; Shimizu, 2021)

Year	Scientific Work Involving Plasmids and Vectors
1943	Luria, S. E., & Delbrück, M. (1943). Mutations of bacteria from virus sensitivity to virus resistance. *Genetics*, 28, 491.
1945	Tatum, E. L. (1945). X-ray induced mutant strains of Escherichia coli. *Proc. Nat. Acad. Sci.*, 31, 215.
1953	Crick, F.H.C., & Watson, J. D. (1953). Molecular structure of nucleic acids: A structure for deoxyribose nucleic acid. *Nature*, 171, 737–738.
1958	Meselson, M., & Stahl, F. W. (1958, July). The replication of DNA in Escherichia coli. Proceedings of the National Academy of Sciences of the United States of America, 44(7), 671–682.
1968	Hayes, W. (1968). *The genetics of bacteria and their viruses* (2nd ed.). New York: John Wiley & Sons Inc.
1981	Baskin, F., Rosenberg, R. N., & Dev, V. (1981). Correlation of double-minute chromosomes with unstable multidrug cross-resistance in uptake mutants of neuroblastoma cells. *Proc Natl Acad Sci USA*, 78(6), 3654–3658.
1982	Barker, P. E. 1982. Double minutes in human tumor cells. *Cancer Genetics and Cytogenetics*, 5(1), 81–94.
	Cowell, J. K. (1982.) Double minutes and homogenously staining regions: Gene amplification in mammalian cells. *Annu. Rev. Genet.*, 16, 21–59.
	Varmus, H. E. (1982). Form and function of retroviral proviruses. *Science*, 216, 812–820.
1987	Murray, A., & Szostak, J. (1987). Artificial chromosomes. *Scientific American*, 257(5), 62–68
1989	Iman, D. S., & Shay, J. W. (1989, August 15). Modification of myc gene amplification in human somatic cell hybrids. *Cancer Research*, 49(16), 4417–4422.
	Wahl, G. M. (1989). The importance of circular DNA in mammalian gene amplification. *Cancer Res.*, 49, 1333–1340.
1990	VanDevanter, D. R., Piaskowski, V. D., Casper, J. T., Douglass, E.C., & Von Hoff, D. D. (1990). Ability of circular extrachromosomal DNA molecules to carry amplified MYCN proto-oncogenes in human neuroblastomas in vivo. *J. Natl. Cancer Inst.*, 82, 1815–1821
1992	Von Hoff, D. D., McGill, J. R., Forseth, B. J., Davidson, K.K., Bradley, T. P., Van Devanter, D. R., & Wahl, G. M. (1992, September 1). Elimination of extrachromosomally amplified MYC genes from human tumor cells reduces their tumorigenicity. *Proceedings of the National Academy of Sciences of the United States of America*, 89(17), 8165–8169.
1998	del Solar, G., Giraldo, R., Ruiz-Echevarría, M. J., Espinosa, M., & Díaz-Orejas, R. (1998, June). Replication and control of circular bacterial plasmids. *Microbiology and Molecular Biology Reviews*, 62(2), 434–464.
1999	Russell, S. J., & Cosset, F. L. (1999). Modifying the host range properties of retroviral vectors. *J Gene Med.*, 1, 300–311.
2001	de Noronha, C. M., Sherman, M. P., Lin, H. W., Cavrois, M. V., Moir, R. D., Goldman, R. D., & Greene, W. C. (2001). Dynamic disruptions in nuclear envelope architecture and integrity induced by HIV-1 Vpr. *Science*, 294, 1105–1108.

(Continued)

TABLE 4.1 (CONTINUED)
Summary of Scientific Work Involving Plasmids and Vectors (Little & Ivins, 1999; Lederberg & Tatum, 1946; Watanabe, 1963; Fredcricq, 1969; Jacob et al., 1963; Caroll et al., 1988; Van Roy et al., 2006; Stephen et al., 2011; L'Abbate et al., 2014; Elsner & Bohne, 2017; Naso et al., 2017; Shimizu, 2021)

Year	Scientific Work Involving Plasmids and Vectors
2002	Tagomori, K., Iida, T., & Honda, T. (2002). Comparison of genome structures of vibrios, bacteria possessing two chromosomes. *Journal of Bacteriology*, 184(16), 4351–4358.
	Ried, M. U., Girod, A., Leike, K., Buning, H., & Hallek, M. (2002). Adeno-associated virus capsids displaying immunoglobulin-binding domains permit antibody-mediated vector retargeting to specific cell surface receptors. *J Virol*, 76(9), 4559–4566.
2004	Warrington, K. H., Jr, Gorbatyuk, O. S., Harrison, J. K., Opie, S. R., Zolotukhin, S., & Muzyczka, N. (2004). Adeno-associated virus type 2 VP2 capsid protein is nonessential and can tolerate large peptide insertions at its N terminus. *J Virol.*, 78(12), 6595–6609.
2005	Lewin, A., Mayer, M., Chusainow, J., Jacob, D., & Appel, B. (2005, June). Viral promoters can initiate expression of toxin genes introduced into Escherichia coli. *BMC Biotechnology*, 5, 19.
2006	Choi, V. W., McCarty, D. M., & Samulski, R. J. (2006). Host cell DNA repair pathways in adeno-associated viral genome processing. *J Virol.*, 80(21), 10346–10356.
2007	Shimizu, N., Hanada, N., Utani, K., & Sekiguchi, N. (2007). Interconversion of intra- and extra-chromosomal sites of gene amplification by modulation of gene expression and DNA methylation. *Journal of Cellular Biochemistry*, 102(2), 515–529.
2009	Morange, M. (2009). What history tells us XIX. The notion of the episome. *Journal of Biosciences*, 34(6), 845–848.
2010	Oliveira, P. H., Prather, K. J., Prazeres, D. M., & Monteiro, G. A. (2010). Analysis of DNA repeats in bacterial plasmids reveals the potential for recurrent instability events. *Applied Microbiology and Biotechnology*, 87(6), 2157–2167
	Storlazzi, C. T., Lonoce, A., Guastadisegni, M. C., Trombetta, D., D'Addabbo, P., Daniele, G., L' Abbate, A., Macchia, G., Surace, C., Kok, K., Ullmann, R., Purgato, S., Palumbo, O., Carella, M., Ambros, P. F., & Rocchi, M. (2010). Gene amplification as double minutes or homogeneously staining regions in solid tumors: origin and structure. *Genome Research*, 20(9), 1198–1206.
	Suerth, J. D., Maetzig, T., Galla, M., Baum, C., & Schambach, A. (2010). Self-inactivating apharetroviral vectors with a split-packaging design. ASM Journals. *J. of Virology*, 84(13), 6626–6635.
2011	Utani, K., Okamoto, A., & Shimizu, N. (2011). Generation of micronuclei during interphase by coupling between cytoplasmic membrane blebbing and nuclear budding. *PLoS ONE*, 6, e27233.
2013	Munch, R. C., Janicki, H., Volker, I., Rasbach, A., Hallek, M., Buning, H., ... et al. (2013). Displaying high-affinity ligands on adeno-associated viral vectors enables tumor cell-specific and safe gene transfer. *Mol Ther.*, 21(1), 109–118.

(*Continued*)

TABLE 4.1 (CONTINUED)
Summary of Scientific Work Involving Plasmids and Vectors (Little & Ivins, 1999; Lederberg & Tatum, 1946; Watanabe, 1963; Fredcricq, 1969; Jacob et al., 1963; Caroll et al., 1988; Van Roy et al., 2006; Stephen et al., 2011; L'Abbate et al., 2014; Elsner & Bohne, 2017; Naso et al., 2017; Shimizu, 2021)

Year	Scientific Work Involving Plasmids and Vectors
2014	Damdindorj, L., Karnan, S., Ota, A., Hossain, E., Konishi, Y., Hosokawa, Y., & Konishi H. (2014). A comparative analysis of constitutive promoters located in adeno-associated viral vectors. *PLOS ONE*, 9(8), e106472.
	Gonçalves, G. A., Oliveira, P. H., Gomes, A. G., Prather, K. L., Lewis, L. A., Prazeres, D. M., & Monteiro, G. A. (2014). Evidence that the insertion events of IS2 transposition are biased towards abrupt compositional shifts in target DNA and modulated by a diverse set of culture parameters. *Applied Microbiology and Biotechnology*, 98(15), 6609–6619.
	Johnston, C., Martin, B., Fichant, G., Polard, P., & Claverys, J. P. (2014, March). Bacterial transformation: Distribution, shared mechanisms and divergent control. *Nature Reviews. Microbiology*, 12(3), 181–196.
	Tanaka, S. S., Mitsuda, S. H., & Shimizu, N. (2014). How a replication origin and matrix attachment region accelerate gene amplification under replication stress in mammalian cells. *PLoS ONE*, 9, e103439.
2017	diCenzo, G. C., & Finan, T. M. (2017). The divided bacterial genome: Structure, function, and evolution. *Microbiology and Molecular Biology Reviews*, 81(3), e00019-17.
	Shah, P., Wolf, K., & Lammerding, J. (2017). Bursting the bubble—nuclear envelope rupture as a path to genomic instability? *Trends Cell Biol.*, 27, 546–555.
2018	deCarvalho, A. C., Kim, H., Poisson, L. M., Winn, M. E., Mueller, C., Cherba, D., Koeman, J., Seth, S., Protopopov, A., Felicella, M., ... et al. (2018). Discordant inheritance of chromosomal and extrachromosomal DNA elements contributes to dynamic disease evolution in glioblastoma. *Nat. Genet.*, 50, 708–717.
2019	Morton, A. R., Dogan-Artun, N., Faber, Z. J., MacLeod, G., Bartels, C. F., Piazza, M. S., Allan, K. C., Mack, S. C., Wang, X., Gimple, R. C., ... et al. (2019). Functional enhancers shape extrachromosomal oncogene amplifications. *Cell*, 179, 1330–1341.e13.
	Shimizu, N., Kapoor, R., Naniwa, S., Sakamaru, N., Yamada, T., Yamamura, Y.K., & Utani, K.-I. (2019). Generation and maintenance of acentric stable double minutes from chromosome arms in inter-species hybrid cells. *BMC Mol. Cell Biol.*, 20, 2.
2020	Wei, J., Wu, C., Meng, H., Li, M., Niu, W., Zhan, Y., Jin, L., Duan, Y., Zeng, Z., Xiong, W., Li, G., & Zhou, M. (2020). The biogenesis and roles of extrachromosomal oncogene involved in carcinogenesis and evolution. *American Journal of Cancer Research*, 10(11), 3532–3550.

MANUFACTURING PROCESS

The manufacture of plasmids and vectors follows common principles such as aseptic operations or manipulations (e.g., use of biological safety cabinets, bioreactors, cell lines, cleanrooms – Class A, B and C, sterile welding and sealing, aseptic connectors). Also, in general, the manufacturing processes have common elements as described elsewhere (Huber et al., 2008). Figure 4.1 summarizes a typical manufacturing process for plasmids and vectors.

Gene Therapy Products

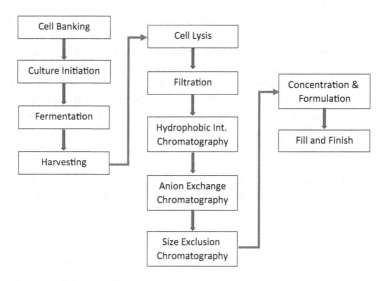

FIGURE 4.1 High-Level Summary of the Manufacturing Process of Plasmids and Vectors.

Each of the steps considered or required for the manufacture of plasmids and vectors are further discussed in the following section:

a) Vector design:
Viral vector design requires the identification of viral genes or elements needed for specific coding for the gene of interest (e.g., transgene delivery and deletion of the remaining sequences; Elsner & Bohne, 2017). Then, the essential genes for vector production are delivered on separate plasmids resulting in single-round infectious viral particles.

b) Cell banking:
Yield and quality of plasmid DNA (pDNA) are significantly influenced by the selected production host (e.g., *E. coli*). Careful selection of an efficient host strain is therefore a critical factor for pDNA production. Strains derived from *E. coli* K12 have been used for pDNA manufacturing. Although many host features are generic, specific plasmid-host combinations may show unexpected effects, and in some cases, host screening is recommended. It is reported for example that *E. coli* JM108 consistently shows superior performance in small- and large-scale cGMP manufacturing (Huber et al., 2008).

c) Fermentation:
Batch fermentation and high cell-density cultivation (HCDC) or fermentation are used in the manufacture of plasmids. A key feature of HCDC is a well-balanced culture medium that supports the predictable production of pDNA. This is achieved by using a synthetically defined culture medium while avoiding complex compounds like yeast extract

or soy peptone associated with varying quality, undefined chemical composition and potential challenging characteristics, such as foaming, dusting and clumping. The metabolic stress of the host strain as well as plasmid stability have been identified as two of the key parameters that greatly influence pDNA yields (Silva et al., 2011). In summary, Silva et al. (2011) reported that exponential feeding profiles, in contrast with those with constant feeding, showed higher biomass and plasmid yields while exhibiting lower plasmid stability and percentage of viable cells.

d) Harvesting:
Centrifugation is one of the preferred methods for harvesting at the lab scale; a large-scale centrifugation process can be cumbersome and provide low yield. Disk stack centrifuges operating at high speed with intermittent ejection gave supercoiled plasmid yields as low as 40% because of shear damage during discharge (Kong et al., 2008). Other harvesting technology such as microfiltration may be used considering 1,000 kDa screen membranes (e.g., tangential flow filtration) at low TMP and permeate control (MilliporeSigma, 2020).

e) Lysis:
The purpose of lysis is the release of pDNA from the biomass. The cells are lysed at alkaline conditions, followed by neutralization. The consequence is a precipitation and flocculation of host proteins and genomic DNA, whereby pDNA remains soluble in the supernatant. To avoid significant loss of plasmids, proper design of the mixing conditions is necessary, as mechanical stress leads to loss of supercoiled plasmids and to an increased level of host-related impurities, which results in reduced yield (Clemson & Kelly, 2003; Huber et al., 2008).

f) Filtration:
Microfiltration can be used to remove spent fermentation media from the host cells (Parker et al., 2020). It is important to keep in mind the shear and high viscosity, especially for high-concentration formulations. Operating parameters can change retention properties. A 100 kD molecular weight membrane can separate plasmid isoforms based solely on filtrate flux, as the sieving coefficient for each isoform is dependent on filtrate flux (Parker et al., 2020).

g) Purification (chromatography steps) and formulation:
Purification processes for plasmids may consider multiple chromatographic steps, such as the following:
- HIC – Hydrophobic interaction chromatography (HIC) is suitable as the initial (capture) step. The main consideration is the optimization of the binding and elution conditions.
- AEC – Anion-exchange chromatography (AEC) may be considered as a second (intermediate) step due to the negative charge of plasmids. In addition, endotoxin clearance, volume reduction and increase in plasmid concentration are attained during the AEC step.

Gene Therapy Products

- SE – Size exclusion (SE) chromatography may be used for buffer exchange and bulk formulation. Ultrafiltration can be used to further concentrate the resulting plasmid concentration from the SE step. It is important to consider the viscosity of the mixture during the concentration step.

h) Formulation and fill:
Plasmids are unstable in solution. The physical and chemical stability of plasmids is influenced by the formulation (e.g., excipients, dosage form) and storage conditions (Walther et al., 2003; Schleef & Schmidt, 2004). Walther et al. (2003) highlight that during storage of the supercoiled plasmids, the topoisoform associated with the highest transfection efficiency of the plasmids is mainly converted to open circular plasmids (single-strand breaks) and linear plasmids (double-stranded breaks), and its topology (and transfection efficiency) can be maintained when solutions are stored at −80°C. Meanwhile, Quaak et al. (2010) demonstrated that lyophilization and the use of lyoprotectants (e.g., disaccharides – sucrose and trehalose) can be used to increase product stability (i.e., transfection efficiency) and extend their shelf life.

i) Other considerations (e.g., raw materials, single-use assemblies):
Elverum and Whitman (2020) have highlighted the need for manufacturers to collaborate in defining consistent standards for incoming material quality and widespread data collection on patient safety within a therapeutic class. The engagement of cross-functional groups from early development will result in fast, cost-efficient and scalable processes.

Issues like single sourcing of critical materials and single-use assemblies represent a challenge to the supply chain. A collaborative effort from across the industry will help reduce the risks to supply, as well as expedite the development of new therapies. Therefore, a robust supply chain is important in addition to outstanding technologies and innovative strategies (Cytiva, 2022).

QUALITY CONTROL

In terms of quality control, the following assays and methods are commonly utilized in the manufacturing process of plasmids and vectors (Creative Biolabs, 2022):

a) Purity – UV spectrophotometry:
To ensure the final product does not contain impurities that jeopardize the functionality and transfection efficacy of the plasmid/vector.

b) Bioburden – Standard microbial testing:
The presence of microbial contamination may result in undesired cell material in the final product.

c) Appearance – Visual inspection:
The final product shall be free of particles, other contaminants or potential damage due to clumping of the plasmids/vector.

d) Concentration – UV spectrophotometry:
   Confirmation of the target amount of plasmid/vector is in the formulated product.
e) DNA homogeneity (densitometry):
   Method to assess an adequate homogeneity of the final product. Specifically, confirming that the molecules exist as the supercoiled covalently closed circular. This is the most compact form where the circular and covalently closed DNA helix is interwoven in itself.
f) Endotoxin – Kinetic turbidimetric limulus amebocyte lysate (LAL):
   Endotoxins are heat-stable toxins associated with the outer membranes of certain gram-negative bacteria (e.g., *Brucella, Neisseria and Vibro* species). Endotoxins are released when the cells are disrupted. This test determines the concentration of endotoxins in a sample.
g) Identity – EtBr – Stained agarose gel electrophoresis (AGE):
   AGE is a method used in biochemistry, molecular biology, genetics and clinical chemistry to separate a mixed population of macromolecules such as DNA or proteins in a matrix of agarose.
h) Mycoplasma – PCR:
   Mycoplasma is a genus of bacteria that, like the other members of the class Mollicutes, lack a cell wall around their cell membranes. Peptidoglycan (murein) is absent. This characteristic makes them naturally resistant to antibiotics that target cell wall synthesis (like the beta-lactam antibiotics). They can be parasitic or saprotrophic. The test determines the presence of mycoplasma in a sample.
i) pH – potentiometric:
   pH is used to express the acidity or alkalinity of a solution on a logarithmic scale on which seven is neutral, lower values are more acid and higher values are more alkaline.
j) Plasmid identity – double-stranded primer-walking sequencing:
   The goal of a diagnostic digest is to cut the plasmid into specific-sized pieces and analyze the resulting fragments by gel electrophoresis. The pattern of the fragments on the gel can indicate if the plasmid contains the expected size insert.
k) Residual host genomic DNA – PCR:
   Residual DNAs (rDNAs) are traces or low quantities of DNA originating from the organisms used in the production process. The rDNAs may be able to transmit viral infections or cause potential risk for oncogenesis or adverse reactions. The test determines the concentration of residual DNA in a sample.
l) Residual host protein – micro BCA:
   Host-cell proteins (HCPs) are process-related impurities derived from the host organism, which affect product quality and safety. The test determines the concentration of residual host proteins in a sample.

Gene Therapy Products 45

m) Residual host RNA – SYBR gold-stained AGE:
Residual host-cell RNA is a major impurity in current large-scale separation processes for the production of clinical-grade plasmid DNA. The test determines the concentration of residual RNA in a sample.

n) Restriction digest – EtBr-stained AGE:
Restriction digestion uses enzymes to cut up the plasmid. After overnight digestion, the reaction is stopped by the addition of a loading buffer. The fragments are separated by electrophoresis. The gel is then stained with an ethidium bromide to visualize the DNA bands. The test is used to visualize the fragments of DNA.

o) Sterility – USP<71>:
Testing to assess the potential presence of microbial contamination in a sample.

p) Volume in container – USP<1>:
Testing to confirm the actual content (e.g., volume) of the product in a container.

q) Particle matter in injections – USP<788>:
Visual test to identify the potential presence of particulate in a sample.

Proper qualification and validation of the methods are key in the characterization and routine testing of plasmids and vectors. Special attention to impurities as the transfection efficiency and overall functionality of plasmids may be affected.

FINAL REMARKS

The chapter summarizes the history of plasmids and vectors starting with the work from Koch (1877) later lined to plasmids by Little and Ivins (1999) and work conducted by multiple researchers throughout the years to develop artificial plasmids used for gene and cell therapies these days. Then, using that historical review, we have focused on the manufacture of vectors starting with the design of the plasmids and selection of a host, fermentation, purification, formulation, fill and general list of analytical methods commonly utilized in the manufacturing process of plasmids. The complexities around the manufacture of plasmids and the challenges associated to supply chains must be considered by manufacturers early in the process of designing and developing plasmids and vectors.

REFERENCES

Anderson, E. S., & Natkin E. (1972). Transduction of resistance determinants and R factors of the transfer systems by phage Plkc. *Mol Gen Genet*, 114, 261–265.

Bukrinsky, M. I., Sharova, N., Dempsey, M. P., Stanwick, T. L., Bukrinskaya, A. G., Haggerty, S., & Stevenson, M. (1992). Active nuclear import of human immunodeficiency virus type 1 preintegration complexes. *Proc Natl Acad Sci USA*, 89(14), 6580–6584.

Buning, H., Huber, A., Zhang, L., Meumann, N., & Hacker, U. (2015). Engineering the AAV capsid to optimize vector-host-interactions. *Curr Opin Pharmacol*, 24, 94–104.

Carroll, S. M., DeRose, M. L., Gaudray, P., Moore, C. M., Needham-Vandevanter, D. R., Von Hoff, D. D., & Wahl, G. M. (1988). Double minute chromosomes can be produced from precursors derived from a chromosomal deletion. *Mol Cell Biol*, 8(4), 1525–1533.

Clemson M., & Kelly, W. J. (2003). Optimizing alkaline lysis for DNA plasmid recovery. *Biotechnol Appl Biochem*, 37(Pt 3), 235–244.

Colosimo, A., Goncz, K. K., Holmes, A. R., Kunzelmann, K., Novelli, G., Malone, R. W., Bennett, M. J., & Gruenert, D. C. 2000. Transfer and expression of foreign genes in mammalian cells. *BioTechnique*, 29(2), 314–318, 320–322, 324.

Couturier, M., Bex, F., Bergquist, P. L., & Maas, W. K. (1988). Identification and classification of bacterial plasmids. *Microbiol Rev*, 52, 375–395.

Creative Biolabs. (2022). GMP plasmid manufacturing – Creative Biolabs. creative-biolabs.com

Cytiva. (2022). Cell therapy manufacturing: the supply chain challenge | Cytiva. cytivalifesciences.com

Dong, B., Nakai, H., & Xiao, W. (2010). Characterization of genome integrity for oversized recombinant AAV vector. *Mol Ther*, 18(1), 87–92.

Elsner, C., & Bohne, J. (2017). The retroviral vector family: Something for everyone. *Virus Genes*, 53, 714–722.

Elverum, K., & Whitman, M. (2020). Delivering cellular and gene therapies to patients: solutions for realizing the potential of the next generation of medicine. *Gene Ther*, 27, 537–544.

Falkow, S., Citarella, R. V., & Wohlhieter, J. A. (1966). The molecular nature of R-factors. *J Mol Biol*, 17, 102–116. PubMed.

Fredcricq, P. (1969). The recombination of colicinogenic factors with other episomes and plasmids. *In* Wolstenholme GE, & O'Connor M (eds), *Bacterial episomes and plasmids* (pp. 163–178). London: J. & A. Churchill Ltd.

Hastie, E., & Samulski, R. J. (2015). Adeno-associated virus at 50: A golden anniversary of discovery, research, and gene therapy success—a personal perspective. *Hum Gene Ther*, 26(5), 257–265.

Harada, K., Kameda, M., Suzuki, M., & Mitsuhashi, S. (1963). Drug resistance of enteric bacteria. IL Transduction of transmissible drug resistance R factors with phage epsilon. *J Bacteriol*, 86, 1332–1338.

Hayes, W. (1969). What are episomes and plasmids? In Wolstenholme GE and O'Connor M (eds.). *Bacterial Episomes and Plasmids* (pp. 4–8). CIBA Foundation Symposium.

Helinski, D. R. (2022). A brief history of plasmids. ECOSalPlus. Cellular and molecular biology of E.coli, Salmonella and the Enterobacteriaceae. *Am Soc Microbiol*. ecosalplus-0028-2021, 1–22.

Huber, H., Buchinger, W., Diewok, J., Ganja, R., Keller, D., Urthaler, J., & Necina R. (2008). Industrial manufacturing of plasmid DNA. *Genet Eng Biotechnol News*, 28(4). genengnews.com.

Jacob, F., & Wollman, E. L. (1958). Les épisomes, elements génétiques ajoutés. *Comptes Rendus de l'Académie des Sciences de Paris*, 247(1), 154–156.

Jacob, F., Brenner, S., & Cuzin, F. (1963). On the regulation of DNA replication in bacteria. *Cold Spring Harbor Symp Quant Bio*, 28, 329–348.

Jacob, F., & Wollman, E. L. (1961). *Sexuality and the genetics of bacteria*. Academic Press Inc., London, United Kingdom.

Kandavelou, K., & Chandrasegaran, S. (2008). Plasmids for gene therapy. *Plasmids: current research and future trends*. Edited by Georg Lipps. Poole, UK: Caister Academic Press.

Koch, R. (1877). The etiology of anthrax, based on the life history of *Bacillus anthracis*. *Beitriige Znr Biologic Dee Pflanzen*, 2, 277–310.

Kong, S., Rock, C. F., Booth, A., Willoughby, N., O'Kennedy, R., Relton, J., Ward, J. M., Hoare, M., & Levy, M. S. (2008). Large-scale plasmid DNA processing: Evidence that cell harvesting, and storage methods affect yield of supercoiled plasmid DNA. *Biotechnol Appl Biochem*, 51(1), 43–51.

Kornberg, A. (1960). Biologic synthesis of deoxyribonucleic acid. *Science*, 131, 1503–1508.

Kotterman, M. A., & Schaffer, D. V. 2014. Engineering adeno-associated viruses for clinical gene therapy. *Nat Rev Genet*, 15(7), 445–451.

L'Abbate, A., Macchio, G., D'Addabbo, P., Lonoce, A., Tolomeo, D., Trombetta, D., Kok, K., Bartenhagen, C., Whelan, C. W., Palumbo, O., Severgnini, M., Cifola, I., Dugas, M., Carella, M., De Bellis, G., Rocchi, M., Carbone, L., & Storlazzi, C. T. (2014, August 18). Genomic organization and evolution of double minutes/homogeneously staining regions with MYC amplification in human cancer. *Nucleic Acids Res*, 42(14), 9131–9145.

Lederberg, J., & Tatum, E. L. (1946). Gene recombination in *E. coli*. *Nature*, 158, 558–562.

Lederberg, J. (1952). Cell genetics and hereditary symbiosis. *Physiol Rev*, 32, 403–430.

Lee, S., & Cobrinik, D. (2020). Improved third-generation lentiviral packaging with pLKO.1C vectors. *BioTechniques*, 68(6), 349–352.

Lewis, P., Hensel, M., & Emerman, M. (1992). Human immunodeficiency virus infection of cells arrested in the cell cycle. *EMBO J*, Aug. 11(8), 3053–3058.

Limberis, M. P., Vandenberghe, L. H., Zhang, L., Pickles, R. J., & Wilson, J. M. (2009). Transduction efficiencies of novel AAV vectors in mouse airway epithelium in vivo and human ciliated airway epithelium in vitro. *Mol Ther*, 17(2), 294–301.

Little, S. F., & Ivins, B. E. (1999). Molecular pathogenesis of *Bacillus anthracis* infection. *Microbes Infect*, 1, 131–139.

MilliporeSigma. (2020). Application Note: Cell Harvest, Lysis, Neutralization & Clarification of Plasmid DNA. Lit. No. MS_AN7049EN Ver. 1.0. Burlington, MA, 018003. MERC200091_PRO_AppNote_Harvest_Clarification_MSIG.indd (sigmaaldrich.com).

Miyoshi, H., Blomer, U., Takahashi, M., Gage, F. H., & Verma, I. M. (1998). Development of a self-inactivating lentivirus vector. *ASM J Virol*, 72(10), 8150–8157.

Moffat, J., Grueneberg, D. A., Yang, X., ... et al. (2006). A lentiviral RNAi library for human and mouse genes applied to an arrayed viral high-content screen. *Cell*, 124(6), 1283–1298.

Munis, A. M. (2020). Gene therapy applications of non-human lentiviral vectors. *Viruses*, 12, 1106–1127. semanticscholar.org.

National Human Genome Research Institute (NHGRI). (2022). Vector (genome.gov). National Institutes of Health. Building 31, Room 4B09. 31 Center Drive, MSC 2152. 9000 Rockville Pike, Bethesda, MD. USA 20892-2152.

Naso, M. F., Tomkowicz, B., Perry-III, W. L., & Strohl, W. R. (2017). Adeno-associated virus (AVV) as a vector for gene therapy. *BioDrugs*, 31(4), 317–334.

Novick, R. P. (1987). Plasmid incompatibility. *Microbiol Rev*, 51, 381–395.

Oliveira, P. H., Prather, K. J., Prazeres, D. M., & Monteiro, G. A. (2009). Structural instability of plasmid biopharmaceuticals: challenges and implications. *Trends Biotechnol*, 27(9), 503–511.

Oliveira, P. H., & Mairhofer, J. (2013, September). Marker-free plasmids for biotechnological applications – implications and perspectives. *Trends Biotechnol*, 31(9), 539–547.

Parker, T., Cherradi, Y., & Mishra, N. (2020). Scalable purification of plasmid DNA: Strategies and Considerations for vaccines and gene therapy manufacturing. Millipore Sigma. Application Note MS-WP7159EN Ver. 1.0

Quaak, S.G.L., Haanen, J.B.A.G., Beijnen, J.H., & Nuijen, B. 2010. Naked plasmid DNA formulation: effect of different disaccharides on stability after lyophilization. *AAPS PharmSciTech*, 11(1), 344–350.

Russell, D. W., & Sambrook, J. (2001). *Molecular cloning: a laboratory manual.* Cold Spring Harbor, NY: Cold Spring Harbor Laboratory.

Rose, J. A., Hoggan, M. D., & Shatkin, A. J. (1966). Nucleic acid from an adeno-associated virus: chemical and physical studies. *Proc Natl Acad Sci USA*, 56(1), 86–92.

Schleef. M., & Schmidt, T. (2004). Animal-free production of ccc-supercoiled plasmids for research and clinical applications. *J Gene Med*, 6(Suppl 1), S45–S53.

Shimizu, N. (2021). Gene amplification and the extrachromosomal circular DNA. *Genes*, 12(10), 1533.

Silva, F., Queiroz, J. A., & Domingues, F. C. (2011). Plasmid DNA fermentation strategies: influence on plasmid stability and cell physiology. *Appl Microbiol Biothechnol*, 93(6), 2571–2580.

Sobecky, P. A., Mincer, T. J., Chang, M. C., Toukdarian, A., & Helinski, D. R. (1998). Isolation of broad-host-range replicons from marine sediment bacteria. *Appl Environ Microbiol*, 64, 2822–2830.

Stephens, P. J., Greenman, C. D., Fu, B., ... et al. (2011). Massive genomic rearrangement acquired in a single catastrophic event during cancer development. *Cell*, 144(1), 27–40.

Van Craenenbroeck, K., Vanhoenacker, P., & Haegeman, G. (2000, September). Episomal vectors for gene expression in mammalian cells. *European J Biochem*, 267(18), 5665–5678.

Van Roy, N., Vandesompele, J., Menten, B., Nilsson, H., De Smet, E., Rocchi, M., De Paepe, A., Påhlman, S., & Speleman, F. (2006, February). Translocation-excision-deletion-amplification mechanism leading to nonsyntenic coamplification ofMYC andATBF1. *Genes, Chromosom Cancer*, 45(2), 107–117.

Walther, W., Stein, U., Voss, C., Schmidt, T., Schleef, M., & Schlag. P. M. (2003). Stability analysis for long-term storage of naked DNA: Impact on nonviral in vivo gene transfer. *Anal Biochem*, 318(2), 230–235.

Watanabe, T. (1963). Infective heredity of multiple drug resistance in bacteria. *Bacteriol Rev*, 27, 87–115.

Watanabe, T, & Fukasawa, T. (1961). Episome-mediated transfer of drug resistance to Enter obacteriacea. II. Elimination of resistance factors with acridine dyes. *J Bacteriol*, 82, 202–209.

Watanabe, T., Nishida, H., Ogata, C., Arai, T., & Sato, S. (1964). Episome-mediated transfer of drug resistance in *Enterobacteriaceae*. VII. Two types of naturally occurring R factors. *J Bacteriol*, 88, 716–726.

Weinberg, J. B., Matthews, T. J., Cullen, B. R., & Malim, M. H. (1991). Productive human immunodeficiency virus type 1 (HIV-1) infection of nonproliferating human monocytes. *J. Exp. Med*, 174(6), 1477–1482.

5 Cell Therapy Products
Basics and Manufacturing Considerations

Humberto Vega-Mercado
Bristol Myers Squibb, Cell Therapy Operations, Global Manufacturing Sciences and Technology, Warren, USA

CONTENTS

Introduction 49
Manufacturing Process 49
Final Remarks 59
References 60

INTRODUCTION

Scientific developmental and clinical work involving the redirection of cytotoxic T cells, the development of the first generation of chimeric antigen receptors (CAR), expansion of T cells ex vivo, eradication of B cell-lineage cells using CD19 CAR T cells, first T cell treatment and approval of CAR T cell treatments by US Food and Drug Administration (US FDA) and the European Union (EU) have been found in the literature since the 1980s, as summarized in Table 5.1 and discussed elsewhere (e.g., Ferreira et al., 2019; Bulaklak & Gersbach, 2020; Bashor et al., 2022). These advances have resulted in the commercialization of cell therapies for the treatment of diseases that were considered terminal a few decades ago.

Summary of scientific work and product approvals involving cell therapies (Ferreira et al., 2019; Bulaklak & Gersbach, 2020; Bashor et al., 2022) are found in Table 5.1.

The goal of this chapter is to provide a high-level summary of the basic considerations for the manufacture of cell therapies by means of a historical review of developmental work for the manufacture of cell therapies and general process considerations, equipment, and steps required for cell therapies.

MANUFACTURING PROCESS

The manufacture of cell therapies (e.g., autologous or allogenic) follows common principles such as aseptic operations or manipulations (e.g., use of biological

TABLE 5.1
Summary of Scientific Work and Product Approvals Involving Cell Therapies (Ferreira et al., 2019; Bulaklak & Gersbach, 2020; Bashor et al., 2022)

Year	T-Cell-Related Work
1986	Redirection of cytotoxic T cells by T-cell receptors (TCRs) (Dembic et al., 1986)
1987	Report of Ig-TCR chimeric receptors (Kuwana et al., 1987)
1988	Report of T-body chimeric receptors (Gross et al., 1989)
1991	First-generation CAR reported (Irving & Weiss, 1991)
1995	Prevention of autoimmunity by T_{reg} cell transfer in mice (Sakaguchi et al., 1995)
1998	In-vitro suppression of T_{eff} Cells by T_{reg} Cells (Thornton & Shevach, 1998) FDA approval of Apligraf™ – Allogenic fibroblasts (Bashor et al., 2022)
2001	Identification of human T_{reg} Cells (Baecher-Allan et al., 2001; Dieckmann et al., 2001; Jonuleit et al., 2001; Levings et al., 2001; Stephens et al., 2001)
2002	Second-generation CAR – CD28 signaling (Maher et al., 2002)
2004	In-vitro and in-vivo expansion of antigen-specific T_{reg} Cells (T1D on NOD mice) (Tang et al., 2004) Large-scale expansion of Human T_{reg} Cells (Hoffmann et al., 2004)
2005	MBP-specific T_{reg} Cells used to cure EAE in mice (Mekala and Geiger, 2005; Mekala et al., 2005)
2008	Mouse TNP CAR T_{reg} Cells improves colitis (Elinav et al., 2008)
2009	Third-generation CAR – CARs with costimulatory domains (Carpenito et al., 2009) Suppression of CEA tumors by Human CEA CAR T_{reg} cells (Hombach et al., 2009) First T_{reg} Cell therapy – treatment of GvHD (Trzonkowski et al., 2009) Suppression of arthritis by TCR redirected mouse T_{reg} cells (Wright et al., 2009) FDA approval of ChondroCelect™ – autologous chondrocytes (Bashor et al., 2022)
2010	Eradication of B-cell lineage cells with CD19 CAR T (Kochenderfer et al., 2010) Autologous dendritic cells therapy (Provenge) approved in the United States (Bashor et al., 2022)
2011	First T_{reg} cells clinical trial (GvHD) (Brunstein et al., 2011) FDA approval of Laviv™ – Autologous bibroblasts (Bashor et al., 2022)
2012	First T_{reg} cells clinical trial (T1D) (Marek-Trzonkowska et al., 2012) Three allogenic HCSs approved in USA – cord blood (Bashor et al., 2022) FDA approval of Gintuit™ – allogenic fibroblasts (Bashor et al., 2022)
2013	Three allogenic HCSs approved in USA – cord blood (Bashor et al., 2022) Autologous dendritic cells therapy (Provenge) approved in EU (Bashor et al., 2022) EU approval of MACI™ – autologous chondrocytes (Bashor et al., 2022)
2015	Confirmation treg cells persisted for up to one year (T1D) (Bluestone et al., 2015) One allogenic HCSs approved in USA – cord blood (Bashor et al., 2022) EU approval of Holoclar™ – autologous limbal stem cells (Bashor et al., 2022)
2016	Human HLA-A2 CAR treg cells prevent GvHD in mice (MacDonald et al., 2016) FDA approval of MACI™ – autologous chondrocytes (Bashor et al., 2022) FDA approval of Exondys 51® (FDA, 2016a; Bulaklak & Gersbach, 2020) FDA approval of Spinraza® (FDA, 2016b; Bulaklak & Gersbach, 2020)
2017	First T_{reg} cell rherapy study for kidney transplant (Chandran et al., 2017) FDA approval of Kymriah™ and Yescarta™ – autologous CAR T (US-FDA, 2019, 2020) EU approval of Spherox™ – autologous chondrocytes (Bashor et al., 2022)

(Continued)

TABLE 5.1 (CONTINUED)
Summary of Scientific Work and Product Approvals Involving Cell Therapies (Ferreira et al., 2019; Bulaklak & Gersbach, 2020; Bashor et al., 2022)

Year	T-Cell-Related Work
2018	Suppression of FVIII specific B cells with human FVIII BAR T_{reg} Cells (Zhang et al., 2018)
	One allogenic HCSs approved in USA – cord blood (Bashor et al., 2022)
	Approval of Kymriah™ and Yescarta™ – CD19 CAR T in EU (Bashor et al., 2022)
	EU approval of Alofisel™ – allogeneic adipose-derived stem cells (Bashor et al., 2022)
	Japan approval of Alofisel™ – allogeneic adipose-derived stem cells (Bashor et al., 2022)
2019	FDA approval of Vyondys 53 (Bulaklak & Gersbach, 2020; FDA, 2019)
2020	FDA approval of Tecartus™ – autologous CAR T (US-FDA, 2020; Bashor et al., 2022)
2021	FDA approval of Abecma™ and Breyanzi™ – autologous CAR T (Bashor et al., 2022)

Notes: T_{reg} cell = Regulatory T cell; BAR = B cell-targeting antibody receptor; CAR = chimeric antigen receptor; CEA = carcinoembryonic antigen; EAE = experimental autoimmune encephalomyelitis; FOXP3 = forkhead box protein P3; FVIII = factor VIII; GvHD = graft-versus-host disease; HLA = human leukocyte antigen; IA = murine major histocompatibility complex class II molecule I-A; Ig = immunoglobulin; MBP = myelin basic protein; NOD = nonobese diabetic; T1D = type 1 diabetes; TCR = T cell receptor; T_{eff} cell = effector T cell; TNP = 2,4,6-trinitrophenyl. HSC = haematopoietic stem cel; HPC = haematopoietic progenitor cell.

---

safety cabinets, cleanrooms – Class A, B and C, sterile welding and sealing, aseptic connectors). Also, in general, the manufacturing processes have common elements, as described elsewhere (e.g., Irving et al., 2017; Tran et al., 2017; Ferreira et al., 2019; Depil et al., 2020; Crees & Ghobadi, 2021). Figure 5.1 summarizes a typical manufacturing process for autologous and allogenic therapies (e.g., CAR T).

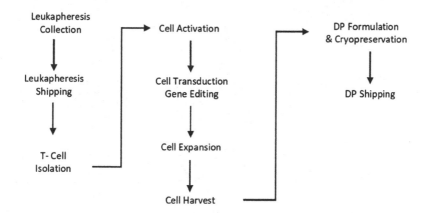

FIGURE 5.1 High-Level Summary of the Manufacturing Process of Autologous and Allogenic Therapies – CAR T (Irving et al., 2017; Tran et al., 2017; Ferreira et al., 2019; Depil et al., 2020; Crees & Ghobadi, 2021).

Each of the steps considered or required for the manufacture of cell therapies (e.g., allogenic or autologous CAR T) is further discussed in the following section:

a) Leukapheresis

The collection of cells from either patients or healthy donors is conducted at a center or hospital where an apheresis device or machine (e.g., COBE Spectra® Apheresis System by Terumo; Spectra Optia® Apheresis System by Terumo BCT; Amicus® Separator by Fresenius Kabi) may be used for the collection of the target cells (T cells) from the patient or donor. Specific protocols defining the collection parameters must be in pace to ensure the correct amount and type of cells are collected. Examples of the equipment used for collection of apheresis are shown in Figure 5.2.

The criticality of patient selection (e.g., clinical trial and commercial production) and apheresis collection criteria to ensure successful manufacture of the therapy are discussed elsewhere (e.g., Elverum & Whitman, 2020; Liu et al., 2022; Roig, 2022). Among the considerations suggested by the different authors are the following:

- History and physical examination of patients
  - Prior treatments
  - Health – viral or bacterial infections
  - Vein assessments
  - Informed concern (specific to the cells to be collected)
  - Education
- Cell handling capabilities
  - Experience and training of the personnel
  - Facilities
  - Equipment

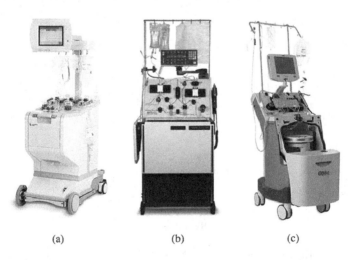

(a)   (b)   (c)

FIGURE 5.2 Example of Apheresis Collection Systems Commonly Used. (Used with permission from Terumo BCT, 2020a, 2020b; Fresenius Kabi, 2020).

Cell Therapy Products

- Detailed apheresis collection protocols
  - Labeling (Compliance with International Society of Blood Transfusion standard ISBT-128, Foundation for the Accreditation of Cellular Therapy (FACT)/Joint Accreditation Committee of the International Society for Cell & Gene Therapy (ISCT) and European Society for Blood and Marrow Transplantation EBMT (Join Accreditation Committee (JACIE)) Standards, Association for the Advance of Blood & Biotherapies (AABB) Standards)
  - Use of anticoagulants (e.g., Anticoagulant Citrate Dextrose Solution A – ACD-A; heparin)
  - Collection parameters (e.g., whole blood to anticoagulation (WB:AC) ratio, AC/citrate infusion rate, flow rates, patient's hematologic parameters and tolerance, target cell dose collection (minimum and optimal), total blood volumes), sampling and testing (calcium, ionized calcium, hypocalcemia), length of time for collection, type of machine used for collection, post collection monitoring of patients for side effects)
  - Storage conditions (e.g., compliant with FACT/JACIE/AABB standards)
  - Chain of Identity (COI) and Chain of Custody (COC)
- Transportation of collected material
  - Monitoring of temperature and humidity conditions per FACT/JACIE/AABB standards

b) Purification of T Cells

The purification step has as its main goal the removal of blood components that are not necessary for the manufacturing process (e.g., red cells, platelet) and to concentrate the type of cells required for the therapy (e.g., T_{reg}, CD4+ T cells, CD8+ T cells) as discussed by Ferreira et al. (2019). Among the methods used to purify cells are magnetic-activated cell sorting (MACS), fluorescence-activated cell sorting (FACS) and centrifugation/filtration applications.

- MACS

  MACS were developed by Miltenyi Biotecs (Miltenyi et al., 1990). MACS are supermagnetic nanoparticles and columns which are used to tag target cells by attaching them to antibodies or ligands on the coating of the magnetic particles (e.g., polymer shell encasing the magnetic material). The cells are then captured in a column placed between magnets. The selection process uses both positive and negative for the selection process. Figure 5.3 summarizes the sorting process using MACS. Besides the Miltenyi Biotec products (MACS® Microbeads), Dynabeads® (Invitrogen™, 2010) are available for cell processing and separation. In terms of equipment to use with the MACS® Microbeads, you will find the autoMACS® Pro Separator and CliniMACS from Miltenyi Biotec (Figures 5.4 and 5.5).

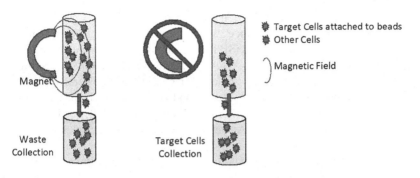

FIGURE 5.3 Summary of the sorting process using MACS® beads.

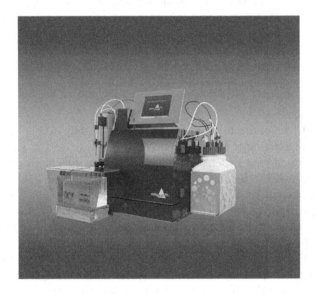

FIGURE 5.4 The autoMACS® Pro Separator from Miltenyi Biotec. (Used with permission from Miltenyi Biotec).

- FACS

  FACS are commonly used for flow cytometry. Specifically, particles and cells suspended in a liquid stream are passed through a laser beam, and the interaction with the light is measured as light scatter and fluorescence intensity by a detector. When a fluorescent label, or fluorochrome, is specifically and stoichiometrically bound to a cellular component, the fluorescence intensity will represent the amount of that particular cell component (Moturu et al., 2019). Figure 5.6 summarizes the sorting process using FACS. An example of a piece of equipment using FACS is the WOLF-G2 from NanoCellect Biomedical, Inc. (Benchtop Microfluidic Cell Sorter: WOLF G2 | NanoCellect) (Figure 5.7).

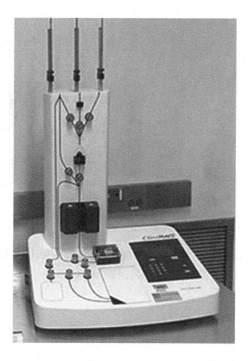

FIGURE 5.5 The CliniMACS Cell Separator from Miltenyi Biotec. (Used with permission from Miltenyi Biotec).

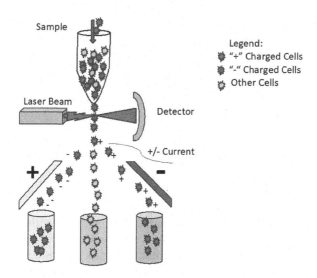

FIGURE 5.6 Summary of the sorting process using FACS. (adapted from Moturu et al., 2019 and NanoCollect Biomedial Inc. 2022).

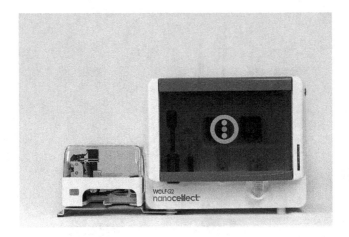

FIGURE 5.7 Picture of the FACS WOLF-G2 from NanoCellect Biomedical, Inc. (2022).

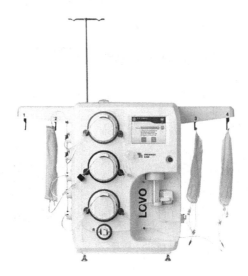

FIGURE 5.8 LOVO® automated cell processing system. (Fresenius Kabi, 2020).

- Centrifugation/filtration

  The LOVO Cell Processing System (Fresenius Kabi, 2014) is a benchtop cell processor that supports unique selection preparation, thaw wash, or harvest wash processes of white blood cell products. The piece of equipment uses a spinning membrane filtration technology, which allows for nonpelletizing cell concentration. Figure 5.8 shows an example of the equipment that may be used for purification and concentration of cells.

  Overall, the method to purify and concentrate the cells shall preserve their viability and functionality. Proper assessment of the intended technology and process parameters shall be conducted to ensure no inactivation of the cells during the purification and concentration steps.

c) Activation

The activation of T cells by means of antibodies (e.g., anti-CD3 and anti-CD28) is required prior to the transduction to ensure the cells continue to be fully functional after their purification and concentration steps and will continue to grow during the expansion step (June et al., 2018; Ferreira et al., 2019; Depil et al., 2020). Preparation of the activation mixture and transfer to the container or bag with the target cells shall be completed without introduction of microbial, particulate and chemical contaminants.

d) Transduction/Gene Editing

Technologies such as viral vector-mediated transgenesis enable the permanent insertion of recombinant DNA coding for a CAR as described by Depil et al. (2020). Preparation of the vector and transfer to the container or bag with the target cells shall be completed without the introduction of microbial, particulate and chemical contaminants.

e) Expansion

The expansion of activated and transduced T cells may be poor in vitro and significant variability in ex vivo. It is common to use stimulants to drive cell proliferation and IL-2 to sustain T-cell expansion. This variability may relate to the activation state of the cells as well as the age of the donor (Ferreira et al., 2019).

A key result reported by Putnam et al. (2009), which studied expansion in vitro in response to anti-CD3-anti-CD28, IL-2, and using rapamycin to maintain lineage purity, is that CD4+CD127(lo/-) and CD4+CD127(lo/-)CD25+ T cells could be expanded and used as T_{regs}. However, expansion of CD4+CD127(lo/-) cells required the addition of rapamycin to maintain lineage purity. Also, expansion of CD4+CD127(lo/-)CD25+ T cells, especially the CD45RA+ subset, resulted in high yield, functional T_{regs} that maintained higher FOXP3 expression in the absence of rapamycin. T_{regs} from type 1 diabetic patients and control subjects expanded similarly and were equally capable of suppressing T-cell proliferation. Regulatory cytokines were produced by T_{regs} after culture; however, a portion of FOXP3+ cells was capable of producing interferon (IFN)-gamma after reactivation. IFN-gamma production was observed from both CD45RO+ and CD45RA+ T_{reg} populations. Meanwhile, Canavan et al. (2016) identified that in vitro expansion enhanced the suppressive ability of CD45RA(+) T_{regs}. These cells also suppressed activation of lamina propria and mesenteric lymph node lymphocytes isolated from inflamed Crohn's mucosa.

Alnabhan et al. (2018) reported that serum promotes better CD19 CAR T-cell expansion and viability in vitro. CD19 CAR T cells produced in TexMACS medium showed a lower CD4/CD8 ratio. They concluded that the choice of culture medium impacts CD19 CAR T-cell end product. Xu et al. (2018) compared six commonly used cultural media for human T cells (e.g., serum-containing media and serum-free media, namely, RPMI 1640, IMDM, Gibco OpTmizer CTS T Cell Expansion SFM, Gibco AIM-V Medium CTS, LONZA X-VIVO 15 and StemSpan SFEM with or without Dynabeads Human T-Activator CD3/CD28) on in vitro T cell expansion, apoptosis and immune phenotype. Their results suggest that serum-free media provide a better proliferation environment for T cells. Among the three serum-free media, Xu et al. (2018) identified OpTmizer and AIM-V as better T cell culture environments compared with X-VIVO, as T cells showed higher viability in the first two media. Also, they reported that in vitro human T cells keep a relatively resting status among non-CD3/CD28 groups, due to weak proliferation and apoptosis abilities. The phenotypes of T cells in different culture media over time indicate T cells maturation during culture duration. Similar observation is reported by Depil et al. (2020) when culture conditions may also be adapted to amplify some specific subsets (e.g., IL-7 and IL-15 increase the frequency of CD8+CD45RA+CCR7+ cells during the ex vivo expansion of CAR T cells.) The culture medium may also strongly influence cell metabolism and the maintenance of a non-terminally differentiated memory T cell phenotype.

The expansion process will use incubators to ensure a constant and optimal temperature, relative humidity and carbon dioxide levels. It may consider gentle mixing (e.g., rocking motion) or not. Different pieces of equipment may be considered that are off-shelf applications (e.g., ThermoFisher Lab incubators; PHCbi Cell Culture Incubators; XURI Cell Expansion Systems from CYTIVA Life Science).

Overall, expansion of T cells must consider, among other factors, the donor, the target health condition, the type of cells to be collected and transduced, phenotypes, their activation status, composition of the culture media and the environmental conditions for an optimal expansion process (Putnam et al., 2009; Canavan et al., 2016; Alnabhan et al., 2018; Xu et al., 2018; Ferreira et al., 2019; Depil et al., 2020).

f) Harvesting and Drug Product Formulation

Harvesting and drug product formulation are the final steps of the manufacturing process prior to cryopreservation (typical preservation technique for cell therapies). The harvesting process has as its main goal to concentrate the cells by removing and replacing most of the culture media required during the expansion process. Keep in mind, the patient will receive an infusion of the formulated drug product. The amount of drug product for the infusion must be reasonable, while satisfying the specifications for a minimum number of cells, to ensure a relatively rapid procedure with minimal discomfort to the patient.

The harvesting step may use the same or similar equipment used for the initial purification step or additional equipment. Execution of the harvesting and subsequent product formulation shall be completed without the introduction of microbial, particulate and chemical contaminants.

Cell Therapy Products

g) Cryopreservation

Cryopreservation is commonly used for the preservation of cell therapies. Freezing of the formulated drug product is carried out following a predefined series of steps to ensure minimal or no damage to the cells. The goal is to stop the metabolic activity of the cells without destroying the cell walls or cell content.

Typically, cryopreservation is initiated by means of a controlled rate freezing process where the temperature of the drug product is gradually reduced from room temperature down to cryo conditions (e.g., −80°C) over several minutes. This step is critical to avoid the formation of large ice crystals inside the cells that may destroy them. Once the controlled rate freezing is completed, the frozen drug product is then transferred to a tank containing liquid nitrogen for storage at −190°C. Proper handling is required to avoid warming and thawing of the material.

The cryopreservation step shall be completed as designed to avoid potential damage to the cells.

h) Other Considerations (e.g., raw materials, single-use assemblies)

Elverum and Whitman (2020) have highlighted the need for manufacturers to collaborate in defining consistent standards for incoming material quality and widespread data collection on patient safety within a therapeutic class. The engagement of cross-functional groups from early development will result in fast, cost-efficient and scalable processes.

Issues like single sourcing of critical materials and single-use assemblies represent a challenge to the supply chain. A collaborative effort from across the industry will help reduce the risks to supply, as well as expedite the development of new therapies. Therefore, a robust supply chain is important in addition to outstanding technologies and innovative strategies (Cytiva, 2022).

FINAL REMARKS

The chapter summarizes the history of cell therapies since the 1980s when significant developments are reported. Then, using that historical review, we have focused on the manufacture of cell therapies starting with the collection of the cells (leukapheresis) to be used in the manufacture of cell therapy (e.g., CAR T). Once the cells are collected, purification and concentration steps are required to enrich the cells in preparation for the activation and transduction steps. Expansion of the cells using an optimal culture media and process conditions (e.g., temperature, relative humidity and CO_2 concentration) are critical for a successful harvesting and drug product formulation. Finally, the formulated drug product or cell therapy is packaged and preserved in preparation for shipping to the patient. The complexities around the manufacture of cell therapies and the challenges associated with supply chains must be considered by manufacturers early in the process of designing and developing such therapies.

REFERENCES

Alnabhan, R., Gaballa, A., Mork, L. M., Mattson, J., Uhlin, M., & Magalhaes, I. (2018). Media evaluation for production and expansion of anti-CD19 chimeric antigen receptor T cells. *Cythotherapy* 20(7), 941–951.

Baecher-Allan, C., Brown, J. A., Freeman, G. J., & Hafler, D. A. (2001). CD4+CD25high regulatory cells in human peripheral blood. *J Immunol* 167,1245–1253.

Bashor, C. J., Hilton, I. B., Bandukwala, H., Smith, D. M., & Veiseh, O. (2022). Engineering the next generation of cell-based therapeutics. *Nat Rev Drug Discov.* 21, 655–675.

Bluestone, J. A., Buckner, J. H., Fitch, M., Gitelman, S. E., Gupta, S., Hellestein, M. K., Herold, K. C., Lares, A., Lee, M. R., Li, K., Liu, W., Long, S. A. Masiello, L. M., Nguyen, V., Putnam, A. L., Rieck, M. Sayre, P. H., & Tang, O. (2015). Type 1 diabetes immunotherapy using polyclonal regulatory T cells. *Sci Transl Med*, 7, 315ra189.

Bulaklak, K., & Gersbach, C. A. (2020). The once and future gene therapy. *Nat Commun*, 11, 5820.

Brunstein, C. G., Miller, J. S., Cao, Q., McKenna, D. H., Hippen, K. L., Curtsinger, J., Defor, T., Levine, B. L., June, C. H., Rubinstein, P., McGlave, P. B., Blazar, B. R., & Wagner, J. E. (2011). Infusion of ex vivo expanded T regulatory cells in adults transplanted with umbilical cord blood: safety profile and detection kinetics. *Blood* 117, 1061–1070.

Canavan, J. B., Scotta, C., Vossenkamper, A., Goldberg, R., Elder, M. J., Shoval, I., Marks, E., Stolarczyk, E., Lo, J. W., Powel, N., Fazekasova, H., Irving, P. M., Sanderson, J. D., Howard, J. K., Yagel, S., Afzali, B., MacDonald, T. T., Hernandez-Fuentes, M. P., Shpigel, N. Y., Lombardi, G., & Lord, G. M. (2016). Developing in vitro expanded CD45RA+ regulatory T cells as an adoptive cell therapy for Crohn's disease. *Gut* 65, 584–594.

Carpenito, C., Milone, M. C., Hassan, R., Simonet, J. C., Lakhal, M., Suhoski, M. M., Varela-Rohena, A., Haines, K. M., Heitjan, D. F., Albelda, S. M., Carroll, R. G., Riley, J. L., Pastan, I., & June, C. H. (2009). Control of large, established tumor xenografts with genetically retargeted human T cells containing CD28 and CD137 domains. *Proc Natl Acad Sci USA*, 106, 3360–3365.

Chandran, S., Tang, Q., Sarwal, M., Laszik, Z. G., Putnam, A. L., Lee, K., Leung, J., Nguyen, V., Sigdel, T., Tavares, E. C., Yang, J.Y.C., Hellerstein, M., Fitch, M., Bluestone, J. A., & Vincenti, F. (2017). Polyclonal regulatory T cell therapy for control of inflammation in kidney transplants. Am J Transpl, 17, 2945–2954.

Crees, Z. D., & Ghobadi, A. (2021). Cellular therapy updates in B-cell lymphoma: The state of the CAR-T. *Cancers (Basel)*, 13, 5181.

Cytiva. (2022). Cell therapy manufacturing: The supply chain challenge. cytivalifesciences.com

Dembic, Z., Haas, W. Weiss, S., McCubrey, J., Kiefer, H., von Boehmer, H., & Steinmetz, M. (1986). Transfer of specificity by murine alpha and beta T-cell receptor genes. *Nature*, 320, 232–238.

Depil, S., Duchateau, P., Grupp, S. A., Mufti, G., & Poriot, L.(2020). Off-the-shelf allogeneic CAR T cells: Development and challenges. *Nat Rev Drug Discov*, 19:185–199.

Dieckmann, D., Plottner, H., Berchtold, S., Berger, T., & Schuler, G. (2001). Ex vivo isolation and characterization of CD4+CD25+ T cells with regulatory properties from human blood. *J Exp Med*, 193, 1303–1310.

Elinav, E., Waks, T., & Eshhar, Z. 2008. Redirection of regulatory T cells with predetermined specificity for the treatment of experimental colitis in mice. *Gastroenterology*, 134, 2014–2024.

Elverum, K., & Whitman, M. (2020). Delivering cellular and gene therapies to patients: solutions for realizing the potential of the next generation of medicine. *Gene Ther*, 27, 537–544.

FDA. (2016a). *FDA grants accelerated approval to first drug for Duchenne muscular dystrophy*.

FDA. (2016b). *FDA approves first drug for spinal muscular atrophy*.

FDA. (2019). *FDA grants accelerated approval to first targeted treatment for rare Duchenne muscular dystrophy mutation*.

Ferreira, L.M.R., Muller, Y. D., Bluestone, J. A., & Tang, Q. (2019). Next-generation regulatory T cell therapy. *Nat Rev Drug Discov*, 18, 749–769.

Fresenius Kabi. (2020). *Amicus separator operator's manual*. Lake Zurich, IL: Fresenius Kabi amicus_brochure_PP_AM_US_0003_2020.pdf (fresenius-kabi.com).

Fresenius Kabi. (2014). LOVO® Automated Cell Processing System – Brochure. Lake Zurich, IL: Fresenius Kabi. US-MD-LOVO-Brochure.pdf.pdf (myfreseniuskabi.com)

Gross, G., Waks, T., & Eshhar, Z. (1989). Expression of immunoglobulin-T-cell receptor chimeric molecules as functional receptors with antibody-type specificity. *Proc Natl Acad Sci USA*, 86, 10024–10028.

Hoffmann, P., Eder, R., Kunz-Schughart, L. A., Andreesen, R., & Edinger, M. (2004). Large-scale in vitro expansion of polyclonal human CD4+CD25high regulatory T cells. *Blood*, 104, 895–903.

Hombach, A. A., Kofler, D., Rappl, G., & Abken, H. (2009). Redirecting human CD4+CD25+ regulatory T cells from the peripheral blood with pre-defined target specificity. *Gene Ther*, 16, 1088–1096.

Invitrogen. (2010). Cell isolation and activation: Dynabeads® magnetic cell separation technology. Document Connect (thermofisher.com).

Irving, M., Vuillefroy de Silly, R., Scholten, K. Dilek, N., & Coukos, G. (2017 Engineering chimeric antigen receptor T-cells for racing in solid tumors: Don't forget the fuel. *Front Immunol*, 8, 267.

Irving, B.A., & Weiss, A. (1991). The cytoplasmic domain of the T cell receptor zeta chain is sufficient to couple to receptor-associated signal transduction pathways. *Cell*, 64, 891–901.

Jonuleit, H., Schmitt, E., Stassen, M., Tuettenberg, A., Knop, J., & Enk, A. H. (2001). Identification and functional characterization of human CD4+CD25+ T cells with regulatory properties isolated from peripheral blood. *J Exp Med*, 193, 1285–1294.

June, C. H., O'Connor, R. S., Kawalekar, O. U., Ghassemi, S., &Milone, M. C. (2018). CAR T cell immunotherapy for human cancer. *Science*, 359, 1361–1365.

Kochenderfer, J. N., Wilson, W. H., Janik, J. E., Dudley, M. E., Stetler-Stevenson, M., Fedman, S. A., Maric, I., Raffeld, M., Nathan, D. N., Lanier, B. J., Morgan, R. A., & Rosenberg, S. A. (2010). Eradication of B-lineage cells and regression of lymphoma in a patient treated with autologous T cells genetically engineered to recognize CD19. *Blood*, 116, 4099–4102.

Kuwana, Y., Asakura, Y., Utsunomiya, N., Nakanishi, M., Arata, Y., Itoh, S., Nagase, F., & Kurosawa, Y. (1987). Expression of chimeric receptor composed of immunoglobulin-derived V regions and T-cell receptor-derived C regions. *Biochem Biophys Res Commun*, 149, 960–968.

Levings, M. K., Sangregorio, R., & Roncarolo, M. G. (2001). Human Cd25+Cd4+ T regulatory cells suppress naïve and memory T cell proliferation and can be expanded in vitro without loss of function. *J Exp Med*, 193, 1295–1302.

Liu, H. D., Su, L., Winters, J. L., Thibodeaux, S. R., Park, Y. A., Wu, Y. Y., Schwartz, J., Zubair, A. C., Schneiderman, J., Gupta, G. K., Ramakrishnan, S., & Aqui, N. A. (2022). *Considerations for immune effector cell therapy collections: A white paper from the American society for apheresis*. Vancouver, British Columbia: Amer Soc. for Aph.

Marek-Trzonkowska, N., Mysliwiec, M., Dobyszuk, A., Grabowska, M., Techmanska, I., Juscinska, J., Wujtewicz, M. A., Witkowski, P., Mlynarski, W., Balcerska, A., Mysliwska, J., & Trzonkowski, P. (2012) Administration of CD4+CD25highCD127 – regulatory T cells preserves beta-cell function in type 1 diabetes in children. *Diabetes Care*, 35, 1817–1820.

MacDonald, K. G., Hoeppli, R. E., Huang, Q., Gillies, J., Luciani, D. S., Orban, P. C., Broady, R., & Levings, M. K. (2016). Alloantigen-specific regulatory T cells generated with a chimeric antigen receptor. *J Clin Invest*, 126, 1413–1424.

Maher, J., Brentjens, R. J., Gunset, G., Riviere, I., & Sadelain, M. (2002). Human T-lymphocyte cytotoxicity and proliferation directed by a single chimeric TCRzeta/CD28 receptor. *Nat Biotechnol*, 20, 70–75.

Mekala, D. J., Alli, R. S., & Geiger, T. L. (2005). IL-10-dependent infectious tolerance after the treatment of experimental allergic encephalomyelitis with redirected CD4+CD25+ T lymphocytes. *Proc Natl Acad Sci USA*, 102, 11817–11822.

Mekala, D. J., & Geiger, T. L. (2005). Immunotherapy of autoimmune encephalomyelitis with redirected CD4+CD25+ T lymphocytes. *Blood*, 105, 2090–2092.

Miltenyi, S., Muller, W., Weichel, W., & Radbruch, A. (1990). High gradient magnetic cell separation with MACS. *Cytometry*, 11(2), 231–238.

Moturu, R., Kamath, S., & Hasan, S. (2019). Fluorescence-activated cell sorting (FACS). (PDF) Fluorescence Activated Cell Sorting (FACS) (researchgate.net).

NanoCellect Biomedical Inc. (2022). Benchtop Microfluidic Cell Sorter: WOLF G2. San Diego, CA.

Putnam, A. L., Brusko, T. M., Lee, M. R., Liu, W., Szot, G. L., Ghosh, T., Atkinson, M. A., & Bluestone, J. (2009). Expansion of human regulatory T-cells from patients with type 1 diabetes. *Diabetes*, 58, 652–662.

Roig, J. (2022). Managing variability in autologous apheresis collection. *Cell & Gene Ther Insig*, 8(2), 189–192. www.insights.bio

Sakaguchi, S., Sakaguchi, N., Asano, M., Itoh, M., & Toda, M. 1995. Immunologic self-tolerance maintained by activated T cells expressing IL-2 receptor alpha-chains (CD25). Breakdown of a single mechanism of self-tolerance causes various autoimmune diseases. *J Immunol*, 155, 1151–1164.

Stephens, L. A., Mottet, C., Mason, D., & Powrie, F. (2001). Human CD4+CD25+ thymocytes and peripheral T cells have immune suppressive activity in vitro. *Eur J Immunol*, 31, 1247–1254.

Tang, Q., Henriksen, K. J., Bi, M., Finger, E. B., Szot, G., Ye, J., Masteller, E.L., McDevitt, H., Bonyhadi, M., & Bluestone, J. A. (2004). In vitro-expanded antigen-specific regulatory T cells suppress autoimmune diabetes. *J Exp Med*, 199, 1455–1465.

Terumo BCT. (2020a). Spectra Optia® Apheresis System. Spectra Optia System overview brochure (terumobct.com). Lakewood, Colorado.

Terumo BCT. (2020b). COBE Spectra® Apheresis System. COBE Spectra Apheresis System (terumobct.com). Lakewood, Colorado.

Thornton, A. M., & Shevach, E. M. (1998). CD4+CD25+ immunoregulatory T cells suppress polyclonal T cell activation in vitro by inhibiting interleukin 2 production. *J Exp Med*, 188, 287–296.

Tran, E., Longo, D. L., & Urba, W. J. (2017). A milestone for CAR T cells. *N Engl J Med*, 377, 2593–2596.

Trzonkowski, P. Bieniaszewska, M., Juscinska, J., Dobyszuk, A., Krzystyniak, A., Marek, N., Mysliwska, J., & Hellmann, A. (2009). First-in-man clinical results of the treatment of patients with graft versus host disease with human ex vivo expanded CD4+CD25+CD127- T regulatory cells. *Clin Immunol*, 133, 22–26.

US-FDA. KYMRIAH (tisagenlecleucel). (2019). https://www.fda.gov/vaccines-blood-biologics/cellular-gene-therapy-produ....

US-FDA YESCARTA (axicabtagene ciloleucel). (2020). https://www.fda.gov/vaccines-blood-biologics/cellular-gene-therapy-produ....

Wright, G., Notley, P., Xue, C. A., Bendle, S. A., Holler, G. M., Schumacher, T.N.A., Ehrenstein, M. R., & Stauss, H. J. 2009. Adoptive therapy with redirected primary regulatory T cells results in antigen-specific suppression of arthritis. *Proc Natl Acad Sci USA*, 106, 19078–19083.

Xu, H., Wang, N., Cao, W., Huang, L., Zhou, J., & Sheng, L. (2018). Influence of various medium environments in vitro human T cell culture. *In Vitro Cell Dev Biol Anim*, 54(8), 559–566.

Zhang, A. H., Yoon, J., Kim, Y. C., & Scott, D. W. (2018). Targeting antigen-specific B cells using antigen-expressing transduced regulatory T cells. *J Immunol*, 201, 1434–1441.

6 Facility and Equipment Considerations

Humberto Vega-Mercado
Bristol Myers Squibb, Cell Therapy Operations, Global Manufacturing Sciences and Technology, Warren, USA

CONTENTS

Introduction 65
Facility Design for Cell and Gene Therapy Products 66
Equipment Design Considerations 73
Contract Manufacturing Organizations 75
Supplier-Sponsor Synergies in Development of Cell and Gene Therapies 76
Final Remarks 78
References 78

INTRODUCTION

The facilities and equipment used for the manufacture of cell and gene therapies will have common requirements with conventional sterile pharmaceuticals as well as very specific and unique requirements due to the nature of the drug products. The facilities for manufacturing gene therapies are closer in design to typical biologic facilities where you consider bioreactors of multiple sizes, utilities such as clean steam, water for injection, either permanent piping or single-use systems, media and buffer formulation suites, compressed gases (e.g., air, carbon dioxide, nitrogen), autoclaves and cleaning stations, among other (ISPE, 2021). The facilities for gene therapies can be scaled-up as batch sizes can be increased based on the product's demand.

Meanwhile, facilities for the manufacture of cell therapies, especially autologous, are different in terms of their set-up. Typically, the facilities are design to have workstations to process one patient at a time with a limitation in the number of patients that can be processed simultaneously across the facility as it is critical to ensure the Chain of Identity (COI), traceability is critical and must be established throughout the product lifecycle (ISPE, 2021). These stations can be either individual rooms (self-contained) or a large ballroom where the stations are distributed throughout the manufacturing space with dedicated equipment in each station (e.g., incubators, mixers, supplies, etc.) with potential for common spaces with non-patient/product specific equipment or tools. The main challenges identified for facilities manufacturing cell therapies, but not limited to, include:

1. Increase number of batches as scale-up is not feasible. Each batch represents a patient and mass production (doses) is not possible.
2. Manufacturing personnel required for manual operations due to the nature of the manufacturing process.
3. Cross-contamination prevention and maintaining Chain of Identity (COI). Since patient materials are used as starting material, each lot must be isolated and managed without compromising and jeopardizing any other lot being processed in the facility. These facilities should establish robust segregation practices end to end.
4. Integrated logistics to manage the end-to-end manufacturing process. Apheresis material is received and requires real-time processing (e.g., limited or no hold times allowed) to ensure the cells remain viable for immediate processing. Once apheresis material is initially processed (e.g., T cells isolated and concentrated), potential for cryopreservation of the cells can be considered. Patients are waiting for their treatments; therefore, drug products shall be timely shipped to the treatment centers once the manufacturing process and quality control testing are completed.
5. Raw materials and supplies sourcing must be evaluated and secured. The utilization of single-use components, common across the industry, as well as other elements in the supply chain (e.g., hospitals and blood banks), creates additional logistic challenges. In the case of cell therapy facilities, the prevention of microbial contamination requires in-house handling and preparation procedures that are not required at other facilities or operations (e.g., multilayer bags to transfer from warehouse to clean rooms).
6. The use of multiple viral vectors for the manufacture in multiproduct facilities requires additional segregation of the products beside patients.

In this chapter, we will discuss those common factors as well as the unique ones: scale-up vs. scale-out, segregation requirements, gowning and lockers, single-use technologies, raw materials, barriers, process flow, material flow, personnel flow and waste flow, equipment, utilities and contact manufacturing organizations.

FACILITY DESIGN FOR CELL AND GENE THERAPY PRODUCTS

i. Common elements from traditional manufacturing facilities and cell and gene therapies (CGT)
Cell and gene therapy manufacturing is required to follow aseptic processing principles and practices. In general, aseptic manipulations in Class A/ISO 5 are expected to have a Grade B/ISO 7 environmental background. With those classifications come, among others, the gowning requirements associated with such conditions; environmental

Facility and Equipment Considerations

controls, cleaning and sanitization programs; environmental monitoring programs; access controls; and the use of both personnel and material airlocks (ISPE, 2021; PDA, 2013).

1. Gowning

   Personnel working on cell and gene therapies will follow the practice of washing hands and arms, changing from street clothes to a plant uniform (e.g., scrubs or clean uniform), replacing their shoes with plant-assigned shoes and/or using shoe covers, applying hair nets, bear covers (as applicable) and sanitized gloves in a locker room prior to entering the controlled-non-classified (CNC) production area.

   Once in the CNC area, based on the activities to be performed, the personnel typically enter specific rooms (Class C room that may have additional classified rooms within the space, Class B and Class A following a pressure cascade) where additional gowning may be required. The access to these rooms is through a personnel airlock that allows additional contamination controls. Gowning in Class C, B and A rooms typically involves the use of sterilized laboratory coats, the use of sterile bunny suits, full sterile gowns (e.g., jumpsuit, boots, head cover), additional sterilized gloves, shoe covers, and clean face masks. In each case, the supplies are staged in the airlock for the convenience of the personnel. The following diagram (Figure 6.1) summarizes the typical configuration for gowning activities.

   It is a common practice to have a single flow design (e.g., entrance to rooms is not the same as the exit from the rooms to avoid potential

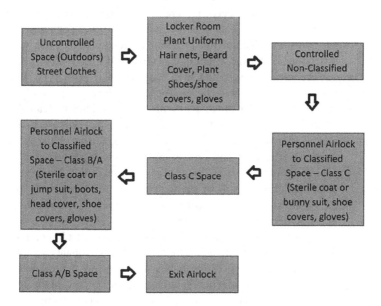

FIGURE 6.1 Example of facility configuration and gowning requirements.

cross-contamination between the personnel entering a room and those leaving the room). Therefore, another airlock is used for the removal of gowning apparel once the employee is heading out of the classified spaces. Appropriate disposal bins are in the exit airlock to collect used apparel for further cleaning and sterilization.

Clean and sterile apparel is commonly sourced from a third-party supplier who manages the delivery and recollection of used pieces (e.g., CINTAS, ARAMARK, STERIS, etc.). These suppliers must be qualified by the facility as part of a supplier qualification program.

2. Environmental Controls

   Environmental controls are critical to maintaining the cleanliness of the processing areas. Among the environmental controls include air handling units, cleaning and disinfection practices, environmental monitoring program, access controls and airlocks.

   a. Air Handling Units and Air Pressure Cascades

      Dedicated air handling units are commonly used across manufacturing facilities to prevent potential airborne cross-contamination. This allows the isolation of processing areas and proper control of air pressure cascades inside the processing areas. In addition, timing controls at the access doors (entrance and exit) are part of the controls to ensure pressure cascades are maintained (e.g., Pressure in Class A/B >> Pressure in Class C >> Pressure in CNC).

   b. Cleaning and Sanitization

      Cleaning and disinfection programs are a must in a manufacturing facility to prevent the spread of chemical, particulate and microbial contamination into classified areas. Qualified cleaning and sanitizing/sporicidal agents must be used as part of a comprehensive cleaning and disinfection program as recommended elsewhere (Marriott, 1999; McLaughlin & Zisman, 2015; PDA, 2012).

      The cleaning activities will have as their primary goal the removal of soils for surfaces. These soils host a variety of contaminants (e.g., chemical, particulate and microbial). Meanwhile, disinfection or sanitization activities have the primary goal of inactivating any viable or dormant microorganisms on the surface of product contact and processing surfaces. Keep in mind that disinfection/sanitization does not replace the initial cleaning step. Also, the effectiveness of sanitizer against microorganisms must be demonstrated (e.g., vegetative vs. spores). Typical cleaning and disinfection agents are summarized in Table 6.1.

      Periodic cleaning and disinfection activities must be conducted in the facility to ensure no proliferation of contaminants within the manufacturing areas. The activities shall consider the following:

      - Rotation of disinfection/sanitizing agents to prevent the development of tolerance to the agents by the microorganisms.
      - Specific timing for cleaning and disinfection activities:

TABLE 6.1
Typical Cleaning and Disinfection Agents (Marriott, 1999)

Cleaning Agents (Not All Inclusive)	Disinfection Agents (Not All Inclusive)
Caustic soda	Iso Propyl Alcohol
Sodium orthosilicate	Ethanol
Trisodium phosphate	Hydrogen peroxide
Tetrasodium pyrophosphate	Peracetic acid
	Quaternary ammonium compounds
	Steam
	Chlorine
	Iodophors

- Before and after equipment usage
- Daily activities
- Weekly activities
- Monthly activities
- Emergency or ad hoc events

c. Environmental Monitoring Program

A comprehensive environmental monitoring program will provide a lagging indication of potential contamination risk from personnel and facilities. The program requires a detailed risk assessment to identify the potential sources of contamination as well as the critical processing steps where contamination may occur (e.g., materials and supplies handling, aseptic manipulations) followed by an initial microbial mapping of the facility before and after major cleaning and disinfection activities. The goal is to confirm the effectiveness of the cleaning and disinfection activities. Once the data from the initial environmental monitoring activities are analyzed, the final locations for routine monitoring shall be defined and periodic monitoring conducted per approved procedure considering the following:

- Before and after specific process steps
- Daily activities
- Weekly activities
- Monthly activities
- Emergency or ad hoc events

Periodic data collection and analysis must be part of the program to identify any potential trend in contamination.

The total amount of personnel allowed in the processing rooms is assessed as part of the initial monitoring studies as well. The goal is to confirm the capability of the air handling unit to maintain the cleanliness of the area as well as the effectiveness of the cleaning/disinfection activities during worst-case occupancy situations.

Monitoring activities shall include air sampling for viable and non-viable particles, settling plates, surface plating and personnel plating (e.g., gowning). Plates and culture media used for environmental monitoring shall be qualified for the intended use. Proper indicator organisms shall be considered for the qualification activities (e.g., molds, spore formers, coliforms).

d. Access Control

Access control is twofold: (a) security and (b) prevent cross-contamination. Security at the manufacturing sites is required to avoid nonauthorized individuals accessing the premises. Meanwhile, authorized personnel (e.g., fully trained) are granted access to the facilities considering their roles and responsibilities. Access to clean rooms will have the most restrictive requirements to keep nontrained personnel from entering and jeopardizing the cleanliness of the area.

e. Personnel and Material Airlocks

Airlocks are transition rooms where personnel increase their cleanliness level by using new clean and/or sterile apparel (e.g., personnel airlock) and allow the cleaning of materials and supplies going into a processing area (e.g., material airlock).

These rooms are cleaned and disinfected per the approved program. Since the rooms serve as a barrier between a clean side and a dirty side, proper steps must be in place to ensure contamination is not carried across the boundary. Also, a pressure cascade is used across the rooms to ensure the cleanest air always flows toward the dirty side of the room.

ii. Differences between traditional manufacturing facilities and CGT

Cell therapies, specifically those with patient-specific products (e.g., autologous therapies), differ from traditional manufacturing facilities in that a strong focus on COI is required among other elements such as scheduling patient-specific apheresis collection, coordinating patient-specific shipping activities of apheresis, preserved mononuclear cells and preserved drug product. In addition, the use of viral vectors requires proper controls and isolation to prevent the potential of cross-contamination or incorrect use during manufacturing operations in multiproduct facilities.

Also, the need to reduce turnaround times for the delivery of drug products has helped the industry to further innovate and use new technologies such as rapid microbiological test methods (e.g., sterility testing), cell counts and vector translation efficacy. Nonconventional test methods have been developed as part of the advances in cell and gene therapy technologies.

Facility and Equipment Considerations

iii. Process-specific requirements

The design and build of the manufacturing facilities must consider the specific process requirements to manufacture the drug products. The receiving and handling of the patient-specific blood material (e.g., apheresis, tissues), purification and concentration of the target cells (e.g., T cells, natural killers), activation of cells, translation using viral vectors, expansion/growth of cells and harvesting, formulation and preservation of the drug products must be considered as part of the design activities to ensure the process is fully supported (e.g., gowning, equipment installation and usage, transporting and transfers from rooms or work stations, environmental conditions). Also, exposure of intermediate materials, raw materials and the final drug product to room temperature conditions as well as requirements for refrigeration and cryopreservation are considered during the design and construction of the facilities.

The facility must be designed to accommodate the process as designed and not to modify the process to fit in the facility.

iv. Modular facilities

The option of modular construction is coming up as an alternative to expedite the construction of manufacturing facilities. Among the benefits are the following:
- Prefabricated rooms with specific design elements (e.g., utilities)
- Prequalified clean rooms
- Relocations capabilities
- Individual commissioning and qualifications

Nevertheless, some of the challenges these modular structures may face are the following:
- Interconnection among modules
- Maximum individual sizes and associated transportation to construction size
- Limited repurposing (structural modification) of rooms

These new prefabricated and prequalified clean room modules are being considered across the biopharmaceutical industry given the fast and reliable delivery time and rapid installation in the manufacturing facility. Examples of these off-site produced clean room infrastructures, among others, from G-CON are presented in Figures 6.2a and 6.2b.

v. Material, people, product and waste flows

Cross-contamination is a major consideration when designing and building facilities to produce cell and gene therapies. Material, people, intermediates and final products and waste flows, among other factors previously discussed, are key to ensuring the quality of raw materials, intermediates and finished drug products. The specific considerations for each of these factors follows:
- Materials

    Raw materials and supplies (e.g., single-use systems and components) shall be received via a warehouse where proper inspection, cleaning and

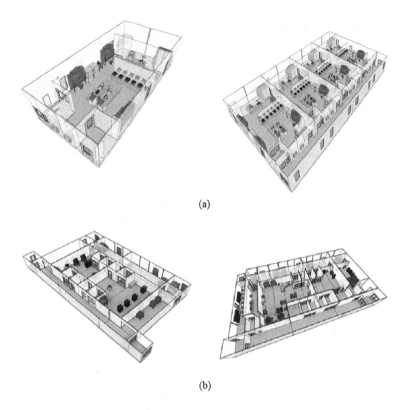

FIGURE 6.2 Examples of G-CON cleanroom PODs for (a) cell therapies and (b) vector production (Courtesy of G-CON Inc., College Station, Texas).

disinfection upon de-boxing are conducted by trained personnel. Once approved for further use by the incoming quality department, the material is transferred for either immediate use or storage nearby the production areas (e.g., kitting room) for further processing. Once required for manufacturing purposes, the materials and supplies are transferred to the processing rooms and transferred via materials airlocks where additional cleaning and disinfections are carried out by trained personnel. The handling and transfer of the materials and supplies are always away from potential contamination (e.g., wastes) or limiting their proximity to actual intermediates or products. Materials will not be transported into a processing room via the personnel airlocks.

- People
Personnel entering the facility will use lockers to wash hands and arms, remove makeup and jewelry and change from street clothes into plant uniforms or scrubs prior to entering the production areas. Visitors to a production facility must be escorted and provided with the necessary short-term training to satisfy the requirements for gowning. Once in plant uniforms or scrubs, personnel will complete

Facility and Equipment Considerations

basic gowning activities (e.g., hair nets, shoe covering, dedicated plant shoes, clean gloves). They will access the CNC area out of the locker and access the processing area via personnel airlocks. These personnel airlocks will be used to get additional gowning components (e.g., sterile lab coats, sterile booties, sterile jumpsuits, head covers, sterile gloves) considering the level of cleanliness in the room the person is ready to access (e.g., CNC to Class C; Class C to Class B/A) and are considered unidirectional. Once a person is ready to leave a room, the person will leave via an exit airlock to prevent potential contamination to those entering the room. Also, personnel cannot use material airlocks to enter a processing room.

- Intermediates and Final Products
Intermediates and final products will be handled within the designated areas (processing rooms) and stored in predesignated storage areas per approved procedures until approved for final packaging and shipping. The transfer of intermediates and finished products must avoid mixing with materials coming into the processing rooms. Clear marking and identification of the intermediates and final products are required all the time (e.g., COI) to avoid potential cross-contamination or mishandling. Once a finished product is ready for transfer to the final storage area, the product will be removed from the processing room via the exit airlock.
- Waste
Waste from the processing rooms will be removed periodically and transferred through a waste airlock. The airlock is commonly dedicated to avoiding the risk of cross-contamination or discarding of intermediates or finished products by accident. Also, personnel shall not exit the processing room through the waste airlock. Waste is neither removed via the material airlocks nor the personnel airlocks.

Figure 6.3 shows a general scheme describing material, personnel, products and waste flows in a theoretical facility.

EQUIPMENT DESIGN CONSIDERATIONS

Pieces of equipment used for cell and gene therapies can be either common to biotechnology applications (e.g., welders, sealers, biological safety cabinets, bioreactors, cell counters) or specific to cell and gene therapies (e.g., flow cytometry, cell isolation and purification/separation units such as SEPAX™, LOVO™ or CliniMacs™). In all cases, the requirements and specifications for the equipment must be aligned with the manufacturing process (e.g., capacity, process conditions, process parameters and outcomes). The following elements shall be considered when selecting or designing pieces of equipment:

i. General requirements (cleaning, maintenance)
The pieces of equipment considered for manufacturing and testing activities shall be easy to clean and maintain. These are basic considerations

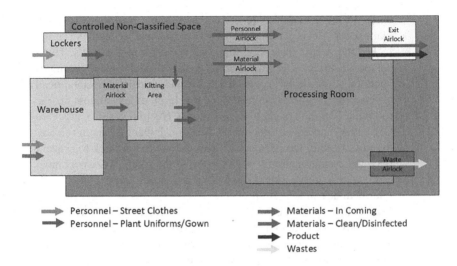

FIGURE 6.3 General material, personnel, products and waste flows in a facility.

expected from Current Good Manufacturing Practices (cGMPs). The utilization of single-use assemblies (e.g., connectors, adapters) and systems (e.g., bioreactor bags, mixing bags), cleaning and maintenance requirements are simplified. Nevertheless, the equipment required for such single-use assemblies still requires proper maintenance and cleaning. Standard operating procedures developed for the operation of the equipment must include specific instructions guiding the users on how to clean and maintain the units. Computerized units will require computer system validation monitoring procedures and tools to ensure proper performance as well.

ii. Process-specific requirements (scale, operation)

The pieces of equipment selected for manufacturing and testing activities must satisfy the user requirements specifications as well as functional requirements specifications. The specifications must be fully aligned with the expected outcomes (e.g., quality attributes) defined for the product as a function of the process conditions (e.g., parameters).

An important consideration when transitioning from the development stage into the clinical and commercial stages is the actual scale and operational conditions considered for the equipment. Some pieces of equipment will be suitable at a laboratory scale level but not suitable for clinical or commercial scales. Also, careful consideration shall be paid for highly customized pieces of equipment, as they may represent a risk of becoming obsolete or difficult to replace in the future.

iii. Supplies: Single-use disposables

The utilization of single-use disposables provides significant flexibility to the manufacturing and testing activities across cell and gene therapies.

Facility and Equipment Considerations

The selection of single-use applications must consider the following factors:
- Extractables and Leachables
  Plasticizers and other plastic-related compounds may interfere with the biological activity of the drug products from gene and cell therapies. Extractables and leachable studies are required to identify those compounds that may be transferred from the plastic and rubbery components of the assemblies into the culture media, intermediate products, and formulated products. Once identified, those compounds are assessed for their toxicity and impact on the manufacturing process and product quality.
- Compatibility
  The compatibility of assemblies with the manufacturing process and testing methods must be evaluated early in the development process to ensure no adverse impact on product quality. Interaction of culture media and buffers with the surface of the single-use components may result in protein aggregation, inactivation of cellular activity, adsorption of chemical components or leaching of toxic compounds.
- Sterilization
  Single-use components are commonly found ready to use in sterile and aseptic operations. Gamma irradiation, ethylene oxide (ETO) and steam sterilization are commonly used by suppliers.

  The suppliers shall confirm the proper qualification of the sterilization process used on their components. The supporting documentation shall be available from the supplier. Also, expiration dating associated with the sterilization process shall be established.
- Container Integrity
  Container closure and boundaries integrity of single-use components and associated packaging components (e.g., bags) must be confirmed to ensure no potential contamination of the buffers, intermediate products, culture media or drug products.
  The manufacturer of the assemblies shall confirm the integrity of their products. Integrity testing may include pressure hold, vacuum decay or leak rates (e.g., helium).
- Connectors, Welding and Seals
  Connectors, welding capability, and seal capability on single-use assemblies shall be confirmed to ensure the functionality of the components satisfies the requirements of the manufacturing process or testing methods.

CONTRACT MANUFACTURING ORGANIZATIONS

Contract manufacturing organizations (CMO) are used by companies to expand manufacturing and testing capabilities outside in-house (brick and mortar) capabilities. The selection of a CMO is not a trivial activity. It requires a thorough assessment of the technical and quality capabilities, among others, of the organization. CMOs are considered an extension of the sponsors' organization, as they will manufacture and/or test the product transferred to the CMO's facility. The following due diligence helps assess the capabilities of CMOs:

    i. Technical capabilities
The due diligence to assess technical capabilities shall include the review of equipment (e.g., tanks, machines, sizes and capacity, qualification), automation (e.g., data collection, controls, operation, qualification), utilities (e.g., compressed gases, cryopreservation, water for injection, clean steam, capacity, qualification), facilities (e.g., lockers, clean rooms, CNC rooms, environmental controls, qualifications, performance), warehousing (e.g., incoming materials, packaging components, shipping rooms) and production capabilities (e.g., scheduling, turnaround times).

A comprehensive checklist shall be developed by the sponsor to consistently assess the technical capabilities of the CMO. The checklist shall capture those items previously mentioned.

    ii. Due diligence – quality systems
The due diligence to assess quality systems shall include the review of systems, procedures and practices to manage compliance-related documentation such as standard operating procedures (SOPs), deviations, change control documents, corrective actions and preventive actions (CAPAs), complaints and documentation required for commissioning, qualification and validation.

A comprehensive checklist shall be developed by the sponsor to consistently assess the quality system capabilities of the CMO. The checklist shall capture those items previously mentioned.

SUPPLIER-SPONSOR SYNERGIES IN DEVELOPMENT OF CELL AND GENE THERAPIES

The manufacture of cell and gene therapies utilizes both common materials (e.g., standard buffers, standard culture media, single-use assemblies – bags, tubing), specialized materials (e.g., plasmids, lentiviral vectors (LVV), adeno-associated virus vectors (AAV), specialized buffers and culture media), off-shelf equipment (e.g., sealers, welders, freezers, refrigerators) and specialized equipment (e.g., cell counters, cell processing/purification equipment, cryopreservation units). The facility must be designed to provide protection for the required materials and supplies and to accommodate the necessary inventories needed during routine operation. The following shall be considered when selecting the supplier for those materials and equipment.

    i. Raw materials
Typical raw materials required for the manufacture of cell and gene therapies include purified amino acids, pharmaceutical grade chemicals

Facility and Equipment Considerations

(e.g., excipients such as sodium chloride, phosphate salts, formulated dimethyl sulfoxide (DMSO)), purified water, heat or gamma inactivated serums (human or bovine) among others.

The following shall be considered when identifying and selecting materials and suppliers:
- Pharmacopeia monographs and testing recommendations/requirements
- Single versus multiple suppliers for the material
- Capacity, capabilities and qualifications of the supplier
- Delivery times and options for the material (e.g., inventories, final package configurations)
- Specifications for the material
- Storage conditions required for the materials
- Use test of the material to ensure it satisfies both process and product requirements

Collaboration with quality control – incoming, supply chain, procurement and manufacturing science and technology groups are recommended to ensure a thorough assessment of the material and the supplier prior to final selection.

Storage of inventoried materials must be under approved conditions (e.g., temperature, relative humidity, light exposure) to ensure the quality of the material. These conditions must be aligned with those of the manufacturers and suppliers/distributors.

ii. Supplies

Typical supplies required for the manufacture of cell and gene therapies include collection bags, tubing assemblies, culture expansion bags or bioreactors, formulation bags, final product storage and shipping bags or vials, among others.

The following shall be considered when identifying and selecting materials and suppliers:
- Extractable and leachable profiles
- Single versus multiple suppliers for the material
- Capacity, capabilities and qualifications of the supplier
- Delivery times and options for the material (e.g., inventories, final package configurations)
- Specifications for the component
- Storage conditions required for the supplies
- Use test of the material to ensure it satisfies both process and product requirements.

Collaboration with quality control – incoming, supply chain, procurement and manufacturing science and technology groups are recommended to ensure a thorough assessment of the assemblies and the supplier prior to final selection.

Storage of inventoried materials must be under approved conditions (e.g., temperature, relative humidity, light exposure) to ensure the quality of the material. These conditions must be aligned with those of the manufacturers and suppliers/distributors.

iii. Equipment

There are multiple pieces of equipment used for the manufacture of cell and gene therapies including cell counters, welders, sealers, flow cytometers, cell purification/separation units, refrigerators, freezers, controlled rate freezers, cryo tanks and cryo shippers among others.

The following shall be considered when identifying and selecting materials and suppliers:
- Single versus multiple suppliers for the equipment
- Capacity, capabilities and qualifications of the supplier
- Delivery times and options for the equipment (e.g., inventories, sizes)
- User Requirement Specifications (URS) and Functional Requirements Specifications (FRS)
- Factory acceptance tests
- Installation Qualification (IQ), Operation Qualification (OQ) and Performance Qualification (PQ)
- Maintenance and repairs
- Spare parts

Collaboration with engineering and manufacturing science and technology groups is recommended to ensure a thorough assessment of the equipment prior to final selection.

The facility must be designed to accommodate the equipment for routine operation (e.g., required utilities) and to facilitate routine maintenance and cleaning activities.

FINAL REMARKS

The design, construction and implementation of facilities and pieces of equipment are not trivial activities. These must be fully aligned with the requirements of the manufacturing process (e.g., environmental conditions and classifications; flow of materials, personnel and waste). It is important to have the proper conditions in the facility for the storage of the materials and supplies (e.g., temperature, humidity, light exposure) by the operation of appropriate HVAC, refrigerators and freezers. Close work with suppliers (e.g., materials, supplies and equipment) to understand those requirements is expected.

REFERENCES

European Commission. (2015). EudraLex Volume 4. EU guidelines for good manufacturing practice for medicinal products for human and veterinary use. Annex 15. Qualification and validation. Brussels, Belgium.

European Medicines Agency (EMA). (n.d.). www.ema.europa.eu/en.

Facility and Equipment Considerations

Food and Drug Administration. (2004). Guidance for Industry. *Sterile drug products produced by aseptic processing – current good manufacturing practices*. Rockville, MD: FDA. www.fda.gov.

Food and Drug Administration. (2011). *Guidance for industry. Process validation: General principles and practices*. Rockville, MD: FDA. www.fda.gov.

G-CON. (2020). *Cell therapy POD layout catalog. Easing decision making reducing design time*. College Station, TX.

G-CON. (2022). *Viral vector POD layout catalog. Easing decision making with trusted certainty*. College Station, TX.

Gibson, M., & Schmitt, S. (2014). *Technology and knowledge transfer. Key to successful implementation and management*. Bethesda, MD: PDA. River Grove, IL: DHI Publishing, LLC.

ISPE®. (2021). *Advanced therapy medical products (ATMPs) – autologous cell therapy*. Tampa, FL: ISPE.

Koo, L. Y. (2019). *FDA perspective on commercial facility design for cell and gene therapy products. PharmaED – Process Validation Summit 2019*. La Joya, CA.

Marriott, N. G. (1999). *Principle of food sanitation* (4th ed.). Gaithersburg, MD: Aspen Publisher, Inc.

McLaughlin, M. A., & Zisman, A. (2015). *The aqueous cleaning handbook. A guide to critical cleaning procedures, techniques, and validation*. White Plains, NY: AI Technical Communications, LLC.

Parenteral Drug Association. (2006). *Technical report no. 28. Process simulation testing for sterile bulk pharmaceutical chemicals*. Bethesda, MD: PDA.

Parenteral Drug Association. (2012). *Technical report no. 29. Points to consider for cleaning validation*. Bethesda, MD: PDA.

Parenteral Drug Association. (2013). *Technical report no. 60. Process validation: A lifecycle approach*. Bethesda, MD: PDA.

Parenteral Drug Association. (2014). *Technical report no. 65. Technology transfer*. Bethesda, MD: PDA.

Parenteral Drug Association. 2019. *Technical report no. 81. Cell-based therapy control strategy*. Bethesda, MD: PDA.

Pharmaceutical Inspection Co-operation Scheme (PIC/S). (n.d.). www.picscheme.org

Vega-Mercado, H. (2019). *Technology transfer and process validation for CAR T products. PharmaED – Process Validation Summit 2019*. La Joya, CA.

7 Analytical Methods for In-Process Testing and Product Release

Neil A. Haig, Jamie E. Valeich, Laura DeMaster, Christopher Do, Razvan Marculescu, Jianbin Wang, Bridget S. Fisher, Xiaoping Wu, Akshata Ijantkar, Cayla N. Rodia, Sean D. Madsen, Mohammed Ibrahim, and Mel Davis-Pickett
Bristol Myers Squibb Cell Therapy-Analytical Development, Seattle, USA

Jennifer F. Hu
Gentibio, Seattle, USA

Jennifer J. Robbins, James Collins Powell, and Divya Shrivastava
Bristol Myers Squibb Cell Therapy-Analytical Development, Warren, USA

Satyakam Singh
Simnova Biotherapeutics, Cambridge, USA

CONTENTS

Introduction 83
General Properties of a Typical CAR T Cell Product 84
    Design 84
    MoA and Mode of Action 85
    CQAs 85
    Novel CQA Discovery 87
General Considerations for the Control of Product Quality 88

Regulatory Expectations for Analytical Procedures 88
   Development of Analytical Procedures 88
   Control Strategy for Analytical Procedures 89
   Documentation of Analytical Procedures in Regulatory Fillings 90
   Types of Performance Attributes 90
   Study Designs for the Evaluation of Performance Attributes 91
      Robustness 91
      Specificity 91
      Accuracy 92
      Precision 92
      Sensitivity 93
      Linearity 93
      Range 94
   Qualifications 94
      Phase I 94
      Phase II/III 95
      Analytical Life Cycle Management during Clinical
      Development 96
   Validations and Verifications 96
      Analytical Life Cycle Management during Commercial 97
Analytical Procedures 100
   Transduction and Characterization of Integration 100
      Control of Vector Dose during the Manufacture of
      a CAR T Cell Product 101
      Control of Transduction 102
   Genome Editing and Characterization of Edited Cells 104
      Genome Editing Technologies 104
      Intended Genome Editing Outcomes 105
      Unintended Genome Editing Outcomes 106
   Identity 108
   Purity and Characterization of Phenotype 109
      General Considerations for Analytical Flow
      Cytometry Methods 109
      QC Flow Cytometry Methods 111
      Characterization Methods That Evaluate T Cell Phenotype
      and Cellular Impurities 112
   Potency 113
      Percent Viable Cells and VCC 114
      Strength 116
      Biological Activities Relevant to the MoA and/or
      Mode of Action 116
   Process-Related Impurities 119
      Impurity Risk Assessment 120

Risk Assessment Approaches at Early Stages of
Development or for Emerging Modalities/Novel
Impurities 121
Impurity Methods Development Considerations and Key
Impurity Method Platforms 121
Microorganisms, Pyrogens and Adventitious Agents 123
Sterility 123
Mycoplasma 124
Endotoxin 126
Replication Competent Retrovirus 127
General CQAs 128
Container Content 128
Uniformity of Dosage Units 129
Container Closure Integrity 129
pH 130
Appearance (Visual Inspection) 130
References 130

INTRODUCTION

Cell therapy products are living non-germline cells of autologous, allogeneic or xenogeneic origin. Cell therapies are distinguished from transfusable blood products by being processed or manipulated by *ex vivo* selection, expansion, propagation, pharmacological treatment, genetic modification and/or other biological alteration. The Food and Drug Administration (FDA) regulates all cell therapies as biologics whether they are intended to be administered as a diagnostic, prophylactic, or therapeutic (section 351(i) of the Public Health Service (PHS) Act (42 U.S.C. 262(i) and Guidance for Human Somatic Cell Therapy and Gene Therapy).

There are multiple cell therapy products that have been commercialized or are under clinical development. This includes those that are propagated, expanded, and/or selected such as hematopoietic progenitor cells, and those that are also genetically modified such as chimeric antigen receptor (CAR) T cells, CAR natural killers (NK) cells, and T-cell receptor (TCR)-modified T cells (El-Karidy et al., 2021). If a cell therapy is genetically modified, it is considered to be both a cell therapy and gene therapy product by the FDA (1998).

This chapter will focus on the analytical procedures used to support the clinical development and commercial life cycle management of a CAR T cell product. This hypothetical product incorporates both lentiviral transduced transgenes and CRISPR-Cas9 gene editing, which are techniques that are widely applicable to both autologous and allogeneic cell therapy products. While there are unique considerations for each type of cell therapy product, much of the information contained in this chapter has been contextualized to be either directly relevant or translatable to other types. However, there will be unique considerations for each particular cell therapy product that is beyond the scope of this chapter.

The section entitled "General Properties of a Typical CAR T Cell Product" will provide a general overview of an archetypal CAR T cell product, mechanism of action (MoA) and critical quality attributes (CQAs). The section entitled *General Considerations for the Control of Product Quality* will summarize general considerations for the development of an integrated control strategy. This includes attributes that are typically controlled through either testing within the manufacturing process or release testing. The third section will describe the regulatory requirements, expectations and industry best practices for analytical method development, qualification, validation, verification, transfer and management of the analytical life cycle. The regulations and guidance for cell therapy products are similar between regulatory jurisdictions. This chapter has cited FDA regulations and guidance, and when divergences are known between other jurisdictions such as the EMA, additional references are included. Finally, the fourth section will provide a detailed overview of unique considerations for each CQA that is typically relevant to a cell therapy product, and the analytical considerations for providing the necessary assurances that these CQAs are sufficiently controlled.

GENERAL PROPERTIES OF A TYPICAL CAR T CELL PRODUCT

DESIGN

A typical CAR is composed of an antigen binding, hinge, transmembrane and intracellular signaling domains of different origins. The antigen binding domain may be derived from a murine, humanized or human antibody. For example, it may be a variable fragment from a murine antibody that is linked into a single chain (murine scFv). There can be more than one antigen binding domain, either as entirely separate CAR constructs or as a single CAR construct with a linker between two or more antigen binding domains. The hinge domain is commonly derived from IgG4 or CD8a, which promotes CAR dimerization through disulfide binding. In addition, the hinge domain may also enhance signaling by adjusting the spacing between the antigen binding domain and the transmembrane domain for optimized antigen binding. The selection of the transmembrane domain influences where CAR proteins become localized in the plasma membrane, such as toward membrane rafts that are enriched in particular membrane proteins (e.g., TCR macromolecular signaling complexes). Finally, the intracellular signaling domain consists of signaling motifs, which includes a stimulatory motif (e.g., CD3z, FcRg chain) and one or more co-stimulatory motifs (e.g., CD28, 4-1BB). The stimulatory and co-stimulatory motifs are specifically selected to have the desired downstream signaling. These signaling motifs are often separated by a spacer to optimize intracellular signaling (Guedan et al., 2018).

The recognition of antigen by the antigen-binding domain of a CAR results in receptor clustering followed by the formation of an immunological synapse between the CAR T cell and the antigen-expressing cell. This cell-to-cell interaction then results in signaling from the intracellular domains of the CAR, thus activating the CAR T cell. The design of the CAR construct impacts its localization within the plasma membrane, affinity and avidity of antigen binding, co-localization with other signaling molecules, strength and duration of signaling in the immunological synapse and the resulting biological outcome of CAR T cell activation (Guedan et al., 2018).

MoA and Mode of Action

The presumed MoA (e.g., the specific interaction of a drug that produces its pharmacological effect) of a typical CAR T cell product is antigen-specific killing of tumor cells. There are two orthogonal pathways of killing. The first is considered "fast-acting" through degranulation of both perforin and granzymes by cytotoxic CAR T cells (i.e., cytolytic activity). Upon activation, perforin is secreted into the immunological synapse between a cytotoxic CAR T cell and tumor cell. Perforin then forms a pore in the plasma membrane that enables granzymes to gain access to the cytoplasm. Once intracellular, the granzymes cause the activation of caspase cascades that induce apoptosis of the tumor cell. The second major pathway of cell killing is considered "slow-acting" and acts through the expression of tumor necrosis factor (TNF) family ligands on the cell surface of CAR T cells. These ligands can induce apoptosis upon engagement with their cognate receptors on tumor cells. One example of this family is Fas ligand, which can induce apoptosis on-target cells that express the death receptor Fas (Benmebarek et al., 2019). The mode of action (e.g., the functional and/or anatomical changes that result from administration of a drug represents an intermediate level of complexity between MoA and physiological outcome) of a typical CAR T cell product is migration to a site of antigen expression (i.e., tumor cells), and upon stimulation by antigen, proliferation into a larger number of cells, differentiation into effector CAR T cells. The effector CAR T cells can either release cytokines that promote anti-tumorigenic responses and/or induce tumor cell killing (Benmebarek et al., 2019)

The active ingredient(s) are the constituents of the drug product that contributes to the intended pharmacological activity. In a CAR T cell product, this includes all of the CAR T cells that contribute to either the MoA and/or mode of action. Importantly, the active ingredient(s) may be present in the drug product in a form that must undergo further modifications *in vivo* to produce the desired effect (21 CFR 210.3(b)(7)). For example, this includes subsets of CAR T cells that, upon antigen-specific activation *in vivo*, proliferate into a larger number of cells (part of the presumed mode of action), differentiate into helper CAR T cells that can secrete anti-tumorigenic cytokines (part of the presumed mode of action), and/or differentiate into cytotoxic CAR T cells that kill tumor cells (MoA).

CQAs

Attributes that are critical to the safety and efficacy of a cell therapy product must be controlled to maintain consistent product quality. These are called CQAs, and they can be "a physical, chemical, biological, or microbiological property or characteristic that should be within an appropriate limit, range, or distribution to ensure the desired product quality." The manufacturer is responsible for proposing and justifying CQAs as well as the strategy for ensuring they are sufficiently controlled, and these must be agreed to by the regulator during the approval process.

It is possible to define CQAs in cellular starting material, process intermediates, drug substance, drug product or excipients in the final formulation (International Conference on Harmonization (ICH), 2009a and 2009b). Since there is a more limited understanding of CQAs in earlier stages of development, the amount of information required on CQAs varies depending on the phase of the clinical trial (21 CFR 312.23(a)(7)). A broad evaluation of product characteristics throughout clinical development should be conducted to better understand manufacturing consistency, the performance of process controls, product stability, and the attributes that are important for ensuring product safety and efficacy (Considerations for the Development of CAR T Cell Products). Consequently, it is not unusual for there to be a revision in the list of CQAs over the course of clinical development, which should be reflected in amendments to the initial information submitted in the IND (21 CFR 312.31(a)(1)).

The focus of regulatory review during clinical trials is product safety and manufacturing controls (Chemistry, Manufacturing and Control (CMC) Information for Human Gene Therapy Investigational New Drug Applications (INDs)). Therefore, in all phases of clinical development, the specification for release of drug substance and drug product must at a minimum include CQAs for identity, purity, strength (quantity) and safety (21 CFR 312.23(a)(7)(iv)(a)) and (21 CFR 312.23(a)(7)(iv)(b)), and the FDA will put a clinical trial on hold if they believe that the controls described in the IND are insufficient to ensure product consistency and safety (21 CFR 312.42(b)(1)(iv)). The FDA recommends that a CQA for potency should be included in specification prior to initiating studies used to evaluate product efficacy (Potency Tests for Cellular and Gene Therapy Products), therefore a measure of potency is not required for initiating Phase I clinical trials. For commercial licensure, the specification must include CQAs for appearance, identity, purity, strength, potency and safety (21 CFR 610(b)):

- ***Appearance*** *is a visual inspection that evaluates the color and clarity of the product.*
- ***Identity*** *confirms the intended product has been manufactured.*
- ***Purity*** *includes assessments of product-related purity (e.g., intended active ingredient(s) and unintended cellular impurities that are either modified or unmodified) and process-related impurities (e.g., residual activation reagent, residual growth factors.), whether or not impurities impact the safety and efficacy of the product (21 CFR 600.3(r)).*
- ***Strength*** *(quantity) is the concentration of the active ingredient(s) in the product, which is used for dosing.*

Analytical Methods for In-Process Testing and Product Release

- **Potency** is a quantitative measure that is indicative of the intended biological activity of the active ingredient(s) (21 CFR 600.3(s) and 21 CFR 610.10).
- **Safety** includes testing for the level of transduction, sterility, mycoplasma, pyrogens such as endotoxin and adventitious agents such as replication competent lentivirus.

Note: Attributes with a known safety risk are described in this chapter as safety CQAs (e.g., pyrogens such as endotoxin), whereas attributes with a theoretical safety risk are described as a purity CQA (e.g., process-related impurities). However, these are all considered to be measures of purity.

NOVEL CQA DISCOVERY

In addition to the characterization of CQAs that are relevant to the presumed MoA and mode of action, the characterization of exploratory attributes provides an opportunity to identify novel attributes that are important for ensuring the safety and efficacy of a cell therapy product. These exploratory attributes can be measured in raw materials, cellular starting materials, process intermediates and lot release. The importance of exploratory attributes can be determined during process development by evaluating if there is impact of varying these exploratory attributes (e.g., percentage of viable cells in the cellular starting material) on CQAs. Alternatively, the importance of exploratory attributes can be evaluated by determining if there is a statistical correlation between a given attribute and clinical outcomes (i.e., safety and efficacy observed during the clinical trial). Finally, correlative analyses can also be performed on attributes that are already considered CQAs due to their importance to the presumed MoA and mode of action.

The outcome of correlative analyses might inform the identification of novel CQAs that need to be controlled to ensure a safe and effective cell therapy product. As such, confidence in the interpretation of correlative analysis is strengthened by using protocol-driven studies that are sufficiently powered, incorporate predefined statistical tests, have predefined thresholds for interpretation (e.g., level of significance required for escalating the criticality of an exploratory attribute) and include appropriate justification for each aspect of design (ICH Q8(R2)).

Correlative analysis can only assess whether there is a correlation between clinical outcomes and the range/distribution of attribute experience that was observed during clinical trials (which collected safety and efficacy data for each lot administered to patients). Therefore, during process performance qualifications and comparability studies, it may be beneficial to ensure that the cell therapy product remains within attribute experience demonstrated to be safe and effective during clinical trials. This would include all CQAs that are relevant to the presumed MoA and presumed mode of action (irrespective of past correlative analyses) and any exploratory attributes that have correlated with clinical outcomes.

GENERAL CONSIDERATIONS FOR THE CONTROL OF PRODUCT QUALITY

The safety and efficacy of a cell therapy product can be impacted by numerous aspects of its manufacture. These include but are not limited to the cellular starting material, selection or depletion of cell types, activation reagents, activation conditions, strategy for transgene delivery, quantity of transgene introduced into each transgenic cell, expansion reagents, expansion conditions, the phenotype and function of the transgenic cells, formulation, storage conditions, transport and product handling at the clinical site. Therefore, it is necessary to identify the factors that impact known or presumed CQAs and then determine a strategy that ensures they are adequately controlled to consistently deliver a safe and effective cell therapy product (ICH Q8(R2)).

There are a variety of strategies that can be employed to provide the necessary assurances of control. These include but are not limited to the development and validation of a robust manufacturing process that has been demonstrated to adequately control a given source of variance, adherence to current good manufacturing practices (cGMP), raw material testing, training, environmental monitoring, procedural controls, validated analytical procedures, in-process testing, specification testing, stability testing and maintaining the manufacturing process and analytical procedures in the validated state of control. One or more of these strategies may provide the necessary assurances of control; a science and risk-based approach should be used to justify the control strategy (ICH Q6B, ICH Q8(R2), ICH Q9, ICH Q10, ICH Q11). Although there are numerous strategies that can be employed to ensure CQAs are adequately controlled to ensure the consistency of a cell therapy product, this chapter focuses on the analytical procedures used for process controls and release testing.

REGULATORY EXPECTATIONS FOR ANALYTICAL PROCEDURES

Development of Analytical Procedures

Analytical procedures are required to monitor CQAs during the development of the manufacturing process, process performance qualifications, comparability studies, stability studies and the release of lots for clinical and commercial use. The analytical procedures should evaluate test samples from the manufacturing step that is most appropriate. For example, mycoplasma and adventitious agents testing is recommended on cell culture harvest material (CMC Information for Human Gene Therapy INDs). The selection of the instrument platform and methodology should be guided by the intended purpose of the analytical procedure, performance requirements, robustness to changes in execution that are relevant to those that are expected in a QC environment and appropriateness for use in a cGMP testing environment. In order to enable selection of an analytical procedure with appropriate performance characteristics, it may be necessary to begin development of several different analytical procedures (Analytical Procedures and Methods Validation for Drugs and Biologics and ICH Q14).

Once the initial conditions of an analytical procedure are developed, a risk assessment should be performed to identify parameters that are high- and medium risk for impacting analytical performance. The outcome of the risk assessment should inform experiments that evaluate the impact of these parameters on the performance of the analytical procedure. A systemic approach to this evaluation is preferred, such as using a design of experiments (DoE) to evaluate method parameters. These experimental studies should identify both the parameters that must be controlled in order to maintain consistent analytical performance and the ranges to which they must be controlled (Analytical Procedures and Methods Validation for Drugs and Biologics and ICH Q14).

CONTROL STRATEGY FOR ANALYTICAL PROCEDURES

Analytical procedures should incorporate a set of controls that ensure the procedure's performance remains consistent throughout its life cycle of use. This control strategy should include assay acceptance criteria (also called system suitability test), sample acceptance criteria (also called sample suitability assessment) and continued verification of analytical performance through control chart trending. The control strategy for analytical procedures should be established before validation and confirmed after validation has been finalized (ICH Q14).

Assay acceptance criteria (AAC) is established to verify the analytical procedure is performing as expected. AAC are established on attributes that are useful in evaluating the performance of the analytical procedure based on prior knowledge, experience gained during the development of the analytical procedure, and a science and risk-based assessment. AAC are typically established against reference standards, positive controls and/or negative controls. The acceptance criteria should be established based on the performance of the analytical procedure (e.g., tolerance intervals). AAC should be designed to verify that all components of the analytical procedure are performing as expected, including reagents, instruments and analysts (ICH Q14). A test session must conform to AAC to report results from test samples (21 CFR 211.165(d)).

Sample acceptance criteria (SAC) is established to verify that the data acquired from a test sample is suitable for reporting. SAC are established on attributes that are useful in evaluating the quality of data that has been acquired from test sample (ICH Q14). For example, SAC ensure the test sample result is within the reportable range of the analytical procedure. In addition, SAC may ensure the precision obtained between replicate determinations, which are subsequently used to calculate the reportable result, are within the performance that is anticipated for the analytical procedure. To report a result from a test sample, its results must conform to SAC (21 CFR 211.165(d)).

Two-tiered sequential assessment of acceptance criteria is recommended by the authors. The first step is to assess the data from a test session against AAC. If a test session does not conform to AAC, the entire assay is invalid and results from test samples cannot be reported. Conforming to AAC allows for the evaluation of test sample data against SAC (each test sample is reviewed independently against the SAC). If the test sample data conforms to SAC, test sample measurement is valid and can be reported. If data from a test sample does not conform to SAC, then that particular test sample is considered invalid and is not reported.

The final component of a control strategy for an analytical procedure is the continued verification of analytical performance through routine control chart trending. This topic is described in greater detail in the following discussion.

DOCUMENTATION OF ANALYTICAL PROCEDURES IN REGULATORY FILLINGS

A description of each analytical procedure used for process controls and release testing must be included in both an IND and BLA (21 CFR 312.23(a)(7)(iv), CFR 601.2(a) and 601.2(c)). There must be sufficient detail to understand how the analytical procedure is performed, the instrumentation used and its operating parameters, any reference standards and/or assay controls that are used, test sample types that can be evaluated (e.g., drug product) and sample preparation steps. In addition, all AAC and SAC that are used to establish the validity of test sessions and test results, respectively, should be described. Finally, the performance of the analytical procedures must be demonstrated to be scientifically suitable for their intended use.

There are several different types of performance evaluations for an analytical procedure. For example, those that are used to demonstrate the suitability of analytical procedures for use in clinical trials are generally called qualifications within industry, whereas those that are used to demonstrate the suitability for use in commercial products are either validations or verifications. The performance attributes that must be demonstrated in each type of evaluation varies depending on the type of reportable result (i.e., qualitative or quantitative), performance characteristics of the analytical procedure (e.g., reportable range is sufficient to report quantitative results from all test samples), and phase of development (i.e., Phase I, pivotal and commercial).

TYPES OF PERFORMANCE ATTRIBUTES

Depending on whether the reported attribute is qualitative or quantitative, there are differences in the performance attributes that must be demonstrated. There are two types of qualitative attributes: identity and impurities with acceptance criteria based on qualitative limits. Analytical procedures that are used to determine identity do not have performance requirements beyond the demonstration of robustness and specificity. However, an analytical procedure used to report an impurity with a qualitative limit must demonstrate robustness, specificity, and limit of detection (LOD). There are two types of quantitative attributes: those that are quantified well within the reportable range and those that quantify near the limit of quantification (LOQ). In the former case, the analytical procedure must be demonstrated to have suitable accuracy, precision, linearity and range, whereas in the latter case the LOQ also must be determined. Examples of quantitative attributes include purity, some impurities, strength and potency. The definition of each performance

Analytical Methods for In-Process Testing and Product Release

attribute (e.g., accuracy, precision) is described in ICH Q2. The requirements of each performance attribute are summarized the following section. Alternative study designs can be used if justified by a science and risk-based approach.

STUDY DESIGNS FOR THE EVALUATION OF PERFORMANCE ATTRIBUTES

The performance of an analytical procedure must be evaluated to demonstrate its suitability for use. Depending on the type of attributes reported by the method, the performance evaluation will consider one or more of the following performance attributes: robustness, specificity, sensitivity (LOD and/or LOQ), accuracy, precision, linearity and range.

Robustness

In contrast to the other performance characteristics, robustness is characterized during development. The robustness of an analytical procedure is determined by assessing the impact of deliberate but relevant variations in its execution. The parameter range that does not meaningfully impact the reportable result should be defined in the analytical procedure as the proven acceptable range. The evaluation of robustness is included as part of development information and must be made available upon request (ICH Q14 Analytical Procedure Development and ICH Q2 Validation of Analytical Procedures).

Specificity

There are three types of specificity that should be demonstrated. The first is the specificity for the reported attribute. Next is the specificity in representative material. Finally, there is the specificity for detecting relevant losses in the stability of a cell therapy product (ICH Q2 (R1)).

Attribute Specificity

The specificity of a reported attribute must be demonstrated. This often requires the specificity of the reagents used to report an attribute to be demonstrated, such as monoclonal antibodies. There are several approaches that can be taken for demonstrating attribute specificity. For example, specificity can be demonstrated if the analytical procedure is unaffected by components that should not impact the reported result but might be present in a test sample. This includes but is not limited to matrix, variants or impurities or a similar product with a different active moiety. Alternatively, specificity can be demonstrated by comparing results to those obtained by a well-characterized orthogonal analytical procedure (ICH Q2). If an analytical procedure does not have sufficient specificity, it might be necessary to include a matrix approach to achieve the necessary level of specificity (ICH Q2). The approach to demonstrating attribute specificity must be justified based on a science and risk-based approach.

Importantly, it may be necessary to redemonstrate attribute specificity for each type of test sample. For example, drug substance and drug product might contain different concentrations and formulations. In addition, changes to the manufacturing process can alter the purity, impurity, potency and other characteristics of the test sample. A science and risk-based approach should be taken to evaluate the potential of each difference to impact the specificity of an analytical procedure, and if risk is identified, it may be necessary to reevaluate attribute specificity (ICH Q1A(R2)).

Representative Material Specificity

An analytical procedure must be demonstrated to be suitable for evaluating each type of test sample. For example, the suitability for evaluating each test sample type can be demonstrated by evaluating multiple lots of representative material (e.g., n = 3), demonstrating their conformance to all AAC and SAC, and obtaining precise results (e.g., n = 6 for each lot) (ICH Q2 (R1)). The strategy should be justified based on a science and risk-based approach.

Stability-Indicating Specificity

There must be at least one analytical procedure on the specification that is sensitive to relevant changes in product stability. The stability-indicating properties of an analytical procedure can be demonstrated by observing a consistent trend in formal (i.e., GMP) stability studies, or by detecting meaningful changes in samples that have been spiked with engineered, forced-stress, accelerated-stability or stability samples (ICH Q1A). The strategy for demonstrating stability-indicating specificity should be justified based on a science and risk-based approach.

Accuracy

An analytical procedure must be demonstrated to have suitable accuracy across its reportable range. This is accomplished by comparing reportable results to the expected values. The sample type used for these studies can include:

1. A well-characterized reference material with a known value
2. A test sample with a known value determined by a well-characterized orthogonal method
3. A dilution series of representative material
4. A dilution series of engineered, purified, forced-stress, accelerated-stability or stability samples spiked into a test sample.

At least three different concentration levels should be evaluated three times each. Accuracy is reported as the mean percent difference between the reported result and expected value together with the 95% confidence interval. The study design and acceptance criteria must be justified (ICH Q2).

Precision

There are three measures of precision. Repeatability is precision under the same operating conditions that might be expected in a short period of time, intermediate precision considers the sources of variance that are expected during routine test in single QC site and reproducibility considers the sources of variance that are expected during routine testing at multiple QC sites. If only one QC testing site is used, reproducibility does not need to be evaluated.

The repeatability of an analytical procedure can be determined by testing three concentration levels that span the reportable range three times each. Alternatively, it can be determined by conducting six tests of a test sample at 100% concentration. Intermediate precision is determined by introducing multiple sources of variance that are relevant to routine testing at a single QC site, such as different analysts, instruments, reagent lots and days, during the evaluation of the same test sample lot. The minimum number of test results is not specifically described, so the authors recommend obtaining at least six reportable results in alignment with the minimum number of results required for repeatability. Since intermediate precision is more representative of the variance expected during routine QC testing than repeatability, the authors recommend for intermediate precision to be evaluated across the intended reportable range. To consolidate testing, this can also be combined with the testing used to evaluate accuracy. Reproducibility usually replicates the intermediate precision study design at multiple sites.

In each type of evaluation, the standard deviation, relative standard deviation and 95% confidence interval should be reported. The study design and acceptance criteria must be justified (ICH Q2).

Sensitivity

The LOD and LOQ can be established based on visual inspection of data (with appropriate justification), the ratio of signal to background noise or attributes of the linear regression used to establish linearity (see ICH Q2 for more information). LOQ can be established for both an upper and lower limit. The lower limit of quantitation (LLOQ) can be directly established based on the lowest concentration level with acceptable accuracy and precision. Similarly, the upper limit of quantitation (ULOQ) can be established based on the highest concentration level with acceptable precision and accuracy. The study design and acceptance criteria must be justified (ICH Q2).

Linearity

An analytical procedure must have a linear relationship between the concentration of the evaluated analyte and the reported result. Linearity can be evaluated using a dilution series of a test sample to generate different concentration levels (i.e., dilutional linearity study), or alternatively, a dilution series of an engineered, purified, forced-stress, accelerated-stability or stability samples spiked into a test sample (i.e., spiking study). Appropriate statistical models should be used to evaluate test results, such as linear regression by least squared. A minimum of five concentration levels that are appropriately distributed across the range must generate a linear response. This should be supported by a plot of the data, y-intercept and slope of the regression, correlation coefficient and residuals. The impact of any nonrandom residual pattern should be assessed. The study design and acceptance criteria must be justified (ICH Q2).

Range

The reportable range of an analytical procedure is the upper and lower limits that have been demonstrated to have acceptable specificity, accuracy, precision and linearity (ICH Q2). The reportable range must exceed the specification by at least 30% for dose-defining attributes and 20% for other quantitative attributes (see ICH Q2 for more information).

QUALIFICATIONS

The performance characteristics of analytical procedures that are used for either process controls or release testing must be understood to be qualified for use in clinical trials. Depending on the type of analytical procedure, this qualification should evaluate specificity, sensitivity (LOD or LOQ, as appropriate), accuracy, intermediate precision and range. In contrast to validations, qualifications do not require predetermined acceptance criteria, can be executed in development (and/or in QC) and can be reviewed and approved by development (and/or by QC and QA). Nevertheless, the results of the evaluation must demonstrate the analytical procedure is suitable for its intended use (Content and Format of INDs for Phase 1 Studies of Drugs, including Well-Characterized, Therapeutic, Biotechnology-Derived Products and CMC Information for Human Gene Therapy INDs).

Phase I

Analytical procedures must be demonstrated to be suitable for use before initiating Phase I clinical trials if they are required to determine dose or ensure patient safety (Content and Format of INDs for Phase 1 Studies of Drugs, including Well-Characterized, Therapeutic, Biotechnology-Derived Products). There are several reasons patient safety would be of concern, including the following (Content and Format of INDs for Phase 1 Studies of Drugs, Including Well-Characterized, Therapeutic, Biotechnology-Derived Products):

1) Unknown, poorly characterized or impure materials are used in the manufacture of the drug product
2) The drug product contains constituents that are either known or highly likely to be toxic
3) Instability of the drug product
4) The drug product contains impurities that have a potential or unknown health hazard

For a typical CAR T cell product, the analytical procedures that should be qualified before initiation of Phase I include identity, control of retroviral transduction (if relevant), control of gene editing (if relevant), strength, product-related impurities (unintended cellular populations), process-related impurities that pose

Analytical Methods for In-Process Testing and Product Release 95

a safety risk, adventitious agents such as replication competent retrovirus (if relevant), sterility, mycoplasma and endotoxin (CMC Information for Human Gene Therapy INDs).

If a compendial method is available for a given attribute, it must be used unless justified otherwise. For example, it might be beneficial to implement alternatives to compendial methods to enable faster release to patients (e.g., rapid sterility testing). Any alternative to a compendial method must be demonstrated to have equivalent or superior performance to be qualified for use in clinical trials (21 CFR 601.12(2)(vii) and USP <1226>, 2019).

The amount of performance characterization that is expected varies depending on the phase of clinical development. In addition, the performance characteristics that are required may also vary by phase (Analytical Procedures and Methods Validation for Drugs and Biologics). Therefore, in alignment with regulations and guidance, a science and risk-based approach to both identifying the analytical performance requirements at each phase of clinical development and justifying the approach to characterizing their performance (21 CFR 312.23(a)(7) and ICH Q14).

Phase II/III

Patient safety remains the focus of regulatory review. Consequently, the IND should be periodically updated (e.g., annual reports, prior to the initiation of Phase II or III) with new information relating to safety (21 CFR 312.31). This could include the identification, quantification and safety assessment of new impurities, changes to the impurity profile, new analytical procedures used for process controls or release testing and/or additional data relating to product stability (INDs for Phase 2 and Phase 3 Studies Chemistry, Manufacturing and Controls Information).

Prior to the initiation of clinical trials that will be used to evaluate product efficacy (the pivotal trial), it becomes necessary to include a measure of potency on the specification. A potency method quantifies a biological activity that is relevant to the intended effect of the cell therapy product (MoA or mode of action), or alternatively, a surrogate that has been demonstrated to have a sufficient correlation to a relevant measure of biological activity. The potency of a cell therapy product can be measured using a single assay or a matrix of assays (e.g., cell viability and cytolytic activity). In either approach, the acceptance criteria for potency should ensure that lots released for efficacy-indicating clinical trials are biologically active and consistent. If the potency of a cell therapy product is inadequately controlled, the clinical trial will be determined to be "deficient in design to meet its stated objectives" and the clinical trial put on hold (21 CFR 312.42(b)(2)(ii)).

The performance requirements of the analytical procedures may increase in later phases of clinical development (Analytical Procedures and Methods Validation for Drugs and Biologics). This might require the optimization or redevelopment of analytical procedures, additional characterization of analytical performance and/or providing additional information in regulatory submissions. A science and risk-based approach should be taken to identify the analytical performance requirements at later phases of clinical development and justifying the strategy that is taken to demonstrate the suitability of their performance (21 CFR 312.23(a)(7), 21 CFR 312.31(a)(1), and ICH Q14).

Analytical Life Cycle Management during Clinical Development

Once analytical procedures are qualified for use in clinical trials, it is beneficial to monitor their performance to ensure they remain suitable for their intended use. This is accomplished by conducting periodic evaluation of data that is relevant to an analytical procedure's performance, such as data used to evaluate against AAC and SAC, instrument, operator, reagent lots, date of testing. Adverse trends should be investigated, root causes identified and corrective action or preventive action (CAPA) implemented. If it is necessary to optimize a method, a scientific and risk-based assessment must be conducted to determine if the optimization causes differences to the performance characteristics of the analytical procedure. If so, it will be necessary to perform a partial or full requalification of the optimized method. Alternatively, if it is necessary to replace a method, the new analytical procedure must be qualified and the relationship to results obtained from the previous analytical procedure understood through a method comparability (also called bridging) exercise (Analytical Procedures and Methods Validation for Drugs and Biologics).

Across the clinical and commercial life cycle of a product, changes to the manufacturing process and analytical procedures are expected. Consequently, an appropriate number of samples should be maintained to enable comparative studies. The number of samples should be determined based on a science and risk-based approach (ICH Q2(R1), USP <1010> (2012), ASTM E2935 (2021)).

VALIDATIONS AND VERIFICATIONS

The validation or verification of an analytical procedure is the process of demonstrating it is suitable for its intended use. As described earlier, the regulatory requirements for the performance evaluation depend on the type of analytical procedure (e.g., identity, potency). A typical validation design evaluates an analytical procedure for all of the necessary performance characteristics at a single site during a single study. However, there are several variations of a validation that can be pursued (ICH Q2), including the following:

- Iterative validation: This strategy can be used to steadily increase the extent of the analytical performance that has been characterized as the phase of the clinical trial progresses. Iterative validation adheres to the full standards of a validation in terms of the performance characteristics that are evaluated, study design and acceptance criteria, but it is performed in consecutive stages. Each stage is protocol-driven with pre-defined acceptance criteria and is conducted under cGMP.

Analytical Methods for In-Process Testing and Product Release

- Co-validation: This strategy can be used to simultaneously evaluate the performance of an analytical procedure at multiple testing sites. A science and risk-based approach should be used to justify the extent of testing at each site.
- Transfer Validation: This strategy transfers an analytical procedure that is already validated to a new testing site. A science and risk-based approach should be used to justify the extent of testing at each site.
- Cross-validation: This strategy aims to enable the use of two or more analytical procedures for the same purpose (i.e., as alternative analytical procedures). The study design must include side-by-side testing and have the same acceptance criteria.
- Revalidation: Performed in response to optimizations to an analytical procedure. A science and risk-based approach should be taken to justify the extent of revalidation.

A verification is a performance evaluation that demonstrates the suitability of a compendial method for a new product under the actual conditions of use (USP <1226>). It is not possible to transfer a compendial method, so they must be verified separately at each QC testing site. They may be circumstances in which it is beneficial to development alternatives to compendial testing. For example, rapid sterility testing might be needed for products with a short half-life. To use an alternative to a compendial method for release of commercial product, it must be validated to have equivalent or superior performance in a side-by-side comparison (Analytical Procedures and Methods Validation for Drugs and Biologics).

The purpose of a validation/verification study, performance parameters that are under evaluation, methodology of the assessment and acceptance criteria must defined and justified in a pre-approved protocol prior to initiating studies. The performance evaluation must be executed under cGMP conditions with trained analysts using qualified critical reagents (if applicable), reference standards and instruments (i.e., IQ, OQ, PQ, etc.). The results of the validation or verification study, including any deviations, must be summarized in a report. The validation/verification study design, results and outcome must be described in the BLA. The FDA must be notified of any changes to an approved analytical procedure (21 CFR 601.2(a), 21 CFR 601.2(c), 21 CFR 601.12(a), Analytical Procedures and Methods Validation for Drugs and Biologics, and ICH Q2).

Analytical Life Cycle Management during Commercial

Analytical procedures must be maintained in their validated/verified state throughout the life cycle of their use. Therefore, a science and risk-based approach should periodically evaluate the need to develop, optimize or revise the strategy for managing instruments, critical reagents, reference standards, assay controls and analyst training.

Managing Critical Instruments

All instruments required for sample processing or data collection should be qualified prior to use. Qualification includes installation qualification (IQ), operational qualification (OQ) and/or performance qualification (PQ). Additionally, this equipment should be placed on a routine preventive maintenance (PM) schedule with the vendor to ensure proper calibration. At least two instruments must be available at each testing site. It is recommended to periodically evaluate the frequency of PM and stagger their maintenance schedules to ensure instrument availability.

Identifying and Managing Critical Reagents

A critical reagent is defined as any material observed throughout the analytical life cycle to have a direct impact to the results of an assay, and/or any material that is difficult to produce, obtain or replace. Critical reagents should be identified as early and clearly documented in analytical procedures. The following attributes can help guide the identification of critical materials for any given method:

- Complexity: Reagents containing inherent variability due to multi-step purification or synthesis, biological activity, chemical conjugation, etc.
- Rarity: Reagents which are difficult to obtain, produce or replace due to limited suppliers, long lead times and/or difficulty manufacturing.
- Stability: Reagents with limited stability due to formulation, concentration and/or temperature sensitivity. These materials will require special supply agreements such that a new lot can be qualified before the current lot is depleted or reaches expiry.

Examples of materials that may qualify as critical reagents are described in Table 7.1; however, each method should be assessed independently.

Critical reagents should be expected to vary between lots from a single manufacturer. For example, a new lot may require a different concentration to have equivalent performance to a previous lot. Therefore, a procedure for qualifying critical reagents should be documented, including a description of the study design, experimental procedure and acceptance criteria for qualifying new critical reagent lots for use in analytical procedures used for cGMP testing. As new information is gained during the life cycle of an analytical procedure, it might be necessary to revise critical reagents and/or their qualification procedures.

The shelf-life of critical reagents should be evaluated as early as possible. Multiple strategies are available for monitoring the stability of critical materials and setting expiry or retest dates. The optimal approach for each critical reagent may differ depending on their usage and inherent physical properties. In addition, a critical reagent that undergoes a change in supplier or manufacturing process may require reevaluation.

Finally, the amount of critical reagent in each vial, lot size, and shelf-life of these materials should be established, and then their inventory managed to ensure an uninterrupted supply (Ligand Binding Assay Critical Reagents and Their Stability: Recommendations and Best Practices from the Global Bioanalysis Consortium Harmonization Team).

TABLE 7.1
Example of Materials – Critical Reagents

Critical Reagent Category	Examples	Rationale
Antibodies	Monoclonal antibodies (mAbs) Polyclonal antibodies (pAbs)	Differences in purity, formulation, post-translational modifications and clonality may vary between clones, lots and suppliers and can impact binding activity and stability.
Engineered Proteins	Soluble receptors Cytokines Ligands Fusion proteins Enzymes	Differences in purity, formulation, post-translational modifications, may vary between lots and suppliers and can impact binding, activity and stability.
Conjugated Proteins	Biotinylated Enzyme-linked (HRP, AP, etc.) Fluorescently labeled Chemiluminescent	Differences in labeling chemistry and degree of labeling may vary between lots and suppliers and can impact signal intensity. Conjugated materials may also exhibit limited stability compared to their unconjugated counterparts.
Chemically Synthesized Reagents	Peptides Nucleic acids Aptamers	Differences in folding, purity and sequence may vary between lots and suppliers and can impact binding and stability.
Complex Biologics	Cells and cell lysates Tissues Viruses Bacteria	Differences between donor, lot, passage, strain and supplier may impact activity, composition, growth, health, infectivity and phenotype.
Binding Supports	Plates Beads Chips	Differences between lots and suppliers may impact binding affinity, avidity, capacity and uniformity.
Biological Matrices	Serum Plasma	Differences between lots, suppliers and primary source may impact composition and function.

Continued Performance Verification of Analytical Procedures

Once an analytical procedure has been validated or verified, it is necessary to provide continual assurances that it remains fit for its intended purpose throughout the life cycle of its use. This is accomplished by performing trend analyses of attributes that are relevant to the performance of the analytical procedure (Analytical Procedures and Methods Validation for Drugs and Biologics and ICH Q14). This can include but is not limited to trending of the performance attributes that are evaluated against AAC and SAC, reference standard lots, control sample lots, critical reagent lots, instruments and analysts. Trend analyses must be performed at periodic intervals (Analytical Procedures and Methods Validation for Drugs and Biologics). However, it might be beneficial to perform trending in real time using automated systems so that changes in method performance can be detected more rapidly (Adhibhatta et al., 2017). The trend analyses should be protocol-driven with predefined 'action limits' (e.g., A limit that is established for a performance attribute. If exceeded, it indicates the analytical procedure is trending outside of expected performance. The response to exceeding an action limit should be predefined, and at a minimum, trigger an investigation) that trigger an investigation for out of trend results. It may be beneficial to also include 'alert limits' (e.g., an established limit for a performance attribute that is more stringent than an action limit, and therefore, it provides an early warning of a potential trend away from normal operating conditions when exceeded. The response to exceeding an alert limit should be predefined but is usually less comprehensive to exceeding an action limit) for situations that are nearing action limits to provide advanced notice of a potential issue and enable the opportunity to take more immediate action (ICH Q2A).

If a meaningful change in analytical performance is detected during trend analysis (e.g., exceeding an action limit), an investigation must be conducted to identify the root cause and CAPA implemented. This might include clarifying the analytical procedure, changes to analyst training, the identification of new critical reagents and/or modifications to the qualification procedures for instruments, reference standards, assay controls and/or critical reagents. If it is necessary to optimize or replace the analytical procedure, it will be necessary to perform a revalidation, a new validation exercise, an analytical method comparability study or combination of these exercises. A science and risk-based approach should be used to justify the approach.

Sample Retention

It is reasonable to expect that during the life cycle of a product there will need to be changes to analytical procedures. For example, it might be necessary to develop and validate new or alternative analytical procedures in response to improved understanding of CQAs, identification of new impurities and/or risk assessments that evaluate the appropriateness of the current analytical procedures. Therefore, it is necessary to maintain an appropriate number of retention samples for use in comparative studies, such as method bridging exercises. The samples held in reserve should include those that are representative of the marketed product and, when possible, pivotal trial material. A science and risk-based approach should be used to determine the sample types and number of samples that are held in reserve.

ANALYTICAL PROCEDURES

TRANSDUCTION AND CHARACTERIZATION OF INTEGRATION

Gene therapy and cell therapy products commonly use third generation lentiviral-derived vectors (LVVs) to deliver transgenic material due to their versatility and efficiency. LVVs are typically based on HIV-1 retroviruses and can deliver transgenic material of up to 10kb. To mitigate the risk of introducing replication competent retrovirus (RCR), the genes that encode the vector components are provided in *trans* via three separate 'helper' plasmids during the production of LVVs. In addition, the sequences within the three helper plasmids may be optimized to reduce homology in order to further mitigate the likelihood that RCR will be generated as a result of recombination events occurring between the helper plasmids.

The tropism of a lentiviral vector is determined by the pseudotyped envelope. Vesicular stomatitis virus G protein (VSV-G) is typically used due to the capacity to facilitate efficient entry into a diversity of cell types. Once LVV is intracellular, the single-stranded positive sense RNA genome is reverse transcribed into dsDNA. Similarly to HIV-1, LVVs are capable of transporting components that are critical to transduction across the nuclear pore, and therefore, are capable of transducing non-dividing cells. Typically, LVVs integrate throughout the genome in areas of active transcription and are not observed to be enriched in regions associated with oncogenesis (Bushman et al., 2005). However, the integration site specificity should be evaluated for each LVV construct as part of its safety assessment. Self-inactivating LVV constructs reduce the risk of vector-mediated transactivation of adjacent genes by eliminating the promoter/enhancer activity of the viral long terminal repeat (LTR) region (Zufferey et al., 1998).

To increase the efficiency of transduction, an activation step is typically included prior to transduction of cells. This increases expression of the receptor required for vector entry (LDL-R in the case of VSV-G pseudotyped LVVs), reduces the impact of host restriction factors, and overcomes the limited availability of dNTP in resting T cells (Amirache et al., 2014; Milone & O'Doherty, 2018). Consequently, the activation step will result in an increased frequency of transduced cells.

While the integration profile of HIV-1 has been well-characterized and exhibits low risk of insertional oncogenesis, the integration of genetic material into a target cell has a theoretical risk of transformation caused by either the knock-out, modulation or over-expression of adjacent genes at the site of integration. Therefore, there must be assurances that the amount of transduction is controlled within the capability of the manufacturing process. This is accomplished by two orthogonal controls: (1) the amount of vector input used in the production of each cellular therapy lot is controlled, and (2) the specification includes an upper limit for transgene integrations per transduced cell – typically called "Vector Copy Number" (VCN). These analytical controls are described in detail in the following section using a hypothetical CAR T cell product as an example.

Control of Vector Dose during the Manufacture of a CAR T Cell Product

The amount of LVV that is added into the manufacturing process of a CAR T cell product impacts the frequency of transduction, the amount of impurities originating from the LVV dose and the theoretical risk of insertional oncogenesis. Consequently, the input of LVV must be controlled to provide the necessary assurances of a safe, effective and consistent cell therapy product throughout product development. Consequently, the FDA typically expects this analytical control to be implemented prior to the initiation of Phase I clinical trials. LVV input to is controlled using a measure of strength (quantity) that is both biologically relevant and can be precisely quantified. Therefore, the measure of strength that is selected and the analytical procedure that measures the attribute is critical. The starting material for LVV infectivity method may be a suitable cell line or healthy donor cells (Considerations for the Development of Chimeric Antigen Receptor CAR T Cell Products, March 2022). Typically, diluted vector is added to cells and a readout of transduction (qPCR or flow cytometry) at harvest is used to calculate the number of functional LVV particles per unit volume.

Control of Transduction

The analytical procedure used to quantify VCN is typically expected to be implemented ahead of initiating Phase I clinical trials due to the safety risk associated with insertional oncogenesis. The analytical procedure is usually based on quantitative polymerase chain reaction (qPCR) technology. Briefly, the analytical procedure is typically a multiplexed reaction that incorporates two sets of primers and probes. The first set of primers and probe is specific for the transgenic sequence, while the second set is specific for a reference gene contained within the unmodified host genome. The number of amplicons measured at a specific quantification cycle (C_q) is calculated based on a reference standard, which is prepared at predetermined concentrations of target and reference gene copies. The quantification of vector and genomic reference gene enables the determination of the average number of vector integrations per cell. Next, the frequency of CAR T cells (i.e., %CD3+CAR+) in the test sample is used as an input to calculate the number of vector integrations per transduced cell.

qPCR-based Analytical Procedures

General considerations in the development of a robust qPCR-based analytical procedure include:

- Primer/probe design (Borah, 2011):
  - Amplicon size should be 75–200 base pairs long
  - Avoiding regions with secondary structures
  - Avoiding G and C repeats longer than three consecutive bases
  - Genome Copies (GC) content between 40%–60%, ideally 50%
  - Primer melting temperature (Tm) should be between 50°C–60°C
  - Ensuring that there is no 3' complementarity between forward and reverse primers to reduce primer-dimer products
  - Probes should have a Tm 5°C–10°C higher than the primers
  - Probes should be less than 30 nucleotides and not contain a G at the 5' end
  - GC content of the probes should be between 30%–80%
- Evaluation of multiple reference genes during development is recommended, and should demonstrate that the selected reference gene has consistent recovery across dilutional linearity experiments using a titration of gDNA to ensure it has suitable performance throughout development (Freitas et al., 2019).

Analytical Methods for In-Process Testing and Product Release

- Accurate detection of the target and reference genes depend on the specificity of primers and probe set. During their design, a high degree of specificity must be demonstrated to mitigate the risk of nonspecific binding and amplification. In addition, the specificity of the primer and probe set must be verified. For example, gel electrophoresis of the qPCR amplicon is commonly used to confirm the intended amplicon size is present in the absence of unanticipated amplicon sizes. Alternatively, the amplicons can be sequenced to verify the specificity of the primer and probe set (Borah, 2011; Kralik & Ricchi, 2017; Taylor et al., 2019)
- Matrix interference must be thoroughly evaluated during assay development, as qPCR reactions can be inhibited by impurities that are present in genomic DNA (gDNA) samples. This type of matrix interference can impact robustness, accuracy, precision linearity and range. Consequently, qPCR inhibition should be evaluated using dilutional linearity experiments. This procedure aims to identify a range of gDNA input concentration that is both suitable for the test sample type and an optimal recovery is observed. For example, a gDNA sample that is serially diluted tenfold is expected to have a difference in Cq of 3.232 (100% amplification efficiency). Amplification efficiencies between 90%–110% are considered acceptable for qPCR reactions (Borah, 2011; Kralik & Ricchi, 2017; Taylor et al., 2019).
- The DNA Polymerase, dNTPs, Passive Reference dye and other buffer components can be combined into a "master mix." Whether these reagents are added individually or together in a master mix, their respective concentrations should be evaluated and optimized during development to maximize analytical robustness (Borah, 2011; Kralik & Ricchi, 2017; Taylor et al., 2019).

Additional information on the development of robust, sensitive, accurate, precise and linear qPCR assays are described in the Minimum Information for Publication of Real Time – PCR Experiments (MIQE) (Bustin et al., 2009) and the Ultimate qPCR Experiment: producing publication quality, reproducible data the first time (Taylor et al., 2019).

Alternative Technology Platforms

qPCR is currently the industry standard for evaluating VCN. In addition to being a proven QC platform, other advantages of this technology include low operational costs and availability of qPCR instruments that are GMP compliant. However, significant drawbacks to qPCR assays include sensitivity to PCR inhibitors present in the sample (e.g., matrix interference), relatively inferior sensitivity compared to alternative approaches, and a requirement for a reference standard curve to calculate the final reportable value. Alternatively, droplet digital PCR (ddPCR) is a promising technology platform that offers several advantages over qPCR. For example, ddPCR is less effected by the matrix interference (e.g., PCR inhibitors).

In addition, it is claimed that ddPCR quantification does not require a standard curve due to absolute quantification of positive droplets. Considering that the ddPCR reaction occurs in thousands of individual droplets, the assay sensitivity is theoretically lower and can be applied to targets that are extremely low in abundance in mixed matrices (Taylor et al., 2019). The authors recommend robust evaluation of available ddPCR instruments for GMP compliance and the usability of ddPCR-based analytical procedures in a QC environment.

GENOME EDITING AND CHARACTERIZATION OF EDITED CELLS

Genome Editing Technologies

Genome editing enables the generation of engineered cells with therapeutic advantages over other approaches. For example, genome editing can be utilized to engineer off-the-shelf universal cells for allogeneic cell therapy by disrupting TCR and HLA-related genes to avoid graft-versus-host and host-versus-graft immune reactions, respectively, whereas knocking out immune checkpoint genes such as PD1 can potentially overcome part of the challenges associated with the inhibitory tumor microenvironment for the treatment of solid tumors.

The diversity of genome editing tools have rapidly expanded in the past few years. Besides the synthetic triplex-forming peptide nucleic acids (PNA)-based system (Ricciardi et al., 2014) and the less target-specific transposase-based system (Sandoval-Villegas et al., 2021), a few programmable endonucleases have been developed for targeted genome editing purposes. These include meganucleases (Silva et al., 2011), zinc finger nucleases (ZFNs) (Carroll, 2011) transcription activator like effector nucleases (TALENs) (Bhardwaj and Nain, 2021) and the clustered regularly interspaced short palindromic repeats (CRISPR) – associated protein (Cas) system (Katti et al., 2022). In addition, several modified nucleases have also been developed, such as high-fidelity (HiFi), cleavage-inactivated, single-strand nicking, base-editing or prime-editing versions. Each genome editing nuclease has unique properties that should be considered when selecting them for either gene disruption, gene addition or transcriptional/epigenetic regulation of gene expression. Among the few genome editing systems, the CRISPR/Cas system is the most widely used system by scientists in various academic and pharmaceutical fields due to its simplicity in guide RNA (gRNA) design. Among the various types of CRISPR/Cas system, the Cas9 and Cas12a-based systems are the most used endonucleases for genomic editing of T cells.

To engineer T cells, genome editing nucleases are typically delivered *ex vivo*. CRISPR/Cas9 system is used as an example in this section, which requires Cas9, guide RNA and the transgenic sequence. Cas9 can be delivered in the form of DNA, mRNA, protein, ribonucleoprotein (RNP), nonviral nanoparticle or viral particle. Guide RNA can be delivered as a single piece (sgRNA) or as two pieces (crRNA and tracrRNA). In either situation, the guide RNA can be delivered in the form of DNA, RNA, RNP, nonviral nanoparticle or viral particle. One of the most widely used approaches delivers both Cas9 and gRNA together as RNP by electroporation. For targeted integration through homology directed repair (HDR),

Analytical Methods for In-Process Testing and Product Release

transgenic sequences can be delivered as DNA, nonviral nanoparticle or viral particle. Adeno-associated virus type 6 (AAV6) is one of the most frequently employed choices for delivery of transgenic sequences.

The quality of edited T cells is impacted by the choice of both the genome editing tools and the delivery method for transgenic sequences. To ensure the safety and efficacy of the drug product, it is necessary to control the genome editing nuclease components and the editing process. For example, to ensure quality of sgRNA and recombinant Cas9 (or the quality of sgRNA and Cas9 pre-combined as RNP), product-specific and phase-appropriate analytical procedures can be developed to evaluate identity, purity, process-related impurities, potency, pyrogens such as endotoxin, sterility, mycoplasma and the general attributes such as appearance, pH and moisture content. The risk of residual genome editing components in cell therapy products should be evaluated with analytical procedures that are scientifically justified as suitable for use.

Intended Genome Editing Outcomes

Targeted nucleases promote genome editing through the creation of a site-specific DNA double-strand break (DSB) at a predefined site in the genome. Repair of DSBs can occur through either the non-homologous end joining (NHEJ) or HDR pathway. NHEJ causes insertions and deletions (indels) at the cleavage site and is often utilized for gene disruption/knock-out (KO) applications. HDR uses a homologous transgenic sequence as a template for high fidelity DNA repair to either correct a target gene or to knock-in (KI) a transgene at the intended cleavage site in a predetermined fashion. HDR is often utilized for gene correction, gene addition or gene disruption.

The efficiency of genome editing should be evaluated based on the frequency of the intended outcome. This CQA should be assessed during development of the manufacturing process, and there should be process controls and/or release testing to provide the necessary assurances of control. The control strategy that is employed should be justified using a science and risk-based approach.

There are a variety of analytical procedures that can be developed to quantify the intended genome editing frequency at the genomic level. The selection of an optimal technology platform is dependent on multiple factors such as availability of instruments, cost of consumables, usability in a QC environment and compliance with GMP regulations (e.g., IQ, OQ, PQ, 21 CFR part 11). The frequency of KIs can be quantified by quantitative PCR (qPCR), droplet digital PCR (ddPCR) or other technology platforms such as a PCR-restriction fragment length polymorphism (PCR-RFLP) assay, which is more challenging to be implemented in a QC environment (Ota et al., 2007). However, due to the large diversity of potential indels in edited cells, some of which may have only small differences compared to the unedited sequence, direct quantitation of indels using qPCR is impractical. Multiple methods have been explored to quantitate frequencies of indels in edited cells, such as:

- Topo cloning of target region–specific PCR amplicons and Sanger sequencing (Bennett et al., 2020)

- Direct Sanger sequencing of target region–specific PCR amplicons followed with bioinformatic data deconvolution (Bennett et al., 2020)
- Digestion of re-annealed target region–specific PCR amplicons with mismatch-targeting enzyme (such as T7 endonuclease 1 or surveyor nuclease) followed by separation with gel or capillary electrophoresis (Vouillot et al., 2015)
- Indel detection by amplicon analysis (IDAA) utilizes triple-primer PCR amplification to fluorescently tag the target-region amplicons followed by separation with capillary electrophoresis (Yang et al., 2015)
- Genome editing test PCR (getPCR) by combining the sensitivity of Taq polymerase to mismatch at primer 3' end with real-time PCR technique (Li et al., 2019)
- Quantitative Evaluation of CRISPR/Cas9-mediated editing (qEva-CRISPR), a ligation-based dosage-sensitive method utilizing two short half-probes (5' and 3' half-probes) for contiguous target-specificity (Dabrowska et al., 2018)
- Gene-editing frequency digital PCR (GEF-dPCR) or drop-off droplet digital PCR (ddPCR) utilizes mechanical emulsification to partition PCR solution into thousands of nanoliter droplets/reactions for PCR amplification, followed by sequentially reading signals from individual droplets and performing quantitation based on Poisson distribution (Mock et al., 2016)
- Target-specific, next-generation sequencing (NGS) utilizes high throughput deep sequencing technologies to sequence massive target-region amplicons in parallel (Bennett et al., 2020)

Among the various approaches to monitoring indels, analytical procedures based on ddPCR are one of the most promising analytical approaches due to their robustness, sensitivity, precision, accuracy and absolute quantification that does not require standard curves.

It is also possible to monitor the efficiency of genome editing by evaluating its downstream effects, such as measuring changes in protein expression or T cell function. For example, an antibody against a protein may be added to a flow cytometry-based method to quantitate a target gene KO and/or transgene KI. At later stages of clinical development (Phase III/Pivotal), functional assays should be developed to (1) demonstrate the structure-function relationship between the intended genomic editing and functional outcomes, and (2) demonstrate process control during PPQ and comparability studies. The analytical approach for monitoring genome editing should be justified by a science and risk-based approach.

Unintended Genome Editing Outcomes

In addition to the intended outcomes of genome editing, there can be unintended outcomes that are a risk to safety and efficacy, such as indels at off-target sites and chromosome rearrangements, including inversion, duplication and translocation. Consequently, the presence of unintended modifications should be evaluated prior to the initiation of clinical trials, monitored during development of the manufacturing process and included in characterization testing that supports process PQ and comparability studies. If an unintended modification is identified in a cell therapy product and verified in multiple lots, it is a nonrandom event that may necessitate a control strategy to be implemented as soon as possible to ensure the safety risk remains adequately controlled. The control strategy may include, but is not limited to, the optimization of the manufacturing process to control the unintended modification, routine monitoring during manufacturing and/or release testing.

The identification of unintended modifications is challenging due to their inherently unpredictable nature. There are multiple different analytical approaches that can identify these off-target sites, such as:

- *In silico* prediction based on genome-wide search for genomic sequences that are similar to the target sequence
- Biochemical methods involving *in vitro* digestion of purified cell-free non-edited genomic DNA with a RNP before NGS library preparation, such as:
  - *in vitro* Cas9-Digested whole Genome Sequencing (Digenome-Seq) (Kim et al., 2015)
  - Digenome-Seq using cell-free chromatin DNA (DIG-Seq) (Kim & Kim, 2018), Selective enrichment and Identification of Tagged genomic DNA Ends by Sequencing (SITE-Seq) (Cameron et al., 2017)
  - Circularization for *In vitro* Reporting of Cleavage Effects by Sequencing (CIRCLE-Seq) (Tsai et al., 2017)
  - Circularization for High-Throughput Analysis of Nuclease Genome-Wide Effects by Sequencing (CHANGE-seq) (Lazzarotto et al., 2020)
- Cellular methods involve genome editing of cells before genomic DNA extraction and NGS library preparation, such as:
  - Genome-wide Unbiased Identification of DSBs Enabled by Sequencing (GUIDE-Seq) (Tsai et al., 2015)
  - *In situ* Breaks Labeling, enrichment on Streptavidin and next-generation Sequencing (BLESS; Crosetto et al., 2013)
  - Breaks Labeling *In Situ* and Sequencing (BLISS) (Yan et al., 2017)
  - Discovery of In Situ Cas Off-Targets and VERification by Sequencing (Discover-Seq) (Wienert et al., 2019)
  - Whole genome sequencing (Kaeuferle et al., 2022)

Translocation is one of the most well-known chromosome rearrangements. If multiple DSBs occurred simultaneously during the genome editing process, chromosome translocation can potentially occur between two on-target sites, between an on-target and an off-target site or between two off-target sites. However, the process to form translocations is usually much less efficient than the process to form indels. Genome-wide chromosome rearrangements can be evaluated by karyotyping or NGS-based methods such as:

- High Throughput, Genome-Wide Translocation Sequencing (HTGTS) (Chiarle et al., 2011)

- Uni-Directional Targeted Sequencing methodology (UDiTas) (Giannoukos et al., 2018)
- Chromosomal Aberrations Analysis by Single Targeted Linker-Mediated PCR Sequencing (CAST-Seq) (Turchiano et al., 2021).

Since each analytical approach has limitations, it may be beneficial to employ complementary analytical approaches to both overcome limitations and verify findings. For example, once candidate off-target sites have been identified, they may be verified by a target-specific NGS method.

A verified unintended modification of genome editing should be routinely monitored using an appropriate quantitative analytical procedure, such as a ddPCR-based method. Furthermore, the biological impact of off-target modifications should be evaluated. However, the functional impact of off-target editing is highly dependent on the exact location and sequence of the unintended editing. For example, editing at an intergenic or intronic region is generally considered low safety risk whereas editing at an exonic region of a critical gene such as a tumor suppressor gene is generally considered high safety risk. Due to the presence of large diversity of potential indels at either on-target or off-target sites, the functional impact of each indel is impractical or impossible to be fully evaluated. As a result, the predicted safety risk should be assessed based on existing knowledge for each verified off-target site. In addition, a tumorigenicity assay (e.g., an IL-2 independent growth assay) should be performed to evaluate whether the genome edited cells (DPs) contain transformed cells. Prior to the initiation of efficacy-indicating clinical trials (pivotal trials), the impact of any verified off-target modification to the intended biological activities of the cell therapy should be assessed using appropriate analytical procedures for measuring these functional characteristics. To facilitate the development and execution of these analytical procedures, it might be beneficial to generate material that is highly enriched in the verified off-target modification (e.g., via purification, modified manufacturing processes or deliberate engineering).

IDENTITY

In all phases of clinical development and commercial, the specification must include an analytical procedure and acceptance criteria that confirms the identity of each lot of drug substance and drug product (21 CFR 312.23(a)(7) and 21 CFR 610.14). Identity testing must be performed on the final labeled product (after all labeling operations have completed) (21 CFR 610.14), and if necessary, can consist of more than one analytical procedure (ICH Q2(R1)). The analytical procedure(s) must be specific for the intended product and must be able to discriminate between other products that are manufactured at the same facility (21 CFR 610.14). The FDA recommends that identity confirm both the presence of the intended transgene required for the MoA and the cellular composition of the final product (CMC Information for Human Gene Therapy INDs). For example, a CAR T cell product may have a matrix approach to identity that includes an analytical procedure that confirms the presence of the transgene (e.g., VCN by qPCR) and another that confirms the intended cellular composition is present (e.g., CD3+CAR+ by flow cytometry).

Identity is a qualitative attribute (ICH Q2(R1)). If a quantitative method is used to establish identity (e.g., qPCR, flow cytometry), a quantitative threshold should be established to define a positive qualitative result. This threshold can be established as the LOD of the analytical procedure, as demonstrated according to guidance described in Section 6 of ICH Q2(R1), although alternative approaches are acceptable if justified. The IND must include information that demonstrates the analytical procedures(s) has sufficient specificity to verify the identity of the product for use in clinical trials (21 CFR 312.23(a)(7) and Considerations for the Development of Chimeric Antigen Receptor (CAR) T Cell Products), and the BLA must include information that demonstrates the specificity has been validated (21 CFR 211.165(e), 211.194(a)(2)), and ICH Q2(R1).

PURITY AND CHARACTERIZATION OF PHENOTYPE

The purity and phenotypic characterization of cell therapies is traditionally measured by analytical procedures based on flow cytometry. Flow cytometry is a technology that allows for the simultaneous analysis of multiple phenotypic characteristics of individual cells in a single assay. This is accomplished through immunostaining test samples with different fluorophore-conjugated antibodies and then their subsequent excitation by lasers, separation of photons based on their emitted wavelength, and measurement by detectors.

In the cell therapy field, analytical procedures based on flow cytometry are used to support many aspects of product development, manufacturing and release. This includes analytical procedures that are used to increase product and process knowledge, demonstrate manufacturing process control and release the product to patients. Flow cytometry is a technology platform that requires special considerations to ensure analytical procedures are accurate, precise, robust and consistent across time, analysts and testing sites. Here, in this section, key topics and recommendations are provided for the development and life cycle management of analytical procedures based on flow cytometry.

General Considerations for Analytical Flow Cytometry Methods
Antibody Staining Panel Design
Antibody-fluorophore panel design for cell therapy products follows the same principles as any other flow cytometry panel. Briefly, developers should consider the cytometer configuration, the expression levels of the cellular markers to be measured, and the fluorophore brightness. It is best practice to pair low expression antigen or antigens that require high sensitivity on bright fluorophores, while dim fluorophores are better suited for highly expressed antigens. Finally, once the antibody-fluorophore format is determined, it is recommended to evaluate the cytometer-specific spillover spreading matrix (SSM) to pre-emptively identify any spectral issues associated with the selected fluorophores, and if needed,

further refine the pairing of antigen-specific antibodies and fluorophore. One final consideration that is distinct to staining panels used to evaluate cell therapy products is the fluorophore selection for the identity reagent, which is used to measure the attributes of drug product identity (e.g., CD3+CAR+) and strength (e.g., %CD3+CAR+ used to determine live CD3+CAR+ cells/mL). This antibody should be placed on a fluorophore that is spectrally separate from other markers in the panel to avoid compensation issues that might impact the quantification of that attribute.

Additional considerations should be taken to consider the fluorochrome pairing with the antibody for drug product identification, if applicable, (i.e., CAR) to ensure that attributes inherent to the cells, such as autofluorescence, do not interfere with the fluorescent signal identifying these important, often dose-defining populations. Some conjugations are simply not suitable between a mAB and a fluorochrome and so stability of the antibody should be confirmed during development prior to moving into qualification or testing clinical drug products.

Critical Reagent Management

All reagents that are critical to the performance of an analytical procedure should be thoroughly evaluated during development to ensure appropriate specificity, volume per vial (preferable single use), number of vials in each lot, consistent performance across lots and acceptable stability (shelf-life). In addition, critical reagent qualification procedures should be established to ensure that all lots used in the analytical procedure are sufficiently equivalent to ensure consistent analytical performance. The most important critical reagents for analytical procedures based on flow cytometry are staining reagents, such as fluorophore-conjugated antibodies and dyes used to exclude (gate out) dead cells. Since many antibodies for flow cytometry are commercially sourced, attributes such as reagent specificity, storage conditions and shelf-life should be provided by the vendor, or alternatively could be demonstrated by the method developer. Due to potential variability between antibody lots, it is recommended that new antibody lots are calibrated to the previous lot through an appropriate critical reagent qualification procedure, such as titrating the volume of a new reagent lot until its mean fluorescent intensity (MFI) is equivalent to the previously qualified critical reagent lot. Frequency of a population should be confirmed in a two-part qualification where the newly qualified antibody is tested in context of a panel and performance is confirmed prior to use. Similarly, exclusion dyes are important for the removal of dead cells that can impact flow cytometry results through autofluorescence and nonspecific staining, and therefore, should be managed as a critical reagent. Although various dyes exist for discrimination of dead cells, amine-reactive dyes are often preferred due to the ability to fix cells stained with these reagents and the wide availability of different fluorescent dyes that allow for easy integration into staining panels.

One critical reagent that warrants special consideration during development is the reagent used to detect expression of the transgene. This is usually an antibody that is specific to the active moiety that confers specificity of the cell therapy, such as the CAR in a CAR T cell, and it is usually used to evaluate identity (presence of CD3+CAR+ cells), CAR T cell purity (%CD3+CAR+) and strength (%CD3+CAR+ is used to calculate live CD3+CAR+ cells/ml). Evaluation of specificity, resolution (separation of positive and negative populations), consistency between lots, stability and supply should be conducted at the earliest stages in the development of this analytical procedure. If this reagent is developed and produced in-house, it may be beneficial to establish a supply agreement with a reputable vendor for later stages of clinical development and commercial manufacture.

Critical Instruments

There can be considerable variability in flow cytometers between vendors, models, individual instruments and across time. The source of this variability is sample injection, fluidics, laser type, laser wavelength, laser power, filters and optics and detectors. For method developers, key parameters to consider when selecting a suitable cytometer platform include the laser and detector configurations, detector stability, dynamic range, electronic noise, fluidics and detector sensitivity, accuracy, precision and linearity.

There are important differences in the needs of a flow cytometry used in an analytical procedure that is implemented for QC release and stability testing versus characterization testing. For example, flow cytometers used in QC must have an IQ, OQ, PQ and be 21 CFR Part 11 compliant to meet regulatory requirements. Furthermore, flow cytometers that support QC release testing benefit from being operator-friendly, highly reliable, accurate, precise and highly reproducible across instruments. Finally, it may be necessary to implement routine instrument calibration and critical instrument PQ procedures to ensure performance across time, instruments and testing sites remains consistent. In contrast to flow cytometry instruments used in QC, those used in development for product characterization may prioritize high parameter configurations to support analyses of a larger number of attributes. These might include both CQAs and exploratory attributes used in correlative analysis to identify new CQAs. Examples might include cell viability, cell health, differentiation, homing/trafficking receptors, activation, exhaustion, anergy and senescence.

QC Flow Cytometry Methods

Purity (e.g., %CD3+ and %CD3+CAR+), identity (e.g., presence of CAR-positive T cells), and strength (e.g., viable CAR T cells/mL are typically reported by analytical procedures based on flow cytometry. Note, T cell Purity (i.e., %CD3+) is typically a CQA used to control the level of non-T cell impurities. These QC methods must be stringently developed, optimized, qualified prior to the initiation of Phase I clinical trials, and validated ahead of process PQ. A brief summary of the development stages is described next.

Method Development

In the initial stages of developing a flow cytometry method for QC, it is important to define the attributes that will be reported by the analytical procedure (e.g., CD3 and CAR), the staining required to report those attributes (e.g., stains to exclude dead cells), and the gating tree necessary for reporting each attribute. This process aims to identify the minimum requirements of the analytical procedure, and minimize the likelihood of unessential stains, gating and/or attributes that would add unnecessary complexity to the QC environment. Method developers should then identify the analytical controls that are necessary to ensure appropriate method performance and objectivity of reported results. For example, this will include at a minimum an assay control lot and fluorescence minus one (FMO). Next, the properties of an easy to perform (e.g., automatable), consistent and robust analytical procedure in QC should be considered to inform subsequent method development activities. Finally, the dead cell stains, antibody clones, and antibody-fluorophore combinations should be selected for optimal staining (see 7.5.4.1.1).

The performance of the staining panel should be evaluated on multiple lots of representative material, which is ideally lots that were manufactured using patient starting material rather than healthy donor material. If the desired resolution has not been obtained between positive and negative cell populations, the panel design should be revisited. The ability to maintain these critical reagents across the analytical life cycle should be carefully considered, and if risk is identified, its mitigation should be prioritized.

Method Optimization

Once the dead cell stain, antibody clones and antibody-fluorophore combinations have been selected, the assay should be thoroughly optimized to ensure consistent performance. The aim of this stage is to determine the reagents, concentrations, procedures and operating conditions that are necessary for a robust analytical procedure that is easy to perform in a QC environment. Optimization experiments may consider the concentration of critical reagents, fixatives and buffers, the length of incubations, temperatures, etc. (Du et al., 2015). Optimization may consider alternative reagents, techniques and/or instruments that increase ease of use, reduce cost of goods or improve analytical performance and consistency.

Method Robustness

The robustness of analytical procedures should be evaluated across the operating ranges that were determined during method optimization to verify their robustness. Examples include the any incubation or temperature ranges the assay can be executed, stability of antibody cocktails, post-thaw stability of test samples and post-fix stability of test samples. This ensures consistent analytical performance across the operating ranges QC analysts require to perform it during routine use. The robustness experiments, results and conclusions should be described in a method development report.

Characterization Methods That Evaluate T Cell Phenotype and Cellular Impurities

Characterization methods can be used to support process development, investigations, process characterization, process PQ and comparability assessments.

Analytical Methods for In-Process Testing and Product Release

Characterization methods can also be used to report exploratory attributes that expand product knowledge through analysis of their correlation with clinical outcomes. Characterization methods should be scientifically sound and have suitable specificity, sensitivity (LOQ), accuracy, precision, linearity and range for their intended use. There may be a reduced scope and stringency for characterization methods used in early clinical development compared to those supporting late clinical development or commercial. For example, characterization methods should be qualified prior to their use in process PQs and comparability assessments.

It might be necessary to modify, optimize, or replace characterization methods over the course of a product life cycle as product knowledge increases and/or limitations in analytical performance are identified. In this event, a comparison of method performance will need to be conducted (i.e., method bridging study) to support the implementation of the revised method by enabling comparison of data across the program life cycle.

POTENCY

Potency is defined by 21 CFR 600.3(s) as the "specific ability or capacity of the product, as indicated by appropriate laboratory tests, to effect a given result." Potency is a CQA and an appropriate analytical procedure should be implemented ahead of efficacy-indicating (pivotal) clinical trials. Potency assays are required to support product conformance, product stability, process PQ and comparability studies. A commercial specification will include acceptance criteria for one or more measures of potency in order to ensure that released lots are within both process capability and experience demonstrated to be safe and effective during clinical trials.

It is well-recognized that cell therapy and gene therapy products can have complex mechanisms of action and modes of action that may not be fully understood by the initiation of clinical trials. Because of this, regulatory authorities recommend a phase-based approach to potency testing. During this early stage of clinical development, the product should be thoroughly characterized with a suite of analytical techniques to improve product understanding. This aims to identify functional attributes that reflect the MoA and mode of action, determine those that are most appropriate for demonstrating manufacturing process control and product safety and efficacy and enable time to develop robust analytical procedures to measure these measures of potency. One or more potency assay(s) should be implemented for QC release and stability testing prior to the initiation of efficacy-indicating clinical trials. These should be well-characterized with specification acceptance criteria that ensure that released lots of the product have consistent biological activity. This ensures that released products are sufficiently controlled for clinical trials design to evaluate the efficacy of a therapeutic.

The exception to the phase-based approach to potency is measures of strength and cell viability. All phases of product development, including early phase clinical trials, require a measure of strength (quantity) for dosing. It is a regulatory expectation that strength quantifies viable cells (e.g., <u>viable</u> CAR T cells/mL), therefore a measure of viable cell concentration (VCC) is also required to initiate Phase I clinical trials. In addition, the percentage of viable cells (PVC) is typically used to control the concentration of nonviable cells that are present in the product administered to patients. Since nonviable cells are an impurity that represents a risk to patient safety, PVC is a CQA that is also required on the specification prior to the initiation of Phase I clinical trials.

Percent Viable Cells and VCC

There are two measures of cell viability that are relevant to the quality of a cell therapy product. The first is the PVC, which ensures that the amount of nonviable cells that would be administered to patients' remains controlled. This CQA may be a stability-indicating attribute that can be useful in increasing process and product understanding, and if verified as a stability-indicating attribute, provide assurances of product stability when justifying shelf-life. The second measure relevant to product quality is the VCC, which is usually during determined alongside the PVC. Strength (quantity) must quantify the active ingredient in terms of viable cells.

PVC and VCC are typically evaluated through the entire manufacturing process to monitor process performance. In addition, PVC and VCC can serve as in process controls (e.g., vector volume required for viral transduction). Finally, PVC and VCC are CQAs that are evaluated as part of QC release and stability testing.

Analytical Performance Requirements

An analytical procedure that reports PVC and/or VCC should be capable of direct cell counting. Direct cell counting identifies a signal from each cell, which is then counted manually or by an instrument. Fluorescence is commonly used as the signal of choice and various staining dyes can identify differential cell subsets that distinguish cells based on labeling of distinct cell attributes (Table 7.2). To report VCC, the analytical procedure should quantify live and dead cells in a defined volume, although other approaches may be acceptable with appropriate controls, demonstrated analytical performance, and justification (e.g., VCC

TABLE 7.2
Example Dyes and Stains for Identifying Live and Dead Cells

Dye	Targets	Identifies
Acridine Orange (AO)	Nuclear DNA, permeable to all cells	Live cells with intact membrane
Calcein AM	Active, cytosolic esterases	Live, metabolically active cells
Propidium Iodide (PI)	Nuclear DNA, permeable to cells with compromised membranes	Dead, necrotic cells
7-Aminoactinomycin (7-AAD)	Nuclear DNA, permeable to cells with compromised membranes	Dead, necrotic cells
Annexin V	Membrane-bound phosphatidylserine	Dying, apoptotic cells

Analytical Methods for In-Process Testing and Product Release

by flow cytometry using counting beads). Analytical procedures that report PVC and/or VCC should be able to identify and quantify both live cells and dead cells unequivocally and objectively. These attributes can then be used to calculate PVC as the percentage of live to total cells and VCC as the number of live cells in a given volume.

Instrumentation and Critical Reagents

Advances in instrumentation and technologies have led to a growth in methods capable of evaluating PVC and VCC. In general, cell viability methods can be categorized into manual and automated assessments. Manual assessments should be avoided to minimize subjectivity and improve reproducibility. Instruments with automated image capture and analysis provide the least biased assessment of PVC and VCC. An instrument should be selected based on capability, throughput, precision and regulatory compliance. The instrument should be able to assess the signal of choice and the settings customized to optimize the identification of live cells, dead cells and debris. Instruments using a plate-based assessment provide higher throughput analysis that enables replicate determinations for each test sample, which increases the accuracy and precision of the assay. As an autologous product moves into commercial, a higher throughput option allows for batch testing multiple test samples in one test session, easing resource and time burdens for analysts. Similarly, to all analytical procedures used in QC release and stability, the instrument will require installation qualification, operation qualification and performance qualification (IQ/OQ/PQ) and all software must be 21 CFR Part 11 compliant.

Analytical Development

The performance of the analytical procedure should be thoroughly optimized during development. For example, any cellular aggregates should be disrupted by mixing prior to staining and quantification. The selection of dyes, concentration and volume should be optimized to ensure robustness. The analytical procedure should include an assay control, which could be representative material or an engineered sample that is sensitive to the same types of sample handling and assay artifacts as the test samples. The assay control should have AAC that establishes the validity of each test result (by demonstrating the analytical procedure is performing as expected). Finally, the analytical procedure will benefit from replicate determinations to improve precision and enable SAC that ensures analytical performance remains within expectations. The results from the assay control and variance between replicate determinations can be monitored using control chart trending to ensure the analytical procedure remains in the state of control throughout its analytical life cycle.

Demonstrating Suitable for Its Intended Use

All analytical procedures used for QC release and stability testing should undergo a performance evaluation that demonstrates it is suitable for its intended use. The general requirements of the performance evaluation are described in more detail in the introduction, including phase-specific considerations. There are several unique considerations to evaluating the performance of analytical procedures that report PVC and VCC, including the design of the dilutional linearity studies, the type of samples used for demonstrating dilutional linearity and the use of orthogonal analytical procedures to verify the accuracy of the reported results.

Dilutional linearity should test relative accuracy since cell counting methods do not have standardized reference materials. Per ISO 20391-2, dilutional linearity should be assessed from a stock sample of a known concentration. Dilutions that span the intended range of the method, including the lower and upper limits of quantitation, should be prepared from the stock sample and assessed over a series of measurements. Samples selected for demonstrating dilutional linearity should be representative of the final drug products. Samples should demonstrate similar biologic properties to ensure staining is identifying characteristics of the final sample and should be sensitive to sample handling artifacts similar to the final product. Because there is a lack of standardized reference materials for cell counting, it is pertinent to assess cell counting accuracy using an orthogonal method (i.e., a manual, dye exclusion-based method to test accuracy of an automated, fluorescence-based method). This assessment gives assurances that the dyes selected is accurately identifying the live and dead cell and that the counting method (automated versus manual) is not introducing any bias.

Strength

Strength is a CQA that is required for dosing, and, consequently, it is required on the specification for QC release and stability during all phases of clinical development and commercial. Strength quantifies the number of viable cells containing the active ingredient (e.g., <u>viable</u> CAR T cells/mL). An example calculation is described in Equation 1. This equation utilizes VCC and CAR T cell Purity (CAR frequency, %CD3+CAR+), which is typically determined by flow cytometry (see section 5).

Equation (1):

$$\text{Strength} = \text{Viable Cell Concentration} \times \left(\frac{\text{cells}}{(\text{mL})}\right)(\text{CAR Frequency})(\%)$$

$$= \frac{(\text{Viable CAR}^+ \text{T cells})}{\text{mL}}$$

Biological Activities Relevant to the MoA and/or Mode of Action

Analytical procedures that measure the biological activity of cell therapy products ensure the potency of the product remains adequately controlled. There may be numerous biological activities that are relevant to the MoA and/or mode of action. Analytical procedures should be developed to quantify each of these biological activities, with one typically utilized for QC release and stability testing and the others for characterization used to support product development, process performance characterization and comparability studies. Ideally, each of these analytical procedures would correlate with clinical efficacy. However, such a correlation might not be possible due to the complexity of biological activities that are relevant to product efficacy and an inability of a limited number of *in vitro* assays to recapitulate the *in vivo* environment of each patient (Carmen et al., 2012; Karanu et al., 2020).

In this circumstance, the analytical procedure used to measure potency for QC release and stability testing should be selected due to its specificity for the product, relevance to the MoA or mode of action, ability to demonstrate manufacturing process control and product stability, performance of the analytical procedure, suitability for application in a QC environment, and ability to manage the analytical procedure across the product life cycle. This method(s) should be developed and its performance demonstrated prior to the initiation of clinical trials used to evaluate efficacy (i.e., Phase 3 or the pivotal trial), whereas the characterization methods should be developed and their performance demonstrated ahead of process PQ, but preferably earlier (e.g., ahead of process characterization or sooner).

Analytical Performance Requirements

All analytical procedures that quantify measures of biological activity must be specific for the engagement of the activity moiety within the active ingredient(s). For example, in a CAR T cell product, this would be the quantification of biological activity that results from antigen-specific stimulation through the CAR. This contrasts with nonspecific biological activities that might result from cell-to-cell interactions that are not specific to the CAR construct.

The biological activities that are generally relevant to CAR T cell products are discussed. The MoA for a typical CAR T cell product is tumor cell killing through cytotoxic activity. The mode of action may include antigen-specific expansion through proliferation and the release of anti-tumorigenic cytokines such as IFNg. The QC release and stability method may be selected based on its relevance to the MoA, mode of action and/or correlations with clinical outcomes (safety and efficacy). If a correlation is not identified, the analytical procedure may be selected based on its analytical performance (robustness, sensitivity, accuracy, precision, linearity and range), ability to provide assurance of manufacturing process control such as product stability, suitability for a QC environment and/or ability to maintain over a program life cycle. Table 7.3 summarizes the potency assays for several FDA approved cellular therapeutics (adapted from Karanu et al., 2020).

Once the analytical procedure(s) is selected for QC release and stability testing, the other analytical procedures that quantify orthogonal and relevant measures of biological activity should be considered for characterization testing. These analytical procedures are used to support process development, process PQs, and comparability studies and/or to expand process and product understanding. Analytical procedures that are used to support process PQs and/or comparability studies should be demonstrated as scientifically suitable (qualified) for their intended use, including an evaluation of their specificity, accuracy, precision, linearity and range.

TABLE 7.3
Potency Assays for FDA Approved Cell Therapies

Product	Cellular Identity	Current Potency Assays
Cord Blood	Hematopoietic Progenitor Cells	Total nucleated cells, CD34+ cells, PVC and strength
Yescarta, Kymriah, Breyanzi	Anti-CD19 CAR T Cells	PVC, CAR T cell Purity, IFNγ Secretion and strength
MACI	Chondrocytes	Expression of Hyaline 1 and strength
Gintuit	Keratinocytes and Fibroblasts	Cytokine-release Assay (unknown MoA) and strength
Laviv	Postauricular Fibroblasts	Strength (unknown MoA)
Provenge	CD54+ Dendritic Cells	Antigen-specific CD54 upregulation and strength

Adapted from Karanu et al. (2020).

Analytical Development

There are usually two stages in an analytical procedure that quantifies the biological activity of CAR T cells. Firstly, CAR T cells must be stimulated by their target antigen and, secondly, the quantification of the resulting biological activity. The first stage is often accomplished using a bioassay, in which CAR T cells are stimulated through co-cultured with antigen-expressing target cells. The stimulation of CAR T cells benefits from optimization, including but not limited to, target antigen expression on cancer cells, ratio of targets to effectors, cell density in the co-culture, and duration of stimulation. The goal of this optimization is to identify conditions that are robust across different lots of drug product, insensitive to minor fluctuations in input or execution and generate the appropriate functional response.

The quantification stage measures the antigen-specific biological activity. For secreted cytokines, an ELISA-based method can be used to quantify the amount of secreted cytokine. Quantification is enabled by interpolation of the measured response in a test sample using a cytokine reference standard curve. Careful considerations should be given to identify a recombinant cytokine reference standard and antibody pairs that equivalently recognize secreted cytokine from test samples to the recombinant cytokine reference standard. This can be evaluated by analyzing the full dose-response relationship using a titration of cytokines from test samples and the recombinant cytokine reference standard. Monoclonal antibodies are preferred for this type of method as they can be more easily controlled across lots.

The analytical procedures should include an assay control, which could be representative material or an engineered sample that is sensitive to the same types of sample handling and assay artifacts as the test samples. The assay control should have AAC that establishes the validity of each test result (by demonstrating the analytical procedure is performing as expected). Finally, the analytical procedure will benefit from replicate determinations to improve precision and enable

Analytical Methods for In-Process Testing and Product Release

SAC that ensures analytical performance remains within expectations. The results from the assay control and variance between replicate determinations can be monitored using control chart trending to ensure the analytical procedure remains in the state of control throughout its analytical life cycle.

Instrumentation and Critical Reagents

All instruments and reagents that affect the performance of the analytical procedure should be identified as critical. All critical reagents should have a qualification procedure developed with predefined acceptance criteria to demonstrate each lot is suitable for use prior to their implementation in QC methods. This can be accomplished by demonstrating equivalent performance between an existing qualified lot and a new lot.

PROCESS-RELATED IMPURITIES

Cell therapy products contain impurities that are categorized as either product- or process-related impurities. Product-related impurities are cellular impurities that originate from patient-derived leukapheresis starting material or arise during gene editing. For example, these may include non-T cells, nonviable T cells and nontransduced, partially gene-edited or incorrectly edited T cells. These were discussed in further detail in the section titled "Purity and Characterization of Phenotype." Process-related impurities, on the other hand, include ancillary materials used during the manufacture of cell therapy products that are not intended to be part of the final product. These include residual ancillary materials, gene editing materials and their associated impurities; examples are described in Table 7.4. For a given cell therapy product, the bill of materials for the vector, gene editing materials and drug product serve as the source document for identifying all potential process-related impurities.

Since process-related impurities are not intended to be part of the final drug product, their potential impact to the safety and efficacy of the product must be understood. First, an impurity risk assessment should be performed to understand

TABLE 7.4
Broad Categories of Process-Related Impurities Are Listed below with Examples

Process-related Impurity Categories	Examples
Residual ancillary materials	Selection reagents, activation reagents, media components, formulation reagents and impurities in these ancillary materials
Residual gene editing materials	Residual viral vector, vector-associated impurities, CRISPR-Cas9 ribonucleoprotein (RNP), RNP-associated impurities and other impurities associated with these materials

the risk associated with each impurity using a science and risk-based approach. The outcome of the impurity risk assessment should be used to guide the development of analytical procedures that can evaluate the presence and quantify each medium- and high-risk impurity. If the impurities are either at levels that are a safety concern or are insufficiently controlled between lots, the analytical procedures can be used to support the optimization of the manufacturing process to deplete and/or control the levels of these impurities. The conclusion of product development should be an control strategy that provides the adequate assurances of impurity control to ensure product safety and efficacy.

Impurity Risk Assessment

For each impurity, an impurity risk assessment should be performed to inform selection of an appropriate control strategy. Broad considerations for assessing impurity risks include:

1) *Toxicity risks of the material* – Such information should be determined by a toxicity assessment conducted by toxicological subject matter experts. When possible, exposure limits (ELs) should be derived based on published and internal nonclinical and clinical data and/or guidance from regulatory agencies following ICH Q3B (R2) and ICH S8.
2) *Quantity expected in the final dose* – This quantity is used to determine patient exposure risk to a particular impurity. This should consider the dosing strategy and infusion rate. Quantification of the impurities can be made in the ancillary materials, in process intermediates or in the final drug product. The quantity of the impurity can be estimated through process modeling or empirically determined through their quantification in the drug product.

The ratio of the exposure limit and quantity present in a final dose is defined as the impurity safety margin (S):

$$S = \frac{Impurity\ toxicity\ exposure\ limit\,(EL)}{Impurity\ quantity\ in\ a\ final\ drug\ product\ dose}$$

Stratification of process-related impurity risks may be made depending on the size of the safety margin as well as additional factors such as the uncertainty around toxicity or exposure risk scoring. The impurity risk assessment is then used to guide elements of both analytical and process decision making, including:

- The requirements of analytical procedures for quantifying potential impurities, including:
  - Testing environment (i.e., whether a method will be used for release or characterization)

Analytical Methods for In-Process Testing and Product Release

- Analytical performance characteristics, such as specificity (including matrix compatibility), sensitivity (LOD, LOQ), accuracy, precision, linearity and range.
- Prioritization of analytical development and optimization
- The types of impurity clearance studies that need to be performed
- Process development to improve depletion of impurities

Risk Assessment Approaches at Early Stages of Development or for Emerging Modalities/Novel Impurities

In early stages of development or for programs using novel cell therapy reagents, impurity risk assessments may be challenging to perform due to insufficient data on the toxicological and/or exposure fronts. In these cases, a general toxicological review combined with platform and scientific knowledge should be leveraged. In the absence of measurement data on final impurity quantities in the drug product, impurity risks may be estimated using the starting quantity of an impurity in conjunction with its modeled process clearance.

Starting impurity levels – In the case of impurities that are direct process inputs, starting impurity levels may be determined by process targets or normal operating ranges. Impurities that are carried over by input viral vectors or gene editing reagents (e.g., host cell substrates) may be estimated based on specification ranges. If quantitative specifications are unavailable, worst case observed values from representative batches may be used.

In silico process clearance modeling – Theoretical process clearance may be estimated by *in silico* assessment by tabulating inputs, outputs, concentrations and volumes at each process step. Depending on analysis goals, past process platform experience, and the physiochemical properties of the impurity, process assumptions may be tuned to reflect scenarios ranging from the "worst-case" (e.g., assumes accumulation of input impurities throughout the process without clearance) or "best case" (e.g., assumes impurities are fully soluble and are diluted and removed with wash and media exchange steps), as applicable and scientifically justified.

Impurity Methods Development Considerations and Key Impurity Method Platforms

As introduced in the two previous sections, an initial impurity risk assessment provides context, rationale and targets for impurity method development. Characterization to understand mechanisms of impurity clearance and process capabilities for depleting and/or controlling impurities may be warranted for process-related impurities that are determined to be medium- and high-risk. Analytical procedures that are developed to support these characterization studies must be evaluated and determined to be suitable for their intended use, i.e., methods must demonstrate sufficient specificity, sensitivity (LOD, LOQ), accuracy, precision, linearity and range for the residual analyte in the intended sample type(s) in order to support release, process PQ and/or comparability studies. Results from impurity clearance characterization studies may be used to revise or refine the impurity risk assessment and to inform an integrated control strategy. Potential control strategies include monitoring impurity levels as part of QC release, routine characterization or extended testing. QC release is used to provide assurances levels are control to acceptable limits. Characterization is used to demonstrate the process depletes the impurities to levels that are no longer a concern. For example, depletion to levels that are an order of magnitude (ten times) lower than established safety limit may be justification for removing the attribute from the specification. This approach is only justified if the process is validated to achieve this level of clearance, typically during process PQ. Depending on the approach that is pursued for a given impurity, as well as development timelines, laboratory capabilities and other programmatic considerations, an additional round of method development may be needed to refine or revise an initially selected impurity method. For example, a mass spectrometry-based impurity characterization method that was initially developed and used to determine process clearance may not be optimal for routine execution as a QC release assay. Instrumentation, analyst capability and/or operational (e.g., throughput) considerations may necessitate the development and implementation of a more QC-friendly analytical procedure, such as an immunoassay.

Process-related impurities in cell therapy tend to fall into several major classes, see Table 7.5. The physiochemical properties of the analyte often helps drive selection of technology platform to base an analytical procedure. For example, protein-based impurities may be effectively detected and quantified by mass spectrometry or an immunoassay. As is the case for all analytical procedures outlined so far in this chapter, the selection of the technology platform, analytical development and optimization must ensure the final assay is phase-appropriate with sufficient analytical performance to be suitable for its intended use.

TABLE 7.5
Typical Compound/Molecule Classes for Process-Related Impurities of Interest with Associated Analytical Approaches

Analyte Class	Example Process-related Impurities	Potential Analytical Approaches
Proteins	Selection reagent, activation reagent, media components (e.g., cytokines), vector, vector-associated impurities (e.g., residual nuclease, residual host cell protein), gene editing reagents	LC-MS, immunoassays
Small molecules	Formulation reagents, media additives	LC-MS, HPLC
Nucleic acids	Vector, vector-associated impurities (e.g., residual plasmid, host cell DNA), gene editing reagents	PCR amplification and/or nucleic acid-dye based approaches
Inert substrate beads	Selection reagent, activation reagent	Imaging- or microscopy-based approaches

Analytical Methods for In-Process Testing and Product Release

MICROORGANISMS, PYROGENS AND ADVENTITIOUS AGENTS

The safety of cell therapy products must be established prior to release for patient infusion, including the evaluation of each cell therapy lot for sterility, mycoplasma, endotoxin and, when lentiviral vectors are employed, RCR. There are compendial analytical procedures for sterility, mycoplasma and endotoxin. However, it may be beneficial to develop alternative analytical procedures that offer performance benefits. For example, noncompendial rapid microbial methods have emerged recently as alternative analytical procedures that mitigate the long turnaround time of growth-based compendial methods (i.e., 14 days for sterility testing and 28 days for mycoplasma), which enables more timely release of products that either have a short shelf-life or to patients that are in immediate need.

Sterility

USP <71> Sterility describes the analytical procedure for the detection of aerobes, anaerobes and fungi after a 14-day culture to promote growth. USP <71> Sterility can use the method of a direct inoculation method or membrane filtration. Both analytical procedures require a minimum 14-day incubation with two separate media types to accommodate for anaerobic (Thioglycolate Broth) and aerobic (Trypticase Soy Broth) growth. The analytical procedure, verification protocol, and specification acceptance criteria are described in USP <71> Sterility, which is harmonized with other regulatory authorities such as European Pharmacopeia (EP 2.6.1) and Japanese Pharmacopeia (JP 4.06.). Alternative analytical methods may be developed with improved performance characteristics, such as reduced turnaround time, if appropriately justified as described in the following section.

Analytical Performance Requirements

The analytical procedure that evaluates sterility should evaluate the final filled container of the cell therapy product (21 CFR 610.12(a)). If the cell therapy product is cryopreserved, the FDA recommends testing the product prior to cryopreservation (CMC Information for Human Gene Therapy INDs). The analytical procedure must be able to detect numerous microorganisms that are spiked at not more than 100 colony forming units (CFU) into representative test samples and evaluated using each applicable growth medium. This includes six model microorganisms including *Staphylococcus aureus*, *Pseudomonas aeruginosa*, *Bacillus subtilis*, *Clostridium sporogenes*, *Candida albicans* and *Aspergillus brasiliensis* and at least two representative microorganisms that have been isolated from the manufacturing site, manufacturing process and/or QC testing site, such as *Micrococcus luteus*, *Staphylococcus epidermidis* and *Burkholderia cepacia*. Additional organisms included in Ph. Eur. 2.6.1, 2.6.27, or relevant facility isolates may also be included. Detection must occur in a representative quantity of sample matrix. The sampling guidelines provided in Ph Eur. 2.6.1 are based on the volume of the final batch (i.e., 1% of the total volume for a product batch, consisting of 10 ml–1,000 ml).

Development of Alternative Analytical Procedures

The compendial analytical procedure may not support the short shelf-life or turnaround time required for needs of some cell therapy products. USP <1071> states that a risk-based approach should be taken when considering alternative analytical procedures. An alternative analytical procedure must be compatible with user defined requirements, such as:

- Detect a wide range of organisms, preferably not more than 100 CFU
- Low rate of false positive and negative results
- Suitable for both manual and automation execution
- Available reference standards necessary for (e.g., qPCR metric will be GC or cycle threshold (Ct)

Analytical technology platforms that may be appropriate for rapid sterility testing include:

1. ATP
2. Flow Cytometry
3. Solid Phase Cytometry
4. Isothermal microcalorimetry
5. Nucleic acid amplification (universal primers and probes targeting 16S rRNA and/or 18S rRNA)
6. Respiration (detection of CO_2 during microbial growth)

Demonstrating Suitability for Its Intended Use

The analytical performance of USP <71> Sterility must be verified for testing each new sample type. In addition, it must be verified for use at each QC testing site. Since sterility is an impurity with a limits-based acceptance criterion, the verification of USP <71> must demonstrate its specificity and LOD. The performance requirements for the verification are defined in USP <71> Sterility and described in "Analytical Performance Requirements." Alternative analytical procedures must be validated to have equivalent or superior specificity and sensitivity (LOD) according to USP <71> Sterility in a side-by-side comparison as described in USP <1223> Validation of Alternative Microbiological Methods.

Mycoplasma

USP <63> *Mycoplasma* Tests describes the analytical procedure for the detection of Mycoplasma. The culture-based analytical procedure is used for evaluating cell therapy products and can include either expansion in culture media and detection of CFU's on nutrient agar plates, or expansion in culture media, followed by fluorescent staining of DNA which may allow for detection of difficult to cultivate *mycoplasma* strains.

The duration of the analytical procedure, up to 28 days, may pose challenges for some cell therapy products, such as those with limited shelf-life stability or when rapid turnaround time is required for patients in need. Consequently, it may

Analytical Methods for In-Process Testing and Product Release

be beneficial to consider alternative analytical procedures that overcome these challenges. The implementation of alternative analytical procedures must be appropriately justified as described in the following section.

Analytical Performance Requirements

The analytical procedure that evaluates *mycoplasma* must be able to detect numerous *mycoplasma* species that are spiked at 10 CFU/ml into representative test samples. The test sample should contain supernatant media and cells to ensure that *mycoplasma* that are enriched in each type of sample could be detected. The ability to detect seven model *mycoplasma* species should be evaluated; six are defined by USP <63> and Ph Eur. 2.6.7 (*Acholeplasma laidlawii, Mycoplasma arginini, Mycoplasma fermentans, Mycoplasma hyorhinis, Mycoplasma orale, Mycoplasma pneumoniae*) and one additional strain (*Mycoplasma salivarium*) is defined by JP 4.06 XVII. Additional mycoplasma species are required if the manufacturing process includes materials derived from avian or insect origin.

Development of Alternative Analytical Procedures

The duration of USP <63> *Mycoplasma* Tests, which includes 28 days in culture, may not be suitable for some cell therapy products. In these situations, an alternative analytical procedure with improved performance characteristics can be developed. For example, an analytical procedure based on a nucleic acid amplification technique (NAT) or enzymatic activity-based may be developed. Details regarding requirements for use of NAT for *mycoplasma* detection is specified in EP 2.6.7.

Several kits for detection of mycoplasma detection by qPCR are commercially available, utilizing TaqMan or SYBR Green chemistry. These may include positive controls, as well as controls to ensure there is not inhibition due to sample matrix. Various qPCR platforms are available that are capable of detection of mycoplasma at the required sensitivity. These instruments must have an IQ, OQ, PQ, and data integrity that meets the 21 CFR part 11 requirement to be compliant for use in a GMP environment.

Demonstrating Suitability for Its Intended Use

Similarly to the requirements of other compendial analytical procedures, the performance of USP <63> *Mycoplasma* Tests must be verified for testing each new sample type. In addition, it must be verified for use at each QC testing site. *Mycoplasma* is an impurity with a limits-based acceptance criterion, so the verification of USP <63> must demonstrate its specificity and LOD. The performance requirements for the verification are defined in USP <63> *Mycoplasma* Tests and described earliers "Analytical Performance Requirements."

Alternative analytical procedures must be validated to have equivalent or superior specificity and sensitivity (LOD) to USP <63> *Mycoplasma* Tests in a side-by-side comparison as described in USP <1223> Validation of Alternative Microbiological Methods. These studies may use either live *mycoplasma* samples for spiking or commercially available *mycoplasma* DNA reference standards.

Endotoxin

USP <85> Bacterial Endotoxins describes the analytical procedure for the detection and quantification of the pyrogen endotoxin from gram negative bacteria. Endotoxin must be tested in the final filled container of cell therapy product (21 CFR 610.13(b)). USP <85> Bacterial Endotoxins defines three alternative analytical procedure that uses the limulus amoebocyte lysate from the horseshoe crab (*Limulus polyphemus or Tachypleus tridentatus*) to detect endotoxin. The first measures gel formation following a gel-clot technique, the second measures turbidity after cleavage of an endogenous substrate, and the third is a chromogenic technique that measures cleavage of a synthetic peptide-chromagen complex. Any of the three techniques can be used. If more than one is used and there is a dispute in the analytical outcome, the final decision should be based on the outcome of the gel-clot technique. These analytical procedures, verification protocol and specification acceptance criteria described in USP <85> Bacterial Endotoxins is harmonized with other regulatory authorities such as European Pharmacopeia (EP 2.6.14) and Japanese Pharmacopeia (JP 4.01). Alternative analytical procedures may be developed with improved performance characteristics, such as reduced turnaround time, if appropriately justified as described in the following section.

Development of Analytical Procedures

USP <85> can be developed internally or commercially available kits can be used. Pyrogen-free labware that is shown to be noninterfering should be used for development and routine testing. In addition, all reagents, buffers and diluents that are used in the analytical procedure should be demonstrated to not cause assay interference. Furthermore, the endotoxin reference standard solution should be calibrated against a national or international reference standard.

An interference screen testing multiple dilutions up to the Maximum Valid Dilution (MVD) should be performed to determine an appropriate dilution for routine testing during development and confirmed during verification. MVD is calculation by endotoxin limit × concentration of Sample Solution) / λ, where λ is the labeled sensitivity of the gel-clot method or lowest concentration of the standard curve for the turbidimetric or chromogenic method. In addition, an interference screen testing multiple dilutions up to the MVD should be performed to determine an appropriate dilution for routine testing during development and confirmed during verification. At least three lots should be tested to show that the chosen dilution exhibits interference free testing.

Demonstrating Suitability for Its Intended Use

The performance of USP <85> Bacterial Endotoxins must be verified for testing each new sample type. In addition, the analytical procedure must be verified for use at each QC testing site. Endotoxin is an impurity that is quantified, so the verification of USP <85> must demonstrate its specificity, accuracy, precision, linearity, LOQ and range. The performance requirements for the verification must be justified based on a science and risk-based approach.

Analytical Methods for In-Process Testing and Product Release

Replication Competent Retrovirus

Generation of RCR in the manufacture of third generation LVV is a theoretical safety concern. RCR has not been detected to date in any clinical trials using a third generation, SIN-inactivated, split plasmid designed LVV-transduced cell product (Cornetta et al., 2018). However, to ensure the safety of LVV, RCR testing of each lot is required. In addition, the FDA also requires cell therapy products to be evaluated for RCR when retroviral vector-based vectors have been used in their manufacture (Guidance for Industry – Supplemental Guidance on Testing for Replication Competent Retrovirus in Retroviral Vector-Based Gene Therapy Products and During Follow-Up of Patients in Clinical Trials Using Retroviral Vectors). Note, this differs from other regulatory jurisdictions such as the EMA. The FDA guidance recommends for RCR to be evaluated throughout the manufacturing process of both LVV and cell therapy products (including release testing), and during follow-up monitoring of patients in clinical trials.

Analytical Performance Requirements

It is recommended that 1% or 10^8 (whichever is less) of *ex vivo* transduced cells are tested by co-culture with a permissive cell line. It is also critical that the method have the necessary sensitivity to detect RCR, and in addition, have a minimal number of false positive and false negative results. FDA guidance requires that the method selected must have the sensitivity to detect < 1 RCR/dose equivalent. An alternative analytical procedure for the RCR detection may be developed that does not require a co-culture, such as through the detection of VSV-G copies. There are various qPCR platforms and fluorescent plate readers that are available for detection of RCL or surrogate markers. A performance assessment must demonstrate that an alternative analytical procedure has the necessary robustness, specificity and sensitivity, and the instrumentation must meet GMP compliance requirements (i.e., IQ, OQ, PQ and data integrity requirements according to 21 CFR part 11).

Analytical Development

The most sensitive analytical procedure for the detection of RCR that is appropriate for a QC testing environment should be developed and implemented. The current analytical procedure of choice is the cell culture-based method. In this analytical procedure, test articles are co-cultured with a highly permissive cell line over at least five passages, followed by endpoint detection of RCR-specific components. This method has a turnaround time of up to six weeks, making it undesirable when the timely release of a cell therapy product is required. In this situation, more rapid alternative analytical procedure may be developed for lot release testing *in lieu* of culture-based methods.

Rapid analytical procedures used for surrogate RCR detection include:

- p24 ELISA that quantify the concentration of p24 antigen. This can be a commercially available kit or an analytical procedure that is developed in-house. The analytical strategy is to demonstrate there is no increase in the amount of p24 antigen across the manufacturing

process. Therefore, the levels of p24 antigen in the cell therapy drug substance or drug product is compared to the levels in in-process samples immediately after transduction. An increased presence of p24 antigen indicates the presence of RCR, whereas similar or reduced levels of p24 antigen indicates the absence of RCR.
- Analytical procedures based on molecular techniques such as qPCR may be used for evaluating RCR by detecting nucleic acid sequences that should not be present in a cell therapy product, such as VSV-G.

RCR testing should be performed on *ex vivo* transduced cells prior to product release. Testing may also be used to monitor levels of the target analyte in process to ensure there is not an increase throughout the manufacturing process. The testing strategy should be justified by a science and risk-based approach.

Demonstrating Suitability for Its Intended Use

Analytical procedures that detect RCR need to have a performance assessment that demonstrates its robustness, specificity, and sensitivity (LOD). The analytical procedure should be sensitive enough to identify RCR at or below the level of sensitivity required by regulatory guidance (< 1 RCR/dose equivalent) and have low levels of false positives. The LOD may be established using a standard curve of the target analyte. FDA guidance recommends that a RCR reference standard should be developed that represents attributes of the LVV for use as a positive control and for characterization of analytical performance. Due to safety concerns with handling replicating lentivirus for use as a positive control, surrogates such as DNA or plasmids with a known copy number of the target analyte (e.g., determined by a well-characterized orthogonal analytical procedure) may be used to establish the LOD and as a positive control in the method. Untransduced cells or cell lines may be used as a RCL negative control.

GENERAL CQAs

Container Content

There must be sufficient volume within each drug product vial to extract (recover) the nominal volume described on the label (21 CFR 201.51(g), USP <1>, and FDA Guidance, Allowable Excess Volume and Labeled Vial Fill Size in Injectable Drug and Biological Products). Therefore, it is necessary to establish the 'excess' volume that is required to ensure the extractable volume meets or exceeds the nominal volume. The nominal volume plus the excess volume is referred to as the 'overfill' volume. Generally, the product label describes only the nominal volume and not the overfill volume (FDA Guidance Allowable Excess Volume and Labeled Vial Fill Size in Injectable Drug and Biological Products).

The analytical procedure, protocol, and acceptance criteria are described in USP <697> Container Content for Injections. USP <1151> Pharmaceutical Dosage Forms provides guidance on recommended excess volumes. Any volume that exceeds the excess volumes recommended in USP <1151> must be justified

(FDA Guidance Allowable Excess Volume and Labeled Vial Fill Size in Injectable Drug and Biological Products). For example, it may be necessary to use a larger excess volume than described in USP <1151> to provide adequate assurance that the extractable volume meets or exceeds the nominal volume described on the label. It may be possible to justify the removal of this CQA from the specification for QC release by demonstrating sufficient control of container content during process development, process characterization and/or process PQ.

Note, 'overfill' should not be confused with 'overage.' An overage is an excess volume that exceeds the overfill required to ensure the extractable volume meets or exceeds the nominal volume. Any overage should be justified to the regulatory authorities, including the amount of overage, reason, rationale for the volume of overage and the impact of the overage to product safety and efficacy. The practice of overage to compensate for degradation that occurs during the manufacture and/or shelf-life of a product is discouraged by regulatory authorities (ICH Q8(R2)).

Uniformity of Dosage Units

There are typically two approaches to dose a cell therapy product. Volume is often used to dose when strength (e.g., viable CAR T cells/mL) is tightly controlled between product lots. Alternatively, strength is used to adjust the dose volume in circumstances where strength is variable between product lots. In either case, the attribute used for dosing must be sufficiently controlled across a narrow range to ensure patient safety and efficacy (USP <1> and USP <905>).

The protocol for evaluating the uniformity of dosage units and acceptance criterion is described in USP <905>. Briefly, the content uniformity of a cell therapy product can be established by evaluated 30 or more test samples. The dosage units must be within ± 25% of the label claim for the given lot. Alternative approaches and acceptance criteria must be justified based on a science and risk-based approach. Sufficient control of content uniformity may be demonstrated during process development, process characterization and/or process PQ to justify the removal of this CQA from the specification for QC release and stability.

Container Closure Integrity

A cell therapy product must be filled into a packaging system (e.g., vial) that prevents both the loss of content and contamination. The integrity of the container must be validated to prevent both the penetration of microbial contamination and the gain or loss of any chemical or physical attribute that is necessary to ensure the safety and efficacy of the product (USP <1>). The integrity of the container closure can be evaluated according to procedures and acceptance criteria described in USP <1207> Sterile Product Packaging – Integrity Evaluation. Alternative approaches and acceptance criteria must be justified based on a science and risk-based approach. Sufficient control of container closure integrity can be demonstrated during process development, process characterization, and/or process PQ to justify the removal of this CQA from the specification for QC release and stability.

pH

pH is a CQA that must be adequately controlled (USP <1>) to ensure product safety and efficacy. pH can assure each lot is in its intended formulation, is sufficiently neutral for safe administration and is within a range that ensures product stability. USP <791> pH describes the analytical procedure for the measurement of pH. Briefly, a pH meter must be calibrated using at least three pH reference standards that are traceable to international or national reference standards, such as the NIST pH reference standards. This calibration should include reference standards that are at least one pH value above and one pH value below the expected pH value of the test sample. Next, the performance of the analytical procedure is verified using a control. If assay acceptance criterion is met for the control, the analytical procedure can be used to evaluate test samples. It may be possible to justify the removal of this CQA from the specification for QC release and stability by demonstrating sufficient control of pH during process development, process characterization and/or process PQ.

Appearance (Visual Inspection)

Appearance (visual inspection) is an assessment of color, clarity and visible particulate matter.

Color and clarity are evaluated to ensure that each released lot is consistent with experience. In addition, these physical characteristics can be used in the clinic to provide assurances that a product mix up or contamination has not occurred prior to administration. Of the product. Color and clarity are generally performed in accordance with European Pharmacopeia (EP) 2.2.2 and 2.2.1.

Particulate matter is extraneous particles that are present in the product that are mobile and/or undissolved (USP <788>). Particulate matter can originate from ancillary materials used in the manufacture of a cell therapy product (e.g., viral vectors), generated during the manufacture of a cell therapy product (e.g., from nonviable cells), or extraneous to the manufacturing process (e.g., fibers, glass, metal, elastomeric materials, and precipitates) (USP <790>).

Particulate matter may pose a risk to patient safety, and consequently, all products that are parenterally administered must be inspected for the presence of particulate matter (USP <1>). All (100%) manufactured containers in a lot must be visually inspected. In addition, each lot must also be sampled at lot release at a frequency that meets or exceeds ANSI/ASQ Z1.4 (or ISO 2859-1) with an AQL of at least 0.65% (USP <790>). The analytical procedure for evaluating visible particulate is described in USP <790> and the lot must be "essentially free" of visible particulates. Essentially free is defined as having no more than the specified number of vials may be observed to contain visible particulates. All containers with visible particulates must be rejected (USP <1>).

REFERENCES

Adhibhatta, S., DiMartino, M., Falcon, R., Haman, E., Legg, K., Payne, R., Pipkins, K.R., & Zamamiri, A. (2017). Continued process verification (cpv) signal responses in biopharma. *Pharm Eng.*, 37, 57–64.

Analytical Methods for In-Process Testing and Product Release 131

American Society Testing Materials (ASTM). (2021). ASTM E2935-20e1: Standard practice for conducting equivalence tests for comparing testing processes. ASTM International, West Conshohocken, PA.

Amirache, F., Lévy, C., Costa, C., Mangeot, P. E., Torbett, B. E., Wang, C. X., Nègre, D., Cosset, F. L., & Verhoeyen, E. (2014). Mystery solved: VSV-G-LVs do not allow efficient gene transfer into unstimulated T cells, B cells, and Hscs because they lack the LDL receptor. *Blood*, 123(9), 1422–1444.

Benmebarek, M. R., Karches, C. H., Cadilha, B. L., Lesch, S., Endres, S., & Kobold, S. (2019). Killing mechanisms of chimeric antigen receptor (CAR) T cells, *Int J Mol Sci*, 20, 1283.

Bennett, E. P., et al. (2020). INDEL detection, the 'Achilles heel' of precise genome editing: A survey of methods for accurate profiling of gene editing induced indels. *Nucleic Acids Res*, 48(21), 11958–11981.

Bhardwaj, A., & V. Nain, (2021). TALENs – An indispensable tool in the era of CRISPR: A Mini Review. *J Genet Eng Biotechnol*, 19(1), 125.

Borah, P. (2011). Primer designing for PCR. *Sci Vis*, 11(3), 134–136.

Bushman, F., Lewinski, M., Ciuffi, A. et al. (2005). Genome-wide analysis of retroviral DNA Integration. *Nat Rev Microbiol*, 3, 848–858. doi:10.1038/nrmicro1263

Bustin, Stephen A., et al. (2009).The MIQE guidelines: Minimum information for publication of quantitative real-time PCR experiments. *Clinical Chemistry*, 55(4), 611–622.

Cameron, P., et al. (2017). Mapping the genomic landscape of CRISPR-Cas9 Cleavage. *Nat Methods*, 14(6):600–606.

Carmen, J., Burger, S. R., McCaman, M., & Rowley, J. A. 2012. Developing assays to address identity, potency, purity and safety: Cell characterization in cell therapy process development. *Regen Med*, 7(1), 85–100.

Carroll, D., (2011). Genome engineering with zinc-finger nucleases. *Genetics*, 188(4), 773–782.

Chiarle, R., et al. (2011). Genome-wide translocation sequencing reveals mechanisms of chromosome breaks and rearrangements in B Cells. *Cell*, 147(1), 107–119.

Cornetta, K., Duffy, L., Feldman, S.A., Mackall, C.L., Davila, M.L., Curran, K.J., Junghans, R.P., Tang, J.Y., Kochenderfer, J.N., O'Cearbhaill, R., Archer, G., Kiem, H-P., Shah, N.N., Delbrook, C., Kaplan, R., Bretjens, R., Riviere, I., Sadelain, M. and Rosenberg, S.A. (2018). Screening Clinical Cell Products for Replication Competent Retrovirus: The National Gene Vector Biorepository Experience. Mol. Ther. Methods Clin. Dev., 10, 371–378.

Crosetto, N., et al. (2013). Nucleotide-resolution DNA double-strand break mapping by next-generation sequencing. *Nat Methods*, 10(4), 361–365.

Dabrowska, M., et al. (2018). qEva-CRISPR: A nethod for quantitative evaluation of CRISPR/Cas-mediated genome editing in target and off-target sites. *Nucleic Acids Res*, 46(17), e101.

Du, L. Gover, A., Ramanan, S. & Litwin, V. (2015). The evolution of guidelines for the validation of flow cytometric methods. *Int J Lab Hematol*, 37(S1), 3–10.

El-Karidy, A.E.H., Rafei, M., & Shammaa, R. (2021). Cell therapy: Types, regulation, and clinical benefits, *Front Med (Lausanne), 8,* doi:10.3389/fmed.2021.756029.

FDA. (1998). Guidance for the industry: Guidance for human somatic cell therapy and gene Therapy. U.S. Department of Health and Human Services Food and Drug Administration Center for Biologics Evaluation and Research. Rockville Pike, Rockville, MD.

Freitas, Flávia C. P., et al. (2019). Evaluation of reference genes for gene expression analysis by real-time quantitative PCR (qPCR) in three stingless bee species (Hymenoptera: Apidae: Meliponini). *Sci Rep* 9(1), 1–13.

Giannoukos, G., et al. (2018). UDiTaS, a genome editing detection method for indels and genome rearrangements. *BMC Genomics*, 19(1), 212.

Guedan, S., Calderon, H., Posey Jr., A.D., & Maus, M. V. (2018). Engineering and design of chimeric antigen receptors. *Mol Ther Methods Clin Dev*, 12, 145–156.

International Conference on Harmonization (ICH). (2003). Stability testing of new drug substances and products. Q1A(R2). www.ich.org.

International Conference on Harmonization (ICH). (2005). Validation of analytical procedures: Text and methodology. Q2(R1). www.ich.org.

International Conference on Harmonization (ICH). (1999). Specifications: Test procedures and acceptance criteria for biotechnological/biological products. Q6B. www.ich.org.

International Conference on Harmonization (ICH). (2006). Quality risk management. Q9. www.ich.org.

International Conference on Harmonization (ICH). (2009a). Pharmaceutical development. Q8(R2). www.ich.org.

International Conference on Harmonization (ICH). (2009b). Pharmaceutical quality system. Q10. www.ich.org.

International Conference on Harmonization (ICH). (2012). Development and manufacture of drug substances (chemical entities and biotechnological/biological entities). Q11. www.ich.org.

International Conference on Harmonization (ICH). (2022). Analytical procedure development. Q14 (Draft). www.ich.org.

Kaeuferle, T., et al. (2022). Genome-wide off-target analyses of CRISPR/Cas9-mediated T-cell receptor engineering in primary human T cells. *Clin Transl Immunol*, 11(1), e1372.

Karanu, F., Ott, L., Webster, D. A., & Stehno-Bittle, L. (2020). Improved harmonization of critical characterization assays across cell therapies. *Regen Med*. 15(5), 1661–1678.

Katti, A., et al. (2022). CRISPR in cancer biology and therapy. *Nat Rev Cancer*, 22(5), 259–279.

Kim, D., et al. (2015). Digenome-seq: Genome-wide profiling of CRISPR-Cas9 off-target effects in human cells. *Nat Methods*, 12(3), 237–243, 1 p following 243.

Kim, D., & Kim, J. S. 2018. DIG-seq: A genome-wide CRISPR off-target profiling method using chromatin DNA. *Genome Res*, 28(12), 1894–1900.

Kralik, P., & Ricchi, M. (2017). A basic guide to real time PCR in microbial diagnostics: Definitions, parameters, and everything. *Front Microbiol*, 8, 108.

Lazzarotto, C. R., et al. (2020). CHANGE-seq reveals genetic and epigenetic effects on CRISPR-Cas9 genome-wide activity. *Nat Biotechnol*, 38(11), 1317–1327.

Li, B., et al., (2019). A qPCR method for genome editing efficiency determination and single-cell clone screening in human cells. *Sci Rep*, 9(1), 18877.

Milone, M. C., & O'Doherty, U. (2018). Clinical use of lentiviral vectors. *Leukemia*, 32(7), 1529–1541.

Mock, U., Hauber, I., & Fehse, B. (2016). Digital PCR to assess gene-editing frequencies (GEF-dPCR) mediated by designer nucleases. *Nat Protoc*, 11(3), 598–615.

Ota, M., Fukushima, H., Kulski, J. K., & Inoko, H. (2007). Single nucleotide polymorphism detection by polymerase chain reaction-restriction fragment length polymorphism. *Nat Protoc*, 2(11), 2857–2864.

Ricciardi, A. S., et al. (2014). Targeted genome modification via triple helix formation. *Methods Mol Biol*, 1176, 89–106.

Sandoval-Villegas, N., et al. (2021). Contemporary transposon tools: A review and guide through mechanisms and applications of sleeping beauty, piggyBac and Tol2 for genome engineering. *Int J Mol Sci*, 22(10), 5084–5113.

Silva, G., et al. (2011). Meganucleases and other tools for targeted genome engineering: Perspectives and challenges for gene therapy. *Curr Gene Ther*, 11(1), 11–27.

Taylor, Sean C., et al. (2019). The ultimate qPCR Experiment: Producing publication quality, reproducible data the first time. *Trends Biotechnol*, 37(7), 761–774.

Tsai, S. Q., et al. (2017). CIRCLE-seq: A highly sensitive in vitro screen for genome-wide CRISPR-Cas9 nuclease off-targets. *Nat Methods*, 14(6), 607–614.

Tsai, S. Q., et al. (2015). GUIDE-seq enables genome-wide profiling of off-target cleavage by CRISPR-Cas nucleases. *Nat Biotechnol*, 33(2), 187–197.

Turchiano, G., et al. (2021). Quantitative evaluation of chromosomal rearrangements in gene-edited human stem cells by CAST-Seq. *Cell Stem Cell*, 28(6), 1136–1147 e5.

US Pharmacopeia (USP). (n.d.). http://online.uspnf.com.

Vouillot, L., Thelie, A., & Pollet, N. (2015). Comparison of T7E1 and surveyor mismatch cleavage assays to detect mutations triggered by engineered nucleases. *G3 (Bethesda)*, 5(3), 407–415.

Wienert, B., et al. (2019). Unbiased detection of CRISPR off-targets in vivo using DISCOVER-Seq. *Science*, 364(6437), 286–289.

Yan, W. X., et al. (2017). BLISS is a versatile and quantitative method for genome-wide profiling of DNA double-strand breaks. *Nat Commun*, 8, 15058.

Yang, Z., et al. (2015). Fast and sensitive detection of indels induced by precise gene targeting. *Nucleic Acids Res*, 43(9), e59.

Zufferey, R., Dull, T., Mandel, R. J., Bukovsky, A., Quiroz, D., Naldini, L., & Trono, D. (1998). Self-inactivating lentivirus for safe and efficient in vivo gene delivery. *J Virol.*, 72(12), 9873–9880.

8 Validation, Verification and Qualification Considerations

Humberto Vega-Mercado
Bristol Myers Squibb, Cell Therapy Operations, Global Manufacturing Sciences and Technology, Warren, USA

CONTENTS

Introduction 135
Facility and Equipment Qualifications 139
Environmental Monitoring and PQ 142
Process, Method, and Facility Changes 146
Process Validation Life Cycle 147
References 150

INTRODUCTION

Process validation, method validation, verifications and qualifications are all focused on demonstrating in-depth knowledge of the manufacturing processes and analytical tests, understanding of the operation and use of pieces of equipment and comprehensive documentation (e.g., protocols, test scripts, reports, changes controls, deviations) required as support to regulatory filings. Cell and gene therapies follow similar, but not identical, paths of conventional drug products when comes to their development and commercialization (e.g., preclinical evaluations, clinical trials, data analysis, facility design and construction, technology transfers, engineering lots, stability and demonstration lots, performance qualification lots and commercial lots). We must consider a life cycle approach when talking about validations, verifications and qualifications, as pharmaceutical products, in the case of this handbook on cell and gene therapies, continue evolving as a function of time as new products and technologies are introduced, new and advanced pieces of equipment are developed and procedures are optimized.

The goal of this chapter is to provide the readers with high-level guidance about qualification, verification and validation activities associated with cell and gene therapies by consolidating available information from across key publications and forums. It is important to recognize that current regulatory guidance continues to evolve as the number of therapies in pipelines continue to grow. Regulatory groups (e.g., US FDA, European Medicines Agency (EMA), Swiss-Medic, ANVIZA, Japan, China, MHRA) continue to work closely with the industry while cell and gene drug products are developed and brought to clinical and commercial settings. Therefore, the guidance in this chapter will be as general and practical as possible considering current expectations and practices across the industry.

We would like to start discussing three key areas that affect validation, verification and qualification activities: Technology Transfers, Method Transfers, Facility Fit Assessments, and Procedures and Methods. These areas or activities require risk assessments to be conducted in preparation for transfer activities, qualification of equipment completed (e.g., Installation and Operation Qualifications) and procedures and methods drafted and considered as "fit" for use in preparation for final validation activities.

i. *Technology transfers and analytical method transfers*

The objectives of technology transfer and analytical method transfers typically include the following (ISPE, 2021; PDA, 2012a, 2012b, 2014):

1. Confirm satisfactory transfer of the manufacturing process from a "Sending Site" to a "Receiving Site" by demonstrating consistency in the process (e.g., execution, documentation, testing) while complying with the in-process and final product specifications, process parameters, yields, regulatory and quality requirements, environmental, health and safety expectations.
2. Confirm satisfactory transfer of analytical methods from a "Sending Site" to a "Receiving Site" following a detailed and thorough risk assessment of the developed/qualified method to be validated (readiness).
3. Demonstration of comparability of the product produced at the receiving site (new facility or existing facility but new product) and the sending site (e.g., product development-clinical or existing site).
4. Process Performance Qualification (PPQ) of the manufacturing process.
5. Receive licensing approval to manufacture and market by correspondent regulatory bodies (e.g., FDA, EMA).

The successful transfer of the process and analytical methods is critical in the development and commercialization of products. These activities are resource intensive (e.g., manpower, cost and time) and companies are focused on optimal ways to complete them. Also, as described by Gibson and Schmitt (2014), an enhanced approach encouraged whereby a scientific, evidence and risk-based approach through Quality by Design (QbD) initiatives are applied including among others the following:

- GMP Guidance for Active Pharmaceutical Ingredients – International Conference on Harmonization (ICH) 7 (ICH, 2001)
- Quality Risk Management – ICH Q9 (ICH, 2006)
- Pharmaceutical Development – ICH Q8 (ICH, 2009a)
- Quality Systems/Knowledge Management – ICH Q10 (ICH, 2009b)

- Development of Manufacture of Drug Substance (Chemical Entities and Biotechnological/Biological Entities) – ICH Q11 (ICH, 2012a)
- Guidance for Industry. Process Validation: General Principles and Practices (FDA, 2011)
- EU Guidelines for Good Manufacturing Practice for Medicinal Products for Human and Veterinary Use. Annex 15: Qualification and Validation (EC, 2015)

In the case of cell and gene therapies, the challenges associated with technology transfers and analytical method transfers are not limited to the actual manufacturing processes and testing methods but the nature of the drug products: Cell Therapies – Personalized drug products (e.g., Autologous CAR T, Allogenic CAR T, etc.) and Gene Therapies – Gene Editing. Successful technology transfers and analytical methods, following the advances in the past 15–20 years in these technologies, have resulted in the recent approval of at least five cell therapies in the United States of America with the following indications:

- Gilead/Kite – Tecartus® is a CD19-directed genetically modified autologous T-cell immunotherapy indicated for the treatment of (a) adult patients with relapsed or refractory mantle cell lymphoma (MCL) and (b) adult patients with relapsed or refractory B-cell precursor acute lymphomblastic leukemia (ALL).
- Novartis – KYMRIAH®: CD19-directed genetically modified autologous T-cell immunotherapy indicated for the treatment of patients up to 25 years of age with B-cell precursor ALL that is refractory or in second or later relapse.
- BMS – Breyanzi™: CD19-directed genetically modified autologous T-cell immunotherapy indicated for the treatment of adult patients with relapsed or refractory large B-cell lymphoma after two or more lines of systemic therapy, including diffuse large B-cell lymphoma (DLBCL) not otherwise specified (including DLBCL arising from indolent lymphoma), high-grade B-cell lymphoma, primary mediastinal large B-cell lymphoma and follicular lymphoma grade 3B. Breyanzi is not indicated for the treatment of patients with primary central nervous system lymphoma.
- BMS – Abecma™: B-cell maturation antigen (BCMA)–directed genetically modified autologous T-cell immunotherapy indicated for the treatment of adult patients with relapsed or refractory multiple myeloma after four or more prior lines of therapy, including an immunomodulatory agent, a proteasome inhibitor and an anti-CD38 monoclonal antibody.
- LLS – Carvykti™: BCMA-directed genetically modified autologous T-cell immunotherapy indicated for the treatment of adult patients with relapsed or refractory multiple myeloma after four or more prior lines of therapy, including a proteasome inhibitor, an immunomodulatory agent and an anti-CD38 monoclonal antibody.

ii. *Facility fit assessments*

Successful transfers and the required qualifications and validations are only accomplished when the facility where the manufacturing process will be performed is designed and built according, among others, with the following elements:

a. process requirements (e.g., environmental controls, materials flow, personnel flows, waste flows, unit operations, gowning requirements)
b. equipment is fully aligned with the intended process requirements within the facility (e.g., process conditions and parameters, equipment requirements, ancillary operations and activities, materials)
c. required procedures and methods are properly documented considering any constraints associated with the facility (e.g., distance between unit operations, aseptic manipulations, handling of in-process materials and samples).

A thorough assessment of the facility allows the identification of potential gaps or deficiencies in the initial design and construction of a facility. Those gaps or deficiencies are identified and eliminated by making required corrections to the facility, pieces of equipment or procedures/methods or by implementing risk mitigation activities that reduce the risk associated with the observed gap or deficiency.

An approach used for facility fit assessments is a comprehensive list of items required for the successful execution of the manufacturing process and analytical testing (e.g., equipment, procedures, samples, materials, training, documentation, environmental controls/requirements). These items are then assessed or quantified based on the nature of the risk. A typical tool used to model the Facility Fit Assessment is the Failure Mode and Effect Analysis – FMEA. Using the FMEA model, the quantification of the risks may include: (a) How difficult is the implementation of the item? By defining a numeric scale to assign a value representing the degree of difficulty (e.g., 1 = easy to 5 = very difficult), the group conducting the assessment can quantify the difficulty levels. (b) Severity of impact if a gap or deficiency is not solved or mitigated. By defining a numeric scale to assign a value representing the degree of impact (e.g., 1 = low to 5 = high impact), the group conducting the assessment can quantify the severity. (c) Define the Total Risk Ranking or Risk Index Number – Multiply the Difficulty Level and the Severity to get a final score. That total score or Total Risk/Risk Index is then used to rank and prioritize remediation and mitigation activities.

The next step in the Fit Assessment is to identify, by means of a side-by-side comparison, the known conditions at the sending site (e.g., Product Development or other Production site) vs. the current status of those items at the receiving site of the items initially identify for the successful transfer. If a gap or deficiency is noted (e.g., Sending site has a procedure for a particular activity vs. Receiving Site has no such procedure), a recommendation is developed considering the level of risk identified for a particular item, and an owner must be identified with a target

Validation, Verification and Qualification Considerations 139

completion date. During the transfer activities (technology and analytical methods) all the items, gaps and remediation/mitigation activities are periodically reviewed to ensure the issues are properly addressed or to update changes on the gaps or deficiencies.

iii. *Procedures and methods*

At the beginning of the transfer activities, a comprehensive list of the procedures (e.g., operation of equipment and facility, cleaning, sanitation, line clearance, equipment set-up, personnel gowning, personnel-material flows, room classifications and operation, business continuity, safety and environmental requirements) and documents (e.g., development reports, characterization experiments, batch records, forms, methods) required to initiate the qualification and validation activities must be developed. The list shall be used to monitor the progress in developing, updating or correcting those documents that must be finalized by the end of the transfer activities. Readiness for qualification and validation activities is assessed, among other factors, by confirming the required documents (procedures and methods) are in place and approved. Typically, a project manager monitors the progress in the development and implementation of these documents.

FACILITY AND EQUIPMENT QUALIFICATIONS

The design and construction of a facility for cell and gene therapies should follow best engineering practices while satisfying the basic regulatory expectations described elsewhere (e.g., EC, 2015; FDA, 2011). This section will discuss general considerations when designing, building, and qualifying the facility and equipment. The concepts apply to both manufacturing and testing facilities and equipment.

i. *General design and construction considerations*

The manufacture of cell and gene therapies must satisfy the basic current Good Manufacturing Practices (cGMP) as these therapies are drug products. Therefore, the following key considerations must be part of the design and construction:

- Facility and Equipment Design and Construction
  - Proper segregation between rooms (e.g., controlled vs. non-controlled) based on the nature of the activities and operations intended across the facility (e.g., warehousing, lockers and bathrooms, gowning areas, personnel and material airlocks, and controlled-non-classified (CNC) areas.)
  - Environmental controls include, but are not limited to, individual air handling units across the facility to prevent potential cross-contamination, room air pressure cascades from clean areas (Class 100 or A) toward CNC spaces with appropriate airlocks and pressure monitoring devices with visual and audio alarms.
  - Construction of floors, walls and ceiling considering routine cleaning and sanitization requirements and the nature of cleaning and sanitizing agents. Specifically, the materials of construction

are nonreactive and ensure easy access to 100% of the surfaces to avoid accumulation of dust and dirt.
- Personnel, material, product and waste flows are properly mapped to avoid potential cross-contamination. Unidirectional flows are highly preferred.
- Easy access and egress options in case of emergencies. There shall be visual and audio alarms to alert the employees in case of emergencies.
- Equipment shall be properly installed, easily cleanable and safe to operate (or have necessary safety guards to prevent accidents).
- Equipment is fit for the intended use and the process conditions are well within the capability of the equipment (avoid borderline operation conditions).
- Required utilities (e.g., electric power, water, compressed gases) are available across the facility (as needed) with enough capacity, including safety margins for operations.
- Emergency power is available to ensure rapid recovery of the operation during power outages.
- Testing lab facilities are strategically localized within the facility to provide easy and safe access when transferring samples (in-process and final drug products).
- Emergency showers and face wash stations are properly installed with appropriate drainage points.
- Access control protocols and technologies are installed to ensure only authorized personnel are allowed into areas.
- Chain of Identity (COI) requirements and controls (e.g., barcode readers and hardware) are fully integrated across the facility to prevent cross-contamination or mislabeling of process materials.

ii. *Installation qualifications*

Installation Qualification (IQ) activities for pieces of equipment must have a well-defined intended use of the equipment linked to the manufacturing process. This is commonly known as the "User Requirement Specifications" or URS. The URS lists the basic elements needed to use and operate the equipment during the manufacturing process (e.g., energy input – voltage/current consumption, dimensions, material of construction, controllers, Human-Machine Interfaces, Displays, Maintenance, Spare Parts). The equipment is ordered based on the URS. The installation process is governed by an IQ protocol where the specifications listed in the URS are confirmed. Table 8.1 summarizes the typical content of IQ protocols.

iii. *Operation qualifications*

Operation Qualification (OQ) activities for pieces of equipment must have a well-defined functional requirement of the equipment linked to the manufacturing process. These are commonly known as the "Functional Requirement Specifications" or FRS. The FRS lists the operational conditions needed to use and operate the equipment during

TABLE 8.1
Summary of Typical Content of IQ Protocols

Content	General Description
Detailed description of the equipment	Name, Serial numbers, Manufacturer, Manufacturing date, Model.
List of user requirement specifications	Define the basic requirements of the equipment with respect to the operation and use during manufacturing or testing activities.
Test script to confirm URS	Detailed tests to confirm the equipment is the one ordered.
Acceptance criteria	Define the conditions to be satisfied as part of the IQ.
Guidance to manage deviations or non-satisfactory results	Define steps to be followed when a deviation or a non-satisfactory result is observed.
User and operations manuals	Copy of the manufacturer documents.
Approval pages	Company-related approvals to execute the IQ.

TABLE 8.2
Summary of Typical Content of OQ Protocols

Content	General Description
Detailed description of the equipment	Name, Serial numbers, Manufacturer, Manufacturing date, Model.
List of functional requirement specifications and ranges	Define the operational requirements of the equipment with respect to parameter ranges.
Test script to confirm FRS	Detailed tests to confirm the equipment operates as designed.
Acceptance criteria	Define the conditions to be satisfied as part of the OQ.
Guidance to manage deviations or nonsatisfactory results	Define steps to be followed when a deviation or a nonsatisfactory result is observed.
User and operations manuals	Copy of the manufacturer documents.
Approval pages	Company-related approvals to execute the OQ.

the manufacturing process (e.g., operating temperature ranges, operating pressure ranges, operating speed ranges). The confirmation that the equipment operates as required for the process is governed by an OQ protocol where the specifications listed in the FRS are confirmed. Table 8.2 summarizes the typical content of OQ protocols.

It is commonly found across the industry that IQ and OQ protocols are combined as a single document to simplify and expedite the execution of the test scripts.

iv. *Performance qualifications*

Performance Qualification (PQ) activities for pieces of equipment focus on the specific process conditions at which the equipment will operate. These conditions are defined during the development stage and later

TABLE 8.3
Summary of Typical Content of PQ Protocols

Content	General Description
Detailed description of the equipment	Name, Serial numbers, Manufacturer, Manufacturing date, Model.
List of process parameters and targets	Define the specific process conditions required for the equipment (targets).
Test script to confirm parameters	Detailed tests to confirm the equipment satisfies the process conditions.
Acceptance criteria	Define the conditions to be satisfied as part of the PQ.
Guidance to manage deviations or nonsatisfactory results	Define steps to be followed when a deviation or a nonsatisfactory result is observed.
Standard operation procedure (SOP)	Copy of the procedure intended to operate the equipment.
Approval pages	Company-related approvals to execute the PQ.

transferred to the production areas during the technology transfer activities. The OQ confirms that these process conditions are within the general operational capabilities of the equipment. Then, PQ confirms the specific conditions can be consistently achieved during routine use and operation of the equipment as part of the manufacturing process (e.g., operating temperature target, operating pressure target, operating speed target). The confirmation that the equipment operates as required for the process is governed by a PQ protocol where the process parameters are confirmed. In some instances, PQ activities are captured as part of the PPQ rather than in an equipment-specific PQ protocol. Table 8.3 summarizes the typical content of PQ protocols.

v. *Verifications and calibrations*

There are ancillary tools that may not require full qualification activities due to the nature of the units (e.g., pH meters, thermometers, temperature sensors, pressure gauges, pressure sensors, torque wrenches, scales). In those cases, either a pre-use and post-use verification procedure using a certified standard may be implemented for routine use of the equipment or a periodic calibration at a predefined frequency based on the risk level associated with the device. The criticality of the tool shall be assessed while defining the scope of the verification activities, as well as the frequency of the calibrations.

ENVIRONMENTAL MONITORING AND PQ

Environmental controls are of critical consideration in cell and gene therapy facilities. The use of aseptic techniques and single-use devices (e.g., bags, tubing) alone cannot guarantee the microbiological quality required for cell and gene therapy products. In the case of gene therapies, the potential use of sterilizing

Validation, Verification and Qualification Considerations 143

grade filters allows the removal of any microbial contamination. In the case of cell therapies, the option of sterilizing filtration is not feasible due to the size of the cells as compared to potential microbial contamination.

In any case, gene or cell therapies, the facilities require strict environmental controls to avoid potential microbial contamination from reaching the culture media, intermediate products or the final drug product. In this section, we will discuss the most relevant elements required in support of environmental control activities as well as required for the environmental monitoring PQ.

i. *Personnel training: Gowning and aseptic techniques*
Personnel working in the facilities must learn and demonstrate proper aseptic skills. Therefore, a well-designed training program must be in place covering basic gowning steps (e.g., scrubs, sterile coats, sterile gowns, gloving, sanitizing techniques), as well as the aseptic techniques to work in clean rooms (e.g., transfer of materials aseptically, aseptic manipulations and connections, sampling and mixing). Initial and periodic qualifications of the personnel are required to ensure such skills are maintained.

ii. *Cleaning and sanitization*
The basis of the environmental controls is proper cleaning and sanitization practices to reduce or eliminate microbial contamination from the equipment and rooms. The personnel must be trained on well-established procedures for cleaning and sanitization activities. These procedures shall include the use of detergents and sanitizing agents tested and approved for use at the facility. Also, these procedures must clearly define the steps to avoid contamination of culture media, intermediate products and final drug products by residues of the detergents and sanitizing agents.

The following general practices shall be considered (PDA, 2006a, 2006b, 2012a, 2012b):
- Periodic rotation of cleaning and sanitizing agents to ensure their effectiveness in controlling microbial contamination in the facility.
- Periodic monitoring of the microbial flora recovered in the facility (e.g., typical human skin microorganisms, spore formers, airborne vs. waterborne, bacteria vs. molds).
- Daily, weekly and monthly cleanings and sanitization schemes.
- Sporicidal compounds for sanitization activities.
- Lint-free wipes and cloths for cleaning surfaces and equipment.
- Periodic replacement of mobs and cleaning tools.
- Periodic use of sanitizing agents for hand/glove sanitization (e.g., isopropyl alcohol).
- Avoid incompatible compounds with the manufacturing process (e.g., interfere with processes, react with components, create fumes).

iii. *Space/room classifications*
Typical manufacturing facilities will be designed considering personnel, material, waste and product flow due to the specific risks associated with the activities associated with each flow. In-depth details for the design of

a production facility can be found elsewhere (ISPE, 2021; Koo, 2019). The most significant risks are microbial contamination and cross-contamination of the drug product. Common assumptions when designing the facility include, but are not limited to, the following:
- Proper segregation of raw materials, in-process materials and final drug products.
- Routine access to gowning materials.
- Unidirectional flows to reduce or eliminate crossing paths during routine activities.
- Dedicated air handling units to avoid potential cross-contamination between adjacent or common serviced rooms/spaces.
- Proper classification of the rooms and spaces within the facility based on the activities intended in each area.
- Increased pressure differentials (pressure cascade) as cleanliness level among rooms is increased:

    Class A/ISO 5 > Barriers/Air Lock > Class B/ISO 7 > Barriers/Air Lock > Class C/ISO 8 > Barriers/Air Lock > Controlled Not Classified > Lockers > Not Controlled

- Use of biological safety cabinets (BSC) when performing open operations or use of isolators/Restricted Access Barrier Systems (RABS) when the operator's hands are required for aseptic manipulations.
- Increased gowning requirements for personnel working in the areas (e.g., hair and beard covers, sterile scrubs, sterile coats, sterile lab coats, sterile bunny suits, sterile gloves, sterile goggles).
- Single-use waste collection containers (nonbiohazardous and biohazardous) consider both solid and liquid wastes.
- Management of waste gowning, laundry, chemical wastes, warehouse, quality control and municipal wastes (e.g., food, office).
- Separated path for collection and removal of wastes generated during the manufacturing process.
- Separated access or airlocks for personnel and equipment/supplies with specific sanitization procedures.
- Routine access to cleaning and sanitizing agents.
- Integration of testing laboratories to the manufacturing areas.
- Housekeeping requirements.

The classification of the rooms and spaces in the facility is based on the activities to be conducted in the areas, as implied by the assumptions listed earlier. Airborne (viable and nonviable particles) environmental requirements are used across the industry to monitor the performance of the environmental controls within a facility. Main guidance on environmental requirements from European and American regulatory groups (e.g., EMA, n.d.; FDA, 2004; Pharmaceutical Inspection Co-Operation Scheme (PIC/S), n.d.) is summarized in Table 8.4.

TABLE 8.4
Environmental Requirements (FDA, 2004; PIC/S and EU)

Source	Designations		ISO 5 Grade A	ISO 6	ISO 7 Grade B	ISO 8 Grade C	Grade D	CNC
US FDA	In Operation. Max. no. particles based on size	Classification (based on 0.5 μm particles)	100	1,000	10,000	100,000	N/A	N/A
		0.5 μm particle/ft^3	3,500	35,200	352,000	3,520,000	N/A	N/A
PIC/S and EU	In Operation. Max. no. particles based on size	0.5 μm particle/m^3	3,520	N/A	352,000	3,520,000	N/A	N/A
		5.0 μm particle/m^3	20	N/A	2,900	29,000	N/A	N/A
	At Rest. Max. no. particles based on size	0.5 μm particle/m^3	3,520	N/A	3,520	352,000	3,520,000	N/A
		5.0 μm particle/m^3	20	N/A	29	2,900	29,000	N/A
US FDA, PIC/S and EU	In Operation. Microbial Active Air Action Limits (CFU/m^3)		<1	7	10	100	200	N/A

TABLE 8.5
Classifications as Function of Activities and Risks (ISPE, 2021)

Classifications	Activities and Risks	Other Considerations
Class A/ISO 5	High-risk products or processes Open processing, aseptic manipulations	Pressurization: Positive Room Filtration: HEPA/ULPA
Class B/ISO 7	Background to Class A/ISO 5	Pressurization: Positive Room Filtration: HEPA/ULPA
Class C/ISO 8	Background to Class A/ISO 5 in exceptional cases with approval from a health authority	Pressurization: Positive Room Filtration: HEPA/ULPA
Class C/ISO 8	Background for closed processing or low-risk products	Pressurization: Positive Room Filtration: HEPA/ULPA
Class D/CNC	Support areas or corridors with access to Class C/ISO 8 room via airlocks	Pressurization: Facility design Room Filtration: MERV, HEPA/ULPA

The potential impact of microbial contamination in the process is commonly used when defining the classification of the rooms/spaces. Table 8.5 summarizes the classifications as a function of activities and risks commonly considered across the industry.

  iv. *Environmental monitoring performance qualification*

The Environmental Monitoring Performance Qualification (EMPQ) is governed by a protocol where the environmental requirements are confirmed across the different rooms/spaces. This qualification is different from aseptic process simulations (APS) or media challenges where the process-related manipulations and operations are the subject of testing (PDA, 2006a, 2006b). During the EMPQ, the environmental conditions (e.g., viable and nonviable particles) are the focus and are closely monitored. Typically, the EMPQ is executed following a thorough cleaning of the facility and prior to any other qualification activities (e.g., APS or PPQ) with the goal of confirming the facility is ready for operation. Once the EMPQ is completed, a subset of tests is defined for routine monitoring (e.g., daily, weekly and monthly) of the environmental conditions in the facility. This routine monitoring is the basis for the environmental monitoring program required to confirm the state of control of the facility.

PROCESS, METHOD, AND FACILITY CHANGES

Changes to processes, methods and facilities must be managed by means of well-defined change management procedures and systems. The goal is to ensure that the rationale and impact of proposed changes are documented and thoroughly assessed by the different stakeholders (e.g., Quality Assurance, Manufacturing Science and Technology, Operations). Every change has desired consequences (e.g., main goal of the proposed change) when properly developed and implemented. However, a change may result in undesired consequences (e.g., unforeseen situations) when the required assessments, developments or implementations are incomplete or incorrect. The following aspects associated with process, method and facility changes provide the reader with some guidance on the topic of change management and the potential impact on qualifications, verifications and validation.

i. *Management of changes (change controls)*
Changes to processes, methods and facilities shall be properly documented and justified. Once a facility and equipment are fully qualified, any change to the design or intended operational conditions shall be properly assessed to ensure such change does not result in undesirable outcomes or malfunction. Similar consideration shall be made for process changes. A change to a validated process may require at least an abbreviated validation study to ensure the quality of the product is not adversely affected (undesired consequence).

There are multiple automation/computerized solutions in the market to support the compliance aspect of managing changes. The following are examples of computerized solutions available for companies to manage change control documents, in no order or preference: TRACKWISE®, VEEVA™, MONTRIUM™, MasterControl™.

ii. *Impact assessments*
The assessment of any change shall include technical, compliance, regulatory, safety, supply and testing considerations by the correspondent groups and subject matter experts (SMEs). These assessments are used to justify the proposed change as well as to identify the required actions required for a successful implementation of the proposed changes (e.g., updates to SOP, modifications to pieces of equipment, modifications of processing conditions, introduction of new raw materials, required monitoring activities following the implementation of the change, updates or execution of qualification documents, updates or execution of new validation activities).

The risks associated with the manufacture of cell and gene therapies are relatively high when compared to other pharmaceutical processes. Since terminal sterilization is not an option for these therapies, aseptic processing is the only option. Aseptic processing involves protecting the product and contact surfaces from microbiological contamination (PDA, 2004. Impact assessment for proposed changes shall consider risk analyses to reduce or eliminate potential hazards that could be introduced by proposed changes. The explanation and use of risk assessment tools (e.g., FMEA) can be found elsewhere (PDA, 2004).

PROCESS VALIDATION LIFE CYCLE

Process validation is defined as a comprehensive process of data collection and analysis starting at the time a manufacturing process is designed and developed and ending when the process is decommissioned. The life cycle concept links all these elements as described elsewhere (EC, 2015; FDA, 2011; ICH 2012b, Gorsky & Baseman, 2020, Vega-Mercado, 2019). Process validation establishes scientific evidence that a process is capable of consistently delivering a quality product (FDA, 2011). General considerations for process validation include the following:

- Knowledge management through the life cycle.
- Integrated team that includes expertise from a multidisciplinary group (e.g., process engineering, analytical chemistry, microbiology, statistics, quality assurance, manufacturing).
- Planned, executed and properly documented studies to discover, observe, correlate and confirm knowledge about the process and product.
- Use of risk-based decision-making throughout the life cycle to assess the criticality of product attributes and process parameters.

Process validation involves multiple activities over the life cycle of a product and process, which can be divided into the following three stages:

i. *Stage 1: Process design*
   The manufacturing process is defined during this stage. Knowledge is gained through laboratory experiments (development) and scale-up activities. That knowledge will be reflected in the master production and control records. All experiments should be conducted in accordance with sound scientific principles and documented following good documentation practices. The following considerations shall be considered in this stage (EC, 2015; FDA, 2011):
   - The functionality and limitations of commercial equipment.
   - Predicted contribution to variability posed by different components, operators, environmental conditions, and measurement systems.
   - Use of laboratory or pilot-scale models designed to be representative of the commercial process.
   - Use of Design of Experiment (DOE) studies to reveal relationships between variable inputs (e.g., components or process parameters). The result of the DOE studies can provide justifications for ranges of incoming materials, equipment parameters and in-process attributes.
   - Use of risk analysis tools to screen variables for DOE studies.
   - Computer-based simulations may be used to gain additional process understanding.
   - Use of process knowledge as the basis for establishing the strategy for process control. The strategy can be designed to reduce input variation, adjust for input variation during manufacturing or a combination of both approaches.
   - Controls shall include examination of material quality and equipment monitoring. These controls are established in the master production and control records.

Once the commercial production and control records are finalized and contain the operational limits and process control strategy, they are confirmed during the second stage of process validation (PDA, 2019).

ii. *Stage 2: PQ*

In this stage, the process design is evaluated to determine if it is capable of reproducible commercial manufacture. This stage has two elements:

a. Design of facility and qualification of the equipment and utilities.

Qualification of equipment and utilities are covered under individual plans or an overall project plan. The considerations and recommendations for the design of the facility and the qualification of equipment have been discussed previously in this chapter.

b. PPQ

The PPQ combines the facility, utilities, equipment and trained personnel with the commercial manufacturing process, controls and materials to produce commercial lots. The successful completion of the PPQ confirms the process design and demonstrates that the commercial manufacturing process performs as expected. It is a regulatory expectation to have a successful PPQ before commencing commercial distribution of the drug product (FDA, 2011, PDA 2013). The following elements shall be considered during PPQ:

- PPQ should be based on sound science and the overall level of product and process understanding and demonstrable control.
- Cumulative data from designed studies, laboratory, pilot and commercial batches should be used to establish the conditions in the PPQ.
- Previous experience with sufficiently similar products and processes can be helpful.
- Recommended to employ objective measures (e.g., statistical metrics) to achieve adequate assurance.
- Higher level of sampling, additional testing and greater scrutiny of process performance than would be typical of routine production.
- Increased level of scrutiny, testing and sampling should continue through the process verification stage as appropriate to establish levels and frequency of routine sampling and monitoring.
- Usable lifetimes of reusable materials (e.g., chromatography resins, molecular filtration media) should be confirmed by additional PPQ protocols during commercial manufacture.
- A written protocol that specifies the manufacturing conditions (e.g., operating parameters, process limits, raw materials), controls, testing (e.g., in-process, release, characterization) and acceptance criteria are required. Also, the following shall be part of the protocol:
  - Sampling plan, sampling points, number of samples and frequency of sampling.
  - Criteria and performance indicators that allow scientific- and risk-based decisions about the ability of the process to produce

quality products. Detailed description of the statistical methods to be used and provision for assessing deviations from expected conditions and handling of nonconforming data (all data must be included in the PPQ).
- Status of the validation of analytical methods used for the PPQ.
- Review and approval of the protocol by appropriate groups and the quality unit.
- Execution of the protocol should not begin until it is fully reviewed and approved.
  - Any deviations from the protocol shall be properly documented and follow established procedures or provisions in the protocol.
  - The commercial manufacturing process and routine procedures must be followed during PPQ.
  - A final PPQ report should be prepared in a timely manner after completion of the protocol. The report shall discuss and cross-reference all aspects of the protocol. The report shall state a clear conclusion as to whether the process met the conditions defined in the protocol and whether the process is in a state of control.

iii. *Stage 3: Continued process verification*

This stage is for continual assurance that the process remains in a state of control or validated state during routine manufacture. Periodic evaluation of production data and information is expected per cGMPs, which will allow the detection of undesired process variability and the identification of problems. Once an issue is identified, proper action must be taken to correct, anticipate or prevent problems and to maintain the process in control. The Continued Process Verification requires the establishment of an ongoing program to collect and analyze product and process data. The program shall consider the following elements:
- Data collected should include relevant trends and quality of incoming materials, in-process.
- Data should be trended and reviewed.
- Information should verify that the quality attributes are being controlled throughout the process.
- A statistician develops the data collection plan and statistical methods and procedures.
- Production data should be collected to evaluate process stability and capability.
- Process variability should be periodically assessed and monitoring adjusted accordingly.

REFERENCES

European Commission. (2015). EudraLex volume 4. EU guidelines for good manufacturing practice for medicinal products for human and veterinary use. Annex 15. Qualification and validation. Brussels, Belgium.

Validation, Verification and Qualification Considerations 151

European Medicines Agency (EMA). (n.d.). www.ema.europa.eu/en.
Food and Drug Administration. (2004). *Guidance for industry. Sterile drug products produced by aseptic processing – current good manufacturing practices*. Rockville, MD: FDA. www.fda.gov.
Food and Drug Administration. (2011). *Guidance for industry. Process validation: General principles and practices*. Rockville, MD: FDA. www.fda.gov.
Gorsky, I. and Baseman, H.S. 2020. Principles of parenteral solution validation. A proactical lifecycle approach. Academic Press. San Diego, CA. USA.
Gibson, M., & Schmitt, S. (2014). *Technology and knowledge transfer. Key to successful implementation and management*. Bethesda, MD: PDA. River Grove, IL: DHI Publishing, LLC.
International Conference on Harmonization (ICH). (2001). *Good manufacturing practice guidance for active pharmaceutical ingredients*. Q7A. www.ich.org.
International Conference on Harmonization (ICH). (2006). *Quality risk management*. Q9. www.ich.org.
International Conference on Harmonization (ICH). (2009a). *Pharmaceutical development*. Q8(R2). www.ich.org.
International Conference on Harmonization (ICH). (2009b). *Pharmaceutical quality system*. Q10. www.ich.org.
International Conference on Harmonization (ICH). (2012a). *Development and manufacture of drug substances (chemical entities and biotechnological/biological entities)*. Q11. www.ich.org.
International Conference on Harmonization (ICH). (2012b). *Technical and regulatory considerations for pharmaceutical product life cycle management approach*. Q12. www.ich.org.
ISPE®. (2021). *Advanced therapy medical products (ATMPs) – autologous cell therapy*. Tampa, FL: ISPE.
Koo, L. Y.(2019). *FDA perspective on commercial facility design for cell and gene therapy products. PharmaED – Process Validation Summit 2019*. La Joya, CA.
Parenteral Drug Association. (2006a). *Technical report no. 28. process simulation testing for sterile bulk pharmaceutical chemicals*. Bethesda, MD: PDA.
Parenteral Drug Association. (2006b). *Technical report no. 44. Quality risk management for aseptic processes*. Bethesda, MD: PDA.
Parenteral Drug Association. (2012a). *Technical report no. 57. Analytical method validation and transfer for biotechnology products*. Bethesda, MD: PDA.
Parenteral Drug Association. (2012b). *Technical report no. 29. Points to consider for cleaning validation*. Bethesda, MD: PDA.
Parenteral Drug Association. (2013). *Technical report no. 60. Process validation: A life cycle approach*. Bethesda, MD: PDA.
Parenteral Drug Association. (2014). *Technical report no. 65. Technology transfer*. Bethesda, MD: PDA.
Parenteral Drug Association. (2019). *Technical report no. 81. Cell-based therapy control strategy*. Bethesda, MD: PDA.
Pharmaceutical Inspection Co-operation Scheme (PIC/S). (n.d.). www.picscheme.org
Vega-Mercado, H. (2019). *Technology transfer and process validation for CAR T products. PharmaED – Process Validation Summit 2019*. La Joya, CA.

9 Adventitious Agent Contamination Considerations during the Manufacture of Cell and Gene Therapy Products

Hazel Aranha
GAEA Resources Inc., Northport, USA

CONTENTS

Introduction 153
Risk Assessment and Quality Considerations in Adventitious Agent
Contamination Control 156
Regulatory Approach to Adventitious Agent Contamination Control 160
Considerations in Adventitious Agent Contamination Control 161
    Source and Raw Materials Controls 161
        Input Materials 163
        Raw Materials, Reagents and Components 165
    Manufacturing Considerations 168
    In-Process Testing 171
Conclusions 171
References 172

INTRODUCTION

Contamination control during the manufacture of cell and gene therapy (CGT) products takes center stage when we recognize that our usual safeguards and guardrails, to ensure a safe product from a minimal contamination standpoint, are not always applicable. Living cells are both the input and the final product administered to the recipient of the CGT therapeutic; consequently, terminal sterilization of the drug product is not an option. Process operations involving aseptic manual manipulations over multiple days increase the risk of microbial ingress, especially when

TABLE 9.1
Points to Consider from a Contamination Control Standpoint When Designing a Strategy for CGT Products

- Source (patient) material collection and administration (initial and final steps) involves bedside manipulation, increasing risk of contamination
- Manufacturing process involves multiple aseptic manipulations, also many manual operations have potential risk of microbial ingress
- Dedicated viral inactivation and removal steps used in the manufacture of biopharmaceuticals for virus safety assurance cannot be applied
- Final product is not amenable to terminal sterilization by conventional methods, e.g., sterile filtration through a 0.2 μm-rated or 0.1 μm-rated filter
- Short manufacturing times as compared with other biopharmaceuticals
- Raw materials, including ancillary materials, are nutrient rich, which is conducive to microbial contamination. Compendial and reference standards are not available
- Detection methods, e.g., growth-based compendial methods, cannot be directly extrapolated from conventional manufacturing operations. Uptake of validated rapid detection methods has lagged; newer methods will need validation
- Product release testing may be conducted concurrent with product administration – this places the onus of contamination control on a robust quality management program

using systems with the potential for high microbial ingress (open system, benchtop and bedside manipulations). When viral vectors are used to deliver the genetic payload, it is necessary to distinguish between the vector virus and any adventitious viruses that could potentially contaminate the product. Incorporation and validation of virus clearance methods, routinely employed for virus safety assurance with traditional biopharmaceuticals, cannot be incorporated. Table 9.1 highlights points to consider when designing a contamination control strategy for CGT products.

Microbial contamination has a huge impact on product manufacture; these contaminants introduce product variability, can potentially cause loss of potency due to degradation or modification of the product by microbial enzymes, result in changes in impurity profiles and increase levels of bacterial endotoxins. While contamination events are low probability, they are catastrophic when they occur. Consequences of contamination events in biomanufacturing range the spectrum from the impact on patient safety and drug shortages to legal, regulatory and financial implications (Shukla & Aranha, 2015). Follow-up on a contamination event includes investigation management, decontamination, regulatory scrutiny and multiple corrective actions. Just over a decade ago, manufacturing-environment contamination events at a biopharmaceutical company resulted in plant shutdown and drug shortages (Plavsic et al., 2016; Qiu et al., 2013). Marketing of the enzyme replacement therapeutic that required lifelong administration had to be halted due to Vesivirus viral contamination of a production facility. Patients were granted access to alternative treatments, including those that had completed phase III clinical trials but had not yet received licensure, in order to address the critical drug shortage (Hollak et al., 2010).

Microbial contamination (bacteria, fungi) during pharmaceuticals and biopharmaceuticals manufacture has been extensively studied. Even though these market sectors are relatively mature, multiple compliance infractions related to this aspect have continued to occur. Jimenez (2019) summarized data from Food and Drug Administration (FDA) enforcement reports and reported that gram-negative bacteria were the most common microbial contaminants; unidentified microbial contaminant species accounted for 77% and 87% of nonsterile and sterile drug recalls, respectively, which was indicative of poor microbiology practices. Microbial contaminants of CGT products, specifically bacteria and fungi, have been addressed in multiple reviews (Cundell et al., 2020; Golay et al., 2018; Mahmood & Ali, 2017.) Most common contaminants reported were mycoplasma, bacteria and fungi. The most common bacteria reported were gram-positive, spore-forming rods such as *Bacillus cereus, B coagulans* and *B brevis*; gram-positive cocci like *Enterococcus malodoratus, E. casseliflavus* and *Staphyloccus epidermidis*; and gram-negative bacteria such as *Escherichia coli*. While the exact origin of these contaminants may not have been determined, they most likely originated during cell collection or during manual manipulations during cell culture.

The Consortium on Adventitious Agent Contamination in Biomanufacturing (CAACB), a consortium of participants from industry and academia collects comprehensive data (voluntary and confidential) on virus contamination events (Barone et al., 2020). Considering that Chinese Hamster Ovary (CHO) cell lines, a workhorse of the biopharmaceutical industry, are used for production of monoclonal antibodies and recombinant products, most contamination incidents reported were with Murine minute virus (MMV), and the potential source was ascribed to the raw material (though in most cases, not conclusively proven). In contrast, for the human and primate cell lines used in biopharmaceuticals manufacturing, operators or the cell line used in manufacturing were suspected to be the source. Four of the five viruses (herpesvirus, human adenovirus type 1, parainfluenza virus type 3 and reovirus type 3) reported to contaminate the human and primate cell lines used for product manufacture are known to be pathogenic in humans. These data highlight that the viral contamination of products produced in human or primate cell lines poses a higher safety risk to patients due to the lack of the species barrier when human cell lines are used. Contamination during the manufacturing process can occur even in a GMP (good manufacturing practices) environment; contamination with Rhinovirus with the HEK-293 cell line was reported in spite of the inbuilt precautions in manufacturing; this was attributed to an operator-associated event – breach in GMP (Rubino, 2010). Although viruses have been detected in bulk harvests and manufacturing environments, to date, there have been no incidents of iatrogenic infectious virus transmission through cell line-derived biopharmaceuticals (Aranha, 2011; Barone et al., 2020). Noteworthy here is that the contamination events reported were detected early in manufacturing and did not pose a safety concern to the patient.

Contamination control requires a holistic program that encompasses concepts within the context of the entire manufacturing life cycle and process. The end product in gene therapy is often a virus particle (that delivers the genetic payload), which brings with it the obvious challenge of differentiating the product of interest from other undesirable viruses.

This chapter addresses risk considerations, discusses the regulatory issues and provides insights into the manufacturing considerations to be addressed when discussing adventitious agent contamination of CGT products. Many issues and concerns that are being raised with CGT products were expressed when the biopharmaceutical therapeutics landscape was just starting to surface on our horizon in the last century. As an industry, we have demonstrated virus safety assurance and have been able to lay claim to decades of safe monoclonal antibodies and recombinant products. While we are not able to directly extrapolate the strategies applied in the past, they offer useful guideposts in our quest to achieve a similar safety profile for CGT products.

RISK ASSESSMENT AND QUALITY CONSIDERATIONS IN ADVENTITIOUS AGENT CONTAMINATION CONTROL

A holistic risk-based approach is essential when addressing contamination control, especially with CGT products. A contamination control strategy should consider all aspects of the product life cycle with ongoing and periodic review and assessment to update the strategy as appropriate. This assessment should lead to corrective and preventive actions as necessary to reduce the risk of microbial and viral contamination.

The aims of the risk-based approach are twofold: (i) to proactively identify risks associated with the input raw materials and the process and (ii) to eliminate, ideally, the risk and failing which to incorporate strategies to minimize and manage the risk. With CGT products, risk-based manufacturing process design and testing regimes, of necessity, cannot lose sight of the short product life cycle to ensure patients have access to these lifesaving therapies.

In any discussion of potential for virus contamination during the manufacture of CGT products, several considerations related to viral contamination need to be acknowledged. Viruses are obligate intracellular parasites, and, while any overt manifestation of viral infection (manifestation of cytopathogenic effects (CPE) in the cell line or a change in the physiological markers monitored during manufacturing) will result in appropriate remediation (cell culture/bioreactor shutdown), not all viruses exhibit CPE behavior. CHO cell lines used in the manufacture of biopharmaceuticals are known to express endogenous retroviral-like particles (RVLPs) that are noninfectious in humans. In spite of high retroviral loads in the CHO cell lines that have been extensively characterized, these cell lines are deemed acceptable for manufacture. Regulations mandate that appropriate virus clearance steps for these endogenous retroviruses and any adventitious viruses that could potentially enter the manufacturing stream be incorporated to ensure a safe product (ICH Harmonized Tripartite Guideline, 1999).

Both regulatory groups and industry manufacturers recognize and acknowledge the unique aspects associated with virus presence, which could go undetected, sometimes, for decades. Based on our ability to detect them and our understanding, viral contaminants may be (i) 'known-known,' where the potential for the virus presence is known and recognized, as, for example, a contaminant e.g., bovine-derived virus, if bovine serum albumin is used; in this situation, elimination of that raw material is required, and even mandated depending on the criticality of the raw material; (ii) known-unknowns, as, for example, endogenous retroviruses, where CHO cell lines harboring these viruses are still considered safe and used in the manufacture of marketed products provided that appropriate virus clearance strategies are incorporated during manufacture; and (iii) 'unknown-unknowns' where the presence of the adventitious virus is unknown; however, with improvements in detection methods or other advances, their presence is demonstrated decades later even after the product is marketed. Two examples here: transmission of hepatitis B virus upon administration of yellow fever virus during World War II was traced to the formulation with contaminated human serum albumin. While the contaminated excipient was implicated in HBV transmission, it was not until the availability of polymerase chain reaction (PCR) methodologies decades later that it was conclusively proven (Aranha, 2012). A more recent example is the detection of PCR in marketed rotavirus vaccines. Methodologies such as massive parallel sequencing, degenerative PCR and pan microbial microarrays have resulted in detection of viral sequences in cell substrates and virus seed stocks. While evaluating their advanced detection method, massive parallel sequencing, researchers used several marketed viral vaccines and demonstrated the presence of porcine circovirus (PCV) in the tested samples (Victoria et al., 2010). In this situation, a risk-benefit evaluation determined that the balance here was highly in favor of the significant benefit offered by the vaccine (Dubin et al., 2013; Ranucci et al., 2011). Additionally, it must be noted that (i) steps routinely used (low pH, chromatography) in the manufacture of the vaccine provided viral clearance of PCV (Yang et al., 2013, 2020), and (ii) the novel detection method detected nucleic acid sequences and not infectious virus. Discussions with the regulatory authorities have since then resulted in reevaluation of the cell line and efforts to replace it.

A key aspect of risk mitigation during biopharmaceutical manufacture is to conduct viral clearance spiking studies for several downstream purification steps to demonstrate the capacity and capability of the process to remove or inactivate known and unknown viruses. Multiple removal and inactivation unit operations are incorporated to ensure that the processes afford adequate virus clearance and also incorporate an adequate safety factor (Shukla & Aranha, 2015.) The effectiveness of this strategy is corroborated by the evidence that there has been no reported transmission of a contaminating virus to a patient from a therapeutic protein produced using recombinant DNA technology. The CAACB data corroborate the excellent safety record for over 35 years (Barone et al., 2020). Additionally, real-world data analysis from Biological License Application (BLA) submissions from 1995 through January 2021 on the viral clearance capabilities of all unit operations validated for virus clearance provide documentation that the unit operations we have relied upon for decades provide safe and effective removal of viral contaminants during manufacturing (Ajayi et al., 2022). Table 9.2 provides a comparison of similarities and differences between monoclonal antibodies/

TABLE 9.2
Comparison of Monoclonal Antibodies and Recombinant Products and Cell and Gene Therapy Products and Factors That Could Potentially Impact Contamination Levels

	Monoclonal Antibodies/ Recombinant Products	CGT Products
Source and raw materials	• Tiered cell banking system; extensively characterized cell banks	• Autologous products – patient-sourced (aspirate/biopsy/blood sample), which gives rise to variability in starting materials • Allogeneic products – ideally, established and characterized banks should be used • Ongoing efforts to progress toward a tiered cell banking system
Cell culture media	• Ideally chemically defined, animal-derived materials excluded	• Complex media containing multiple supplements; while some supplements may be low risk (Tier I) in some cases medium risk (Tier II or III) may be used when cell growth and differentiation require these supplements • In some cases, in early stages of establishment of cell banks for allogeneic products it may be necessary to use low levels of serum
Manufacturing process	• Well understood • Mature processes • Many downstream operations have been platformed and provide a comprehensive toolbox for purification and product recovery	• Processes for these operations are not easy to standardize in view of their inherent significant variability • Recovery and purification operations include volume reduction, cell harvest, cell sorting • Multiple manual operation steps (efforts ongoing toward automation)
Equipment and facilities	• Both single-use systems (SUS) and stainless steel at large scale; single-use favored at early (seed) stages • Processes largely automated at commercial scale	• Single-use technologies dominate, as batch sizes are relatively small, also at early stage bench scale systems may be used • Operations may be manual, labor intensive; trend toward semi-automation as processes mature
Key sources of variability	• Well-controlled processes, established design space, operations are conducted within the design space • Some degree of variation in product titer may occur depending on feed strategy and culture conditions; however, this must fall within design space	• Variety of scale-out and scale-up operations are in use • Delays are a serious concern as cell expansion depends on cell source (donor material), cell response during pre-processing and processing steps

(Continued)

TABLE 9.2 (CONTINUED)
Comparison of Monoclonal Antibodies and Recombinant Products and Cell and Gene Therapy Products and Factors That Could Potentially Impact Contamination Levels

	Monoclonal Antibodies/ Recombinant Products	CGT Products
Key sources of adventitious agent ingress	• Extensively characterized cell banks and established supplier management for raw materials decrease the potential for microbial contamination • Potential contamination sources are breach of GMP by personnel, manufacturing-environment excursions	• Entire process is at high risk of microbial ingress. Source materials when directly derived from patients (autologous) are a source of intrinsic and extrinsic contamination (during handling operations) • Multiple manual operations are another source of microbial contamination
Storage and shipping	• Usually ambient or refrigerated, stable over several months or as determined by stability studies	• Cryopreserved or live with limited shelf life
Product nature and dose	• Final product essentially 'inert'; relatively homogenous though glycoforms may be present • Usually high dose, administration periodically over extended time	• Product is 'live,' a virus; difficult to characterize heterogenous nature of product due to impact of different cell types in final population • Typically single treatment or application
Product delivery	• Product vialed and then transfused or injected into patient	• Product injected/implanted into patient (live products) or vialing (cryopreserved) • Appropriate courier and traceability required

recombinant products and CGT products and highlights the areas that require additional diligence with the latter.

Quality risk management (QRM) provides a tool that focuses on making patient safety decisions based on sound scientific knowledge and process understanding (ICH Q9, 2021). The nature and complexity of CGT products compounded with a need to collect and organize data mean that the principles of QRM and Risk Management become essential tools for product and process knowledge and manufacturing control.

There are multiple risk assessment tools that are discussed in detail in ICH Q9. Approaches commonly used with biopharmaceutical products are the Ishikawa (Fishbone) approach and the FMEA (Failure Mode and Effects Analysis). The former provides a qualitative approach and requires the user to define the potential sources of risk. For example, in the Ishikawa-Fishbone approach, risk sources are evaluated depending on, hypothetically, the following factors: Management, Man (Personnel), Process, Machine (Equipment), Measurement, and so on. In FMEA, the risk components severity, occurrence and detectability are considered to quantify the risk of the specific failure and its effect. The risk priority number (RPN) functions as quantification of the risk and is calculated by multiplying the severity (SEV), occurrence (OCC) and detectability (DET) values. Values between 1 and 10 may be used for rating. The assessment and quantitative ratings of the components increase with higher risk relevance.

Severity is a measure of the possible consequences of the hazard. Severity can range from low-medium (minor to moderate) to high (critical). Minor to moderate excursions include those that are in locations that either have no direct product contact or from a source with indirect product contact, respectively. When collected from a site where there is direct product contact, it has the potential to jeopardize the GMP status of the facility and result in compliance nonconformance and would constitute a critical hazard. To conduct the FMEA, the severity of a potential failure needs to be assessed. Because of the patient risk of contamination, product-dependent factors must be considered. After defining the potential root cause, its likelihood of occurring is rated; OCC reflects the probability of its occurrence; contributing factors such as human or technical errors are added, and the final OCC rating is calculated. The rating increases with a decreasing probability of detecting the failure. DET is an integral component of determination of RPN which addresses the probability to detect a failure. When discussing microbial and fungal contamination, growth and culture tests are considered the gold standard. Despite the availability and the recognized benefits, implementation of alternative and rapid methods remains slower than desired. It should be noted that new detection technologies used in the research setting may not be designed to meet the rigor of Current Good Manufacturing Practices (cGMPs) and therefore would need to be evaluated and validated prior to introduction into a production environment.

REGULATORY APPROACH TO ADVENTITIOUS AGENT CONTAMINATION CONTROL

Regulatory requirements include demonstration of a quality control system with a defined workflow, in-process controls and specific release criteria. The lessons of the past have led to current regulatory guidance and industry best practice, which relies on three pillars in terms of control of adventitious agents: the selection of appropriate starting and raw materials with a low risk of containing adventitious virus, testing of cell banks and in-process materials to ensure they are free from detectable viruses and incorporation of in-process steps to ensure a safe product.

Regulatory compliance requires new technologies and a pragmatic mindset to standardize and check the quality of the production process of CGT products. Definition of the quality of CGT therapeutics is challenging given their complexity. CGT products must be developed in designated clean environments; in addition, manual cell culture practices are not able to provide the high throughput needed for the development of new CGT products.

Three complementary approaches have been adopted to prevent possible contamination of biotechnology products from an adventitious agent standpoint: selecting and testing cell lines and other raw materials, including cell culture media for the absence of viruses; viral clearance studies (viral validation studies), which evaluate the capability of the purification process to clear viruses; and testing products at appropriate stages of the production for the absence of contaminating viruses (in-process testing). Viral clearance evaluation studies are a major component of assuring the safety of biopharmaceuticals. These virus inactivation and removal steps that are part of the purification process cannot be incorporated during CGT manufacture because the genetic payload delivered to the patient is a live virus and would not withstand these treatments. An exception here is the manufacture of adeno-associated virus (AAV) and lentivirus viral vectors (Table 9.5). The specific characteristics of the vectors could be exploited to use virus clearance methods during manufacture of these vectors, which are considered a critical raw material for manufacture of CGT products.

The main risks for viral contamination in CGT products are cell sources, materials used in cell culture and exposure of the cell culture process stream to the environment and personnel. Tables 9.3 and 9.4 summarize risk mitigation strategies that can be deployed and the operational challenges associated when discussing risk mitigation and management. The associated risks and challenges are discussed in the following section.

CONSIDERATIONS IN ADVENTITIOUS AGENT CONTAMINATION CONTROL

SOURCE AND RAW MATERIALS CONTROLS

The origin of the cells, the key raw material for Advanced Therapy Medicinal Products (ATMPs) is the donor. The final administered dose is given back to the same patient (autologous) or to multiple patients (allogeneic). Thus, it is mandatory to protect the patient's cells from microbiological and viral contamination through the collection, processing and administration life cycle. Short turnaround times/shelf lives are an expectation with these patients who have life-threatening conditions, so the manufacturing process and concurrent testing are expected to meet these demands.

The clinical site environment and operators (collection, product manipulation at clinical site) are key in ensuring a safe product. Depending on the product, they may undergo further manipulation at the clinical site prior to administration in largely non-cGMP environments. When cells are manipulated post-collection, the microbial contaminant could have been either intrinsic (patient source) or introduced during the handling process.

For cell collection and product administration, kits can be an effective tool to drive consistency in the execution of these processes. Kits ensure that the same materials will be used in each process, that instructions are available for each procedure, the labeling is consistent, and (in the case of collection kits) that a qualified shipping solution is stipulated for transporting critical patient cells

TABLE 9.3
Potential Sources of Contamination and Appropriate Risk Mitigation Strategies Employed

Possible Source of Contamination	Risk Mitigation Strategy
Cell culture (autologous cells, viral vectors used as critical raw materials)	Extensive documentation related to origin of cell line, media, passage number, subculturing frequency; conditions related to culturing temperature and CO_2 requirement; cryopreservation and storage conditions
Continuous cell lines	Use cells from a recognized repository that provides documentation and traceability
Lab personnel	Appropriate gowning depending on work area Work areas must be sanitized/disinfected as appropriate Training to high standards of hygiene and cleanliness Periodic retraining, considering criticality, need for proficiency testing and annual competency assessment
Airborne particulates and aerosols	Work in closed environments wherever possible Bedside manipulations must be conducted in a laminar flow hood Manufacturing environment must be International Standards Organization (ISO) 7 or higher
Cell culture medium and supplements	All additives must be sterile-filtered wherever possible; heat or g-irradiation is not an option Supplier selection is key
Plasticware/disposables and glassware	Single-use bags/systems are preferred Any glassware used must be autoclaved or subject to dry heat sterilization
Cell storage conditions	Buffers, saline, cryoprotectants may be a source of microbial contamination
Compressed gases	Air, oxygen, nitrogen and carbon dioxide are sterile-filtered in-line and/or point-of-use CO_2 incubator gases must be HEPA filtered
Water	Major raw material in the manufacture of biopharmaceuticals. If on-site water processing systems are used, they must be designed, maintained and operated in a GMP-compliant manner. Periodic monitoring for microbial counts and endotoxin is required

following their collection. This approach decreases process variation and ensures that clinical results reflect the efficacy of the given therapy rather than variations in the handling.

Facility requirements for performing on-site preparative steps or administration of gene therapy products depend on the nature of the products, their applications and the manipulations required. The most important determinant of facility features is the level of risk for microbial contamination associated with each step. While there are no specific recommendations, it has been suggested that the definition of low-risk and high-risk conditions can be made according to a framework

TABLE 9.4
Operational Challenges during the Manufacture of CGT Products

	Current State	Comments/Clarification
Batch variation	Variation in source material from different patients even when cell type may be the same for a given therapeutic indication	Source-material variation results in variable duration of cell expansion and impacts overall product manufacturing time frame
Scale-up/scale-out	Scale-up of autologous cell therapy manufacturing usually results in 'scale-out' rather than 'scale-up' of the process to increase manufacturing capacity	This requires introduction of additional stations/suites to support cell therapy manufacturing for an increased number of patients
Equipment	Key process equipment used in cell therapy operations (cell washing, cell separation, cell selection) and controlled rate freezing is usually not standard equipment used in GMP environments	Significant focus by suppliers to make available equipment designed for the CGT market Robust validation of equipment and generation of protocols that can be used in cGMP environments
Operational challenge	No terminal sterilization prior to filling No established playbook regarding manufacturing and purification steps, i.e., no platform methods to fall back on	Onus of 'safe product' falls on risk minimization approaches during patient-material collection and subsequent processing Process operations drive room classification
Product testing prior to release	Standard product release is not possible	Tight environmental and manufacturing controls Establishment of a design space, operation within which would facilitate a high probability of a minimal contamination event

similar to that defined in US Pharmacopeia (USP) <797> Pharmaceutical Compounding of Sterile Preparations, General (US Pharmacopeia, 2019).

Input Materials

The input materials for CGT products are derived from an individual. Autologous cell therapy products originate from the collection of cells from human blood or tissues each time a production process is initiated. Tests to assure that the derived cells are free of adventitious virus generally cannot be completed before initiating cell therapy manufacturing, and the process generally proceeds at risk. For allogeneic therapies in which cells from one donor are used to create therapies for multiple patients, the donor cells should also be characterized to assure they are virus free, per regulatory guidance.

The quality of these starting materials (cells) is determined by multiple factors, including cell source, harvest technique and patient-related factors, such as concomitant disease and age. Possible sources of contamination during the collection

TABLE 9.5
Virus Clearance Methods That Can Be Employed for Virus Safety Assurance during Manufacture of Vectors Used in Gene Therapy

Clearance Method	AAV Vector	Lentivirus Vector
Heat	Heat stable; will inactivate contaminating viruses with minimal impact on AAV	Heat sensitive; not a suitable method for this vector virus
Low pH	Low pH stable; low pH hold will inactivate contaminating viruses with minimal impact on AAV	pH sensitive; not a suitable method for this vector virus
Detergent inactivation	Nonenveloped virus; only enveloped contaminating viruses will be inactivated with minimal impact on AAV	Vector virus is enveloped, detergent sensitive, so not applicable
Chromatography	Surface charge differences may be exploited to separate vector virus from contaminants	Surface charge differences may be exploited to separate vector virus from contaminants
Virus-removal filtration	Vector is ~ 20 nm, so filtration through virus-removal filters with pore size distribution > 25 nm to separate larger viral contaminants may be effective	Vector is relatively large, ~ 80–100 nm, so will be retained by virus-removal filters

process include asymptomatic bacteremia of the patient or donor at time of collection, infected catheters or skin plugs from venous access and improper execution of aseptic technique during collection or processing. Depending on the target cell type desired, the harvested calls may comprise mixtures of different types of cells with varying degrees of purity. The cell populations of interest are subject to separation and enrichment methods. While automated separation devices are available, careful consideration must be given to the software selected. Cells must be identified and characterized in terms of identity, purity, potency, viability and suitability for intended use.

Standard practice is to harvest bone marrow (BM) under sterile conditions in operating rooms, peripheral blood stem cells (PBSCs) via apheresis, and umbilical cord blood (UCB) after birth in hospital obstetric units. The potential for and mechanism of contamination differs for these various collection methods. Once cell products are collected, they are transferred to the cell-processing laboratory where they may undergo further manipulation ranging from minimal ex vivo processing to more extensive cell manipulation, resulting in the increased potential for introducing a contaminant during the manipulations.

For patient-specific source materials (autologous cells), according to 21 CFR 1271.90(a)(1) (10), a donor eligibility determination or donor screening for infectious agents is not required on autologous cells. However, as a pathogen could potentially be present in the donor, it should be determined if bioprocessing will increase the risk to the recipient owing to further propagation of the pathogenic agent present in the donor that on administration could cause infection. For allogeneic cells or tissues, manufacturers must perform donor screening and testing, as required in 21 CFR Part 1271. Donors of all types of cells and tissues must be screened for risk factors and clinical evidence of relevant communicable disease agents and diseases, including human immunodeficiency virus; hepatitis B virus; hepatitis C; human transmissible spongiform encephalopathies, including Creutzfeldt–Jakob disease; and Treponema pallidum (syphilis) (21 CFR 1271.75). If cord blood or other maternally derived tissue is used, screening and testing on the birth mothers, as described in 21 CFR Part 1271.80(a) (8), must be performed.

Following apheresis, the cells are exposed to multiple manipulations that are high risk from a microbial contamination standpoint. Studies have been conducted to determine the timing of potential contaminants, which were subsequently classified based on risk analysis. The early phase is a primary period, which includes operations with extrinsic risk at the cell isolation and the passage in culture. Low levels of contamination were reported at this phase which could be due to both extrinsic factors (manual operators) or intrinsic (associated with the cells); the middle phase is a period for stabilization, which has operations for passage to change vessels from a flask to a culture bag. Due to multiple manipulations involving cell feeding, transfers in this phase had the highest potential for contamination. The late phase is the period for the proliferative process in a culture bag after confirmation of the logarithmic increase of cells. The possibility of contamination is low since operations in this phase involve the addition of a culture medium.

In the case of cell sources, both recombinant biopharmaceutical products and viral vector gene therapy products have a low risk of contaminated starting cell sources, as both manufacturing processes start with exhaustively characterized master cell banks. For allogeneic therapies in which cells from one donor are used to create therapies for multiple patients, the donor cells should also be characterized to assure they are virus free, per regulatory guidance. Conversely, autologous cell therapy products originate from the collection of cells from human blood or tissues each time a production process is initiated. Tests to assure that the derived cells are free of adventitious virus generally cannot be completed before initiating cell therapy manufacturing, and the process generally proceeds at risk.

Raw Materials, Reagents and Components

Raw materials quality assurance constitutes the underpinning of microbial and viral safety in the manufacture of biologicals with an acceptable safety and efficacy profile. Cell culture processes employed for the manufacture of biopharmaceutical products and ATMP products utilize a variety of basal medium formulations consisting of a mixture of more than 50 essential nutrients (for example, amino acids, vitamins and trace elements) and other chemicals. These are filter sterilized before use, typically with 0.1-μm-rated sterilizing-grade filters through which most viruses will pass. If any components of media are contaminated with a virus during their manufacture or handling, they may initiate an infection during the cell culture process. Animal-derived and human-derived components (for example, serum and growth factors), which carry a higher risk of virus contamination than other components, are commonly added to media for ATMP production.

Raw materials qualification is an integral component of adventitious microbial or viral agents; we rely heavily on our processes designed to evaluate, track and document to ensure product quality and safety of the raw materials used. Vendor qualification programs are an essential component of raw materials qualification. 'Fit for use' is a key consideration when evaluating raw materials and reagents. Raw materials must be evaluated to identify risks associated with the procured raw material to be evaluated from a fit-for-function standpoint. If lab or research-grade materials are used at the initial stages, they will not pass muster when the process moves toward commercialization.

In accordance with cGMP requirements, the use of animal-derived materials (ADMs) must be avoided. However, often fetal bovine serum (FBS) may be required at the early stages of cell bank establishment. If FBS is used, it must be from countries classified as Grade B (low to no risk of BSE). The certificate of analysis provided should indicate absence of bluetongue virus, bovine adenovirus, bovine parvovirus, bovine respiratory syncytial virus, bovine viral diarrhea virus, haemadsorbing agents, rabies virus, reovirus and bovine spongiform encephalopathy agent.

Ancillary materials (AMs) are considered a subset of raw materials as per the USP. In and of themselves, AMs are complex biological materials like serum derivatives, cytokines, growth factors; therefore, it is difficult to assess their quality attributes. AMs can play a role in cell growth, differentiation or function. Considering this criticality, qualification of the AM used in making the final product is required; if sourced externally, the supplier will be required to provide information related to its qualification.

AMs can exert an effect on the therapeutic substance – cytokine may activate a population of cells, but they are not intended to be in the final formulation. AMs may be classified based on their risk profile: (i) Tier I are low-risk, highly qualified materials with intended us as therapeutic drugs or biologics or medical devices; examples here include heparin or insulin that are available as GMP-grade materials; (ii) Tier II include low risk but well-characterized materials produced in compliance with GMP; (iii) Tier III are moderate-risk materials not intended for use as AMs, they are reagent grade materials or produced for in vitro diagnostics; (iv) Tier IV are high-risk materials not produced in compliance with GMP and are not intended for use in CGT products.

While raw materials are not 'approved,' their use is regulated under other mechanisms that apply to biologics product manufacture. Additionally, suppliers are often not forthcoming with the information as their manufacturing process and formulations are considered proprietary. Manufacturers must fully understand and investigate claims made by suppliers. For example, GMP-grade raw material is a term commonly used; however, it is absent from any regulatory lexicon. Recognizing that compendial raw materials are not always available, regulations recommend that the materials used be characterized to the fullest extent possible.

It is important to recognize that scientists in academic and research environments may not have full knowledge of regulatory and GMP requirements, and, therefore, reagents used at this stage may be laboratory grade or ADMs may be used; consequently, the process may need to be redeveloped if the product is on the commercialization pathway.

Any qualification program for raw materials must address the following: (i) identification and selection, (ii) suitability for use in manufacturing, (iii) characterization and (iv) quality assurance. Identification and selection include an active program of supplier selection, use of primary sources wherever possible and ensuring consistency, traceability and availability of an uninterrupted source of the component. Suppliers should provide information regarding traceability of each material, especially for human- and ADMs. An assessment of suitability is required to determine whether the raw material may pose a risk to the safety, potency and purity of the final therapeutic product. Characterization of the raw materials may pose a challenge, especially when the raw materials are biological-derived (serum-derived extracts, growth factors) and, consequently, difficult to characterize. The level of testing for each component will depend on the risk assessment profile and the knowledge gained about each component during development. The test panel for each raw material should assess a variety of quality attributes, including purity, functionality and freedom from adventitious viruses or microbial contaminants. Tests for sterility, pyrogenicity, mycoplasma and adventitious or infectious agent viral agents must be conducted. The specific adventitious agent panel will depend on the source of the component and how the component is manufactured. Because biochemical testing may not be reflective of process performance due to their heterogenicity and complexity, functional performance testing may be necessary. For example, when FBS or complex enzymatic digests are used, in vitro culture cytotoxicity tests must be conducted. Quality assurance should reflect those found in a typical GMP manufacturing environment. These activities should include the following systems or programs: (i) incoming receipt, segregation, inspection and release of materials prior to manufacturing; (ii) vendor auditing and certification; (iii) certification of analysis verification testing; (iv) formal procedures and policies for out-of-specification materials; (v) stability testing; and (vi) archival sample storage.

Some human cell lines used for vector production may harbor endogenous viruses. An example here is the HEK-293T cell line, known to express the Simian Virus 40 (SV40) large antigen T. SV40 virus has known potential tumorigenicity and was reported in the cell line used for polio vaccine manufacture in the last century. Lookback studies over decades with the cell line contaminated with SV40 virus used did not demonstrate any evidence of tumorigenicity. The HEK-293T cells are used as packaging cells for GMP-grade viral vector production systems such as AAVs, retrovirus, adenovirus or lentivirus manufacturing in several CART-T cell clinical trials (Castella et al., 2019; van der Loo & Wright, 2016). In general, any cell line used must be obtained from a standardized collection, e.g., American Type Culture Collection (ATCC), and must be accompanied by results of microbiological quality control tests and PCR-based assays for human pathogenic viruses: HIV, HepB, HPV, EBV, CMV.

In addition to reagents and raw materials, bioreactor bags, tubing and other components that constitute the disposable train must be evaluated to ensure that they do not potentially affect cell growth and viability or contribute to the particulate load of the system. In view of the concern over potential introduction of particulates from these materials into the product, vendors of cell culture products are under increasing pressure to use 'low particulate' manufacturing methods for their products. These particulates, while theoretically inert, could interfere with the efficacy of the product and pose a safety hazard to the patient.

Manufacturing Considerations

The following section highlights a few key considerations to keep in mind during the manufacture of CGT products and highlights areas of concern from a contamination control standpoint. Entire chapters in this handbook provide detailed guidance related to CGT manufacturing and associated processes.

ICH Q7 provides guidance related to GMP (ICH Harmonized Tripartite Guideline, 2000). The current GMP (CGMP) relevant regulations, 21 CFR 210 and 211, would also apply, as would the additional biological product regulations in 21 CFR 600-680 and 1271 on human cells, tissues and tissue-based products The recent 2019 Parenteral Drug Association *Technical Report No. 81: Cell-Based Therapy Control Strategy* provides a comprehensive control strategy that mitigates risks related to product quality.

One of the first considerations when designing manufacturing for CGT products is whether an end-to-end processing platform will be used or whether unit operations will be employed. These decisions will drive containment and facility segregation requirements. Understanding the process equipment intricacies and limitations is essential to ensure the facility is designed appropriately.

The cell therapy manufacturing wherever possible should be performed in closed systems (low risk of microbial ingress) using sterile single-use disposable materials with advanced aseptic manipulation occurring in biosafety cabinets, restrictive access barrier systems (RABS), or isolator systems.

Aseptic processing in a controlled, clean facility (per ISO standards) and rigorous in-process sterility sampling paired with an active environmental monitoring program are critical in reducing risk of adventitious agent contamination and can provide investigative data in the event of a positive sterility result that requires antimicrobial intervention. A well-designed trending program that includes critical data analysis will assist in root cause analysis and corrective and preventive analysis (CAPA) will assist with instituting additional training-related issues or the need for additional engineering controls. Most organisms in engineering-controlled facilities appear to arise from personnel commensal flora, which is generally considered nonpathogenic but can cause clinically significant infections in immunocompromised hosts.

Environmental monitoring (EM) is key in any attempt to manage contamination in manufacturing environments. Typically, EM results are not available until after seven days of incubation and, therefore, identification of any recovered

Adventitious Agent Contamination Considerations

organisms is not available prior to release of CGT products. If any deviations occur during processing, they must be triaged and the risk to the patient determined before the investigation is fully completed. Therefore, proactive measures to prevent ingress of microorganisms are paramount. The impact of manufacturing-environment excursions cannot be underestimated. A reported incident of a flood in the manufacturing environment required the immediate remedial action of fixing the leak; what was not considered was the significant impact of microbial contamination that resulted in a facility deviation, an 'alert limit' due to the increase in bacterial and fungal species that resulted in the bioburden excursions during routine EM.

Maintenance of the aseptic environment is critical to preclude extrinsic contamination risks, similar to conventional pharmaceutical manufacturing. However, intrinsic contamination risks exist in all cell manufacturing processes since cells used as raw materials cannot be sterilized, thus giving rise to the primary and secondary risks of cell contamination and cross-contamination, respectively. Table 9.6 provides an overall summary of the quality challenges and recommended solutions associated with CGT manufacture.

TABLE 9.6
Contamination Control Challenges during Manufacture of CGT Products

Input/Patient Materials

Need	Source-material collection to decrease microbial ingress from handling and environment
Challenge	Differences in cell types/source materials collected require different levels of manipulation and pose different levels of risk
Impact	High impact due to failed lots due to overt contamination; could delay in delivery to patient
Potential Solution	• Define and standardize collection procedures • Make kits available for patient-side manipulations • Training of personnel at collection site • Closed systems whenever possible • For autologous products, banked cells that have undergone extensive characterization

Raw Materials

Need	Processes require input of complex raw materials, e.g., ADMs (serum), AMs
Challenge	Complex raw materials; lack of compendial analytical test methods
Impact	High impact due to failed lots; delay in delivery to patient
Potential Solution	• Establish characterized cell and virus bank for viral vectors and allogeneic products • When animal-derived raw materials are used in early stages of the process, work toward decreasing volumes used and dependence of the cells • Use the 'highest grade' raw materials available; using lab grade or research-grade raw materials that were used in early stages of development will not pass GMP muster

(Continued)

TABLE 9.6 (CONTINUED)
Contamination Control Challenges during Manufacture of CGT Products

Cell Line Potential Tumorigenicity

Need	Well-characterized cell lines that do not harbor endogenous viruses with potential for tumorigenicity
Challenge	Some human cell lines used for vector production (HEK-T) cells harbor endogenous viruses (SV40) with known tumorigenicity
Impact	Low-medium
Solution	• Appropriate risk management • Extensive characterization of cell culture systems • Pharmacovigilance

Manufacturing Process

Need	Control of AA bioburden through process design
Challenge	Multiple manual manipulations
Impact	High; contaminated batches have significant impact on patient access to drug; also business implications
Solution	• Design robust processes • Automate steps where possible • Use SUS • Adequately gowned and trained personnel to ensure no GMP breach

In-Process Testing

Need	Ensure product consistency and acceptability to progress to final processing
Challenge	Sampling for in-process testing
Impact	High
Potential solution	• Use closed systems wherever possible • Automate sampling operations

Personnel are an important variable during the manufacturing process. While the trend is toward automation, appropriately trained personnel are foundational to the development and delivery of ATMPs. Starting with cell harvest from patients and through process development, there should be an adequate number of and appropriately trained personnel. The former is sometimes a challenge for companies that are financially strapped and looking to do 'more with less.' The earlier trend in ATMP development was investments in research and academic institutions that were often financially strapped; currently, biopharmaceutical companies have made significant investments in organic growth of their ATMP portfolios.

Pests are an area of concern even in high-tech environments. While pest control is low-tech, it has a high impact in warehousing and manufacturing environments and is an item often audited in a quality audit. Insects and other pests could carry microorganisms such as spores, bacteria, fungi and even viruses. As part of their CAACB initiative, Barone et al. (2020) have reported 21 cases of contamination of bulk harvests (not final product) with MMV.

IN-PROCESS TESTING

Manufacturing processes have well-defined, go-no-go decision criteria that are established for key manufacturing attributes. In-process control tests are conducted to ensure that the in-process material is of sufficient quality and quantity to ensure manufacture of an acceptable final product. Examples of in-process controls include (i) enumeration and viability, (ii) microbiological criteria (endotoxin, mycoplasma), (iii) expression of phenotypic or genotypic markers, (iv) production of the desired bioactive substance, (v) assays for potential process impurities, (vi) monitoring of culture system parameters (% CO_2, % relative humidity, pH) and (vii) functional tests.

One concern related to in-process testing is the very small volumes that constitute the batch size which then makes sample availability for testing a significant challenge for quality control testing of CGT products. Excessive sample volume would consume material intended for the patient and too small volume would not be statistically significant. In such situations, depending on the kind of tests, surrogate samples (e.g., spent media) or conducting multiple tests on a single sample may be done.

A primary reason for establishing in-process tests is to ensure that the correct product with specified quality and yield is obtained. A secondary reason for performing in-process tests is to gather process and process characterization data that could be useful in assessing the impact of the process changes or excursions, i.e., trending the data. Intermediate in-process material that fails to satisfy in-process control criteria should not be used for further manufacturing. The material may be reprocessed if there are procedures in place for such activities. The reprocessed materials must satisfy the original in-process specification, and the effect of the reprocessing on other quality attributes must be defined before the material can undergo further manufacturing.

Production of a safe and efficacious product involves not only lot-release specifications but also specifications designed to maintain control of the manufacturing process and the final product. Acceptance criteria should be developed on the basis of data from clinical studies and other relevant development data. Once specifications have been established, test results should be trended. Results that are out of specification (OOS), or even those that are out of trend, need to be investigated.

CONCLUSIONS

Integral to ensuring patient safety and end-to-end process and product quality is implementation of a robust risk-based contamination control program. In the absence of dedicated virus clearance steps due to the nature of the CGT product, the risk of contamination prevention relies on establishment of a holistic program that focuses on risk minimization of adventitious agent entry at the beginning of the process and maintenance of this risk prevention approach through minimization of open processing steps, decreased operator intervention through automated systems and purpose-built CGT manufacturing suites rather than retrofitting working spaces. Gaining in-depth knowledge of the manufacturing process as it is conducted in real time and how it fits into all the components of the contamination control program is critical to understanding where vulnerabilities to contamination exist. If the information is not available in-house, several companies recognize the need for specialized expertise and use the services of a subject matter expert (SME) in contamination control as the stakes in the contamination control area are too high in terms of patient risk.

Current focus is away from microbial detection for sterility demonstration to a 'sterility by design' modality to reflect the statistical limitations to detection and improved performance of manufacturing technologies (Tidswell et al., 2021). Through application of QRM initiatives and functioning within a 'design space,' the best possible outcomes would be achievable considering that contamination control is a moving target and mandates a proactive and prescient blueprint for contamination control.

REFERENCES

Ajayi, O. O., Johnson, S. A., Faison, T., Azer, N., Cullinan, J. L., Dement-Brown, J., & Lute, S. C. (2022). An updated analysis of viral clearance unit operations for biotechnology manufacturing. *Current Research in Biotechnology*. doi:10.1016/j.crbiot.2022.03.002

Aranha, H. (2011). Virus safety of biopharmaceuticals: Absence of evidence is not evidence of absence. *Contract Pharma, Nov-Dec 2011*, 82–87.

Aranha, H. (2012). Current issues in assuring virological safety of biopharmaceuticals. *Bioprocess International*, *10*(3), 12–18.

Barone, P. W., Wiebe, M. E., Leung, J. C., Hussein, I. T. M., Keumurian, F. J., Bouressa, J., ... Springs, S. L. (2020). Viral contamination in biologic manufacture and implications for emerging therapies. *Nat Biotechnol*, *38*(5), 563–572. doi:10.1038/s41587-020-0507-2

Castella, M., Boronat, A., Martin-Ibanez, R., Rodriguez, V., Sune, G., Caballero, M., ... Juan, M. (2019). Development of a novel anti-CD19 chimeric antigen receptor: A paradigm for an affordable CAR T cell production at academic institutions. *Mol Ther Methods Clin Dev*, *12*, 134–144. doi:10.1016/j.omtm.2018.11.010

Cundell, T., Drummond, S., Ford, I., Reber, D., Singer, D., & members of the Pharmaceutical Microbiology Expert Discussion, G. (2020). Risk assessment approach to microbiological controls of cell therapies. *PDA J Pharm Sci Technol*, *74*(2), 229–248. doi:10.5731/pdajpst.2019.010546

Dubin, G., Toussaint, J. F., Cassart, J. P., Howe, B., Boyce, D., Friedland, L., ... Debrus, S. (2013). Investigation of a regulatory agency enquiry into potential porcine circovirus type 1 contamination of the human rotavirus vaccine, Rotarix: Approach and outcome. *Hum Vaccin Immunother*, *9*(11), 2398–2408. doi:10.4161/hv.25973

Golay, J., Pedrini, O., Capelli, C., Gotti, E., Borleri, G., Magri, M., ... Introna, M. (2018). Utility of routine evaluation of sterility of cellular therapy products with or without extensive manipulation: Best practices and clinical significance. *Cytotherapy*, *20*(2), 262–270. doi:10.1016/j.jcyt.2017.11.009

Adventitious Agent Contamination Considerations 173

ICH Q5A (R2). (2022). Viral safety evaluation of biotechnology products derived from cell lines of human or animal origin draft version endorsed on 29 September 2022. Currently under public consultation.

ICH Q7. (2000). International conference on harmonisation of technical requirements for registration of pharmaceuticals for human use. Good manufacturing practice guide for active pharmaceutical ingredients Q7 Current Step 5 version dated November 2000.

ICH Q9 (R1). (2021). International conference on harmonisation of technical requirements for registration of pharmaceuticals for human use. Guideline on quality risk management Step 2b. December 2021.

Hollak, C. E., vom Dahl, S., Aerts, J. M., Belmatoug, N., Bembi, B., Cohen, Y., ... Cox, T. M. (2010). Force majeure: Therapeutic measures in response to restricted supply of imiglucerase (Cerezyme) for patients with Gaucher disease. *Blood Cells Mol Dis, 44*(1), 41–47. doi:10.1016/j.bcmd.2009.09.006

Jimenez, L. (2019). Analysis of FDA enforcement reports (2012-2019) to determine the microbial diversity in the contaminated non-sterile and sterile drugs. *Am Pharm Rev*, September–October, 48–73.

Mahmood, A., & Ali, S. (2017). Microbial and viral contamination of animal and stem cell cultures: Common contaminants, detection, and elimination. *J Stem Cell Res Ther, 2*(5), 1–8.

Plavsic, M., Shick, K., Bergmann, K. F., & Mallet, L. (2016). Vesivirus 2117: Cell line infectivity range and effectiveness of amplification of a potential adventitious agent in cell culture used for biological production. *Biologicals, 44*(6), 540–545. doi:10.1016/j.biologicals.2016.08.001

Qiu, Y., Jones, N., Busch, M., Pan, P., Keegan, J., Zhou, W., ... Mattaliano, R. J. (2013). Identification and quantitation of Vesivirus 2117 particles in bioreactor fluids from infected Chinese hamster ovary cell cultures. *Biotechnol Bioeng, 110*(5), 1342–1353. doi:10.1002/bit.24791

Ranucci, C. S., Tagmyer, T., & Duncan, P. (2011). Adventitious agent risk assessment case study: Evaluation of RotaTeq(R) for the presence of porcine circovirus. *PDA J Pharm Sci Technol, 65*(6), 589–598. doi:10.5731/pdajpst.2011.00827

Regulations, C. o. F. 21 CFR 210. (n.d.) Current good manufacturing practices in manufacturing, processing, packaging or holding of drugs: General.

Regulations, C. o. F. 21 CFR 211. (n.d.) Current good manufacturing practices for finished pharmaceuticals.

Regulations, C. o. F. (n.d.) 21 CFR 600 Biological products: General.

Regulations, C. o. (n.d.-a) F. 21 CFR 820 Quality systems regulation.

Regulations, C. o. (n.d.-b) F. 21 CFR 1271 Human ells, tissues, and cellular and tissue-based products.

Rubino, M. J. (2010). Experiences with HEK293: A human cell line. *PDA J Pharm Sci Technol, 64*(5), 392–395.

Shukla, A., & Aranha, H. (2015). Viral clearance for biopharmaceutical downstream processes. *Pharm Bioprocess, 3*(2), 127–138.

Tidswell, E. C., Agalloco, J. P., & Tirumalai, R. (2021). Sterility assurance-current & future state. *PDA J Pharm Sci Technol*. doi:10.5731/pdajpst.2020.012526

US Pharmacopeia. (2019). USP <797> Pharmaceutical compounding of sterile preparations, general chapter.

van der Loo, J. C., & Wright, J. F. (2016). Progress and challenges in viral vector manufacturing. *Hum Mol Genet, 25*(R1), R42–R52. doi:10.1093/hmg/ddv451

Victoria, J. G., Wang, C., Jones, M. S., Jaing, C., McLoughlin, K., Gardner, S., & Delwart, E. L. (2010). Viral nucleic acids in live-attenuated vaccines: Detection of minority variants and an adventitious virus. *J Virol*, *84*(12), 6033–6040. doi:10.1128/JVI.02690-09

Yang, B., Wang, H., Ho, C., Lester, P., Chen, Q., Neske, F., ... Blumel, J. (2013). Porcine circovirus (PCV) removal by Q sepharose fast flow chromatography. *Biotechnol Prog*, *29*(6), 1464–1471. doi:10.1002/btpr.1804

Yang, B., Wang, H., Kaleas, K., Butler, M., Franklin, J., Bill, A., ... Blumel, J. (2020). Clearance of porcine circovirus and porcine parvovirus from porcine-derived pepsin by low pH inactivation and cation exchange chromatography. *Biotechnol Prog*, *36*(4), e2968. doi:10.1002/btpr.2968

10 Training Approaches to Build Cell and Gene Therapy Workforce Capacity

Orin Chisholm
Principal Consultant, Pharma Med, UNSW Sydney, Australia

CONTENTS

Cell and Gene Therapy Workforce Needs 175
Formal Education Pathways 175
Accreditation and Competency Frameworks 176
Short Courses 180
Conclusion 180
References 181

CELL AND GENE THERAPY WORKFORCE NEEDS

With the maturing of the cell and gene therapy sector, there is a need to focus on training and education approaches to building workforce capacity in this emerging sector. Indeed, it has been widely reported that manufacturing is being held up by the lack of a skilled workforce (Ausbiotech, 2021; Chakraverty, 2021; Wyles et al., 2021). A coordinated effort between academic institutions, professional societies and industry is needed to increase workforce capacity by providing micro-credentials, short courses and degree programs to facilitate innovation and development of cell and gene therapy products. Ideally, government policies should foster the development of skills to address gaps in this sector and drive the development of cell and gene therapies.

FORMAL EDUCATION PATHWAYS

Formal education pathways in cell and gene therapy traditionally focus on a biomedical science undergraduate qualification followed by an academic research pathway toward a PhD qualification, while for medical graduates, various MD/PhD programs are often pursued to develop the research skills to bring these products to patients. However, these narrow research-focused pathways do not provide graduates with the broad range of skills needed to develop and commercialize products in this area. Specialist interdisciplinary undergraduate and postgraduate courses that focus on all aspects of the commercialization pathway are needed to develop a workforce that can navigate the complexities of bringing these advanced therapies to market. As such, an industry-focused element within the traditional PhD program would greatly benefit people pursuing these pathways. There are increasing opportunities becoming available in this area. For example, the Office of Health and Medical Research within the Department of Health in New South Wales (NSW), Australia, has developed a PhD scholarship program with local universities where students complete their PhD in an area contributing to the development, delivery or commercialization of cell and gene therapy. Successful candidates must undertake skills development including attending the NSW Health Commercialization Training Program as part of their PhD tenure (NSW government, 2020).

Another example is the embedding of a cell and gene therapy curriculum in medical school programs. As an example, a new transdisciplinary training endeavor has begun at the Mayo Clinic to ensure that healthcare professionals have the requisite skills to develop novel therapies in the cell and gene therapy space (Wyles et al., 2021). The regenerative medicine curriculum is delivered to medical trainees and embedded in the first year of their training program. In later years of their study, students may elect to take further specialty-tailored clinical electives in regenerative medicine to develop skills in clinical trial development and investigational new drug applications. Outcomes of this program are positive with, student self-perceived knowledge and skills in regenerative medicine increasing after completion of the course (Wyles et al., 2021).

Alternatively, people wanting to work in this field may pursue specialist masters-level degrees in various aspects of drug development commercialization, which may contain individual subjects in the development and commercialization of cell and gene therapy products. Such courses may lead to a master's degree in regulatory science, pharmacovigilance, health economics or medical affairs, or they may encompass all these aspects of drug and/or medical device development. Alternatively, subjects in cell and gene therapy may be offered in master's programs focused on genetics and genomics. More specialist master's degrees in cell and gene therapy are also emerging. Some available English-language master's degrees available in North America, Europe and the Asia-Pacific regions are provided in Table 10.1 (graduate certificates and diplomas are not included in this list).

ACCREDITATION AND COMPETENCY FRAMEWORKS

The subjects offered within a degree are continually reviewed and changed in response to the external environment. With the growth in new cell and gene therapies coming onto the market (ARM, 2021), we can expect that curricula will be updated to include more focus on cell and gene therapy product development,

TABLE 10.1
Masters Coursework Degrees in Either Cell and Gene Therapy or Medical Product Development (English Language, Not Exhaustive)

Region	Name of Degree/University	Website
North America	Master of Science in Clinical Research Management (Regulatory Science)/Arizona State University	https://asuonline.asu.edu/online-degree-programs/graduate/master-science-clinical-research-management-regulatory-science/
	Master of Science in Health Sciences in Regulatory Affairs/George Washington University	https://healthsciencesprograms.gwu.edu/program/the-george-washington-university-regulatory-affairs-master-master-of-science-in-health-sciences-1527569596782
	Master of Science in Regulatory Science/Johns Hopkins University	https://advanced.jhu.edu/academics/graduate-degree-programs/regulatory-science/
	Master of Business and Science (Clinical and Regulatory Affairs)/Keck Graduate Institute	https://www.kgi.edu/academics/henry-e-riggs-school-of-applied-life-sciences/academic-programs/master-of-business-and-science/concentrations/
	Master of Science in Drug Regulatory Affairs/Long Island University	http://www.liu.edu/Pharmacy/Academic-Programs/Graduate/MS-Drug-Regulatory-Affairs
	Master of Science in Regulatory Affairs and Health Policy/ Massachusetts College of Pharmacy and Health Science	https://www.mcphs.edu/academics/school-of-pharmacy/regaff-pharmeco-healthpolicy/regulatory-affairs-and-health-policy-ms
	Master of Science in Pharmaceutical Chemistry (Cell and Gene Therapy)/New Jersey Institute of Technology	https://catalog.njit.edu/graduate/science-liberal-arts/chemistry-environmental-science/pharmaceutical-chemistry-ms/
	Master of Science in Biotechnology (Regulatory Science Concentration)/Northeastern University	https://cos.northeastern.edu/biotech/academics/ms-biotechnology/concentrations/
	Master of Science in Regulatory Compliance/ Northwestern University	https://sps.northwestern.edu/masters/regulatory-compliance/index.php#Regulatory%20Compliance%20&%20Program%20Tracks
	Master of Clinical Research/Ohio State University	http://gpadmissions.osu.edu/programs/program.aspx?prog=0229
	Master of Science in Biotechnology Innovation and Regulatory Science/Purdue University	https://polytechnic.purdue.edu/degrees/ms-biotechnology-innovation-regulatory-science

(*Continued*)

TABLE 10.1 (CONTINUED)
Masters Coursework Degrees in Either Cell and Gene Therapy or Medical Product Development (English Language, Not Exhaustive)

Region	Name of Degree/University	Website
	Master of Business and Science (Drug Discovery and Development)/Rutgers, State University of New Jersey	https://mbs.rutgers.edu/program/drug-discovery-development-online
	Master of Science in Regulatory Affairs/San Diego State University	https://regsci.sdsu.edu/
	Master of Science in Regulatory Affairs and Services/St. Cloud State University	https://www.stcloudstate.edu/graduate/regulatory-affairs/default.aspx?utm_source=website&utm_medium=redirect
	Master of Science in Pharmacy Administration/St. John's University	https://www.stjohns.edu/academics/programs/pharmacy-administration-master-science
	Master of Science in Regulatory Affairs and Quality Assurance/Temple University	https://pharmacy.temple.edu/admissions/admissions-requirements/qara-non-thesis-ms
	Master of Biotechnology/ University of Alabama, Birmingham	https://www.uab.edu/shp/cds/biotechnology
	Master of Science in Pharmaceutical Outcomes and Policy (Pharmaceutical Regulation specialty)/ University of Florida	https://onlinepop.pharmacy.ufl.edu/programs/degrees-and-certificates/masters-degree/
	Master of Science in Regulatory Science /University of Maryland	https://www.pharmacy.umaryland.edu/academics/regulatoryscience/
	Master of Regulatory Affairs, Master of Regulatory Science/ University of Pennsylvania	https://catalog.upenn.edu/graduate/programs/regulatory-science-msrs/; https://www.itmat.upenn.edu/itmat/education-and-training/master-of-regulatory-affairs/
	Master of Science in Regulatory Science, Management of Drug Development, Medical Product Quality, or Regulatory Management. Doctor of Regulatory Science (DRSc)/University of Southern California	https://regulatory.usc.edu/programs/
	Master of Science in Biomedical Regulatory Affairs/ University of Washington	https://www.regulatoryaffairs.uw.edu/
	Master of Science in Regulatory Affairs /Northeastern University, Toronto, Canada	https://toronto.northeastern.edu/academic_program/regulatory-affairs/
Europe	Master of Science in Pharmaceutical Medicine/ University of Claude Bernard Lyon, Eudipharm, France	https://www.pharmatrain.eu/diploma-master.php?id=8
	Master of Science in Pharmaceutical Medicine/ pme institute, University of Duisburg-essen, Germany	https://www.uni-due.de/studienangebote/studiengang.php?id=82

Master of Science Cell and Gene Therapies/University of Patras and University of West Attica, Greece	https://mastercgt.com/course/
Master of Science in Cellular Manufacturing and Therapy/NUI Galway, Ireland	https://www.nuigalway.ie/courses/taught-postgraduate-courses/cellular-Manufacturing-therapy.html#
Master of Science in Pharmaceutical Medicine/ Trinity College Dublin, Ireland	https://www.tcd.ie/medicine/pharmacology-therapeutic/postgraduate/msc/
Master of Science in Drug Sciences, Master of Advanced Studies in Medicines Development/ University of Basel, Switzerland	https://www.unibas.ch/en/Studies/Degree-Programs/Degree-Programs/Drug-Sciences.html, http://web.ecpm.ch/master
Master of Science (Genes, Drugs and Stem Cells-Novel Therapies)/Imperial College London, UK	https://www.imperial.ac.uk/study/pg/medicine/genes-drugs-stem-cells/
Master of Science in Drug Development Science/Clinical Pharmacology/Medical Affairs/King's College London, UK	https://www.kcl.ac.uk/scps/postgraduate-study/pharmaceutical-medicine
Master of Science in Cell and Gene Therapy/University College London, UK	https://www.ucl.ac.uk/prospective-students/graduate/taught-degrees/cell-and-gene-therapy-msc
Master of Science in Pharmacovigilance/PIPA and University of Hertfordshire, UK	https://www.herts.ac.uk/courses/postgraduate-masters/pharmacovigilance
Master of Science in Regulatory Affairs/TOPRA/ validated by University of Hertfordshire, UK (TOPRA: The Organization for Professionals in Regulatory Affairs)	https://www.topra.org/TOPRA/TOPRA_Member/mscra/MSc_Regulatory_Affairs.aspx
Master of Science in Advanced Therapy Medicinal Products/University of Manchester, UK	https://www.manchester.ac.uk/study/masters/courses/list/12672/msc-advanced-therapy-medicinal-products/course-details/
Master of Science in Advanced Cell and Gene Therapies/University of Sheffield, UK	https://www.sheffield.ac.uk/postgraduate/taught/courses/2022/advanced-cell-and-gene-therapies-msc
Master of Biomedical Engineering (Molecular and Cellular Biotherapies)/BME Paris Consortium	https://www.bme-paris.com/molecular-cellular-biotherapies-mcb/
Master of Pharmacovigilance (PV) and Pharmacoepidemiology/Eu2P joint – several European universities	https://www.pharmatrain.eu/diploma-master.php?id=7
Asia-Pacific	
Master of Science in Medicine (Pharmaceutical and Medical Device Development)/University of Sydney, Australia	https://www.sydney.edu.au/courses/courses/pc/master-of-science-in-medicine-pharmaceutical-and-medical-device-development.html
Master of Clinical Trials Research/University of Sydney, Australia	https://www.sydney.edu.au/courses/courses/pc/master-of-clinical-trials-research.html
Master of Pharmaceutical Industry Practice/University of Queensland, Australia	https://future-students.uq.edu.au/study/program/Master-of-Pharmaceutical-Industry-Practice-5703

manufacturing and commercialization. However, accreditation of university courses against competency frameworks developed by cell and gene therapy bodies would greatly accelerate the development of new courses and subjects to address the skills gaps in the cell and gene therapy workforce. The development of professional competency frameworks and accreditation standards have been very successfully employed to tailor university degrees such as those in medicine, engineering, nursing and allied health disciplines (Australian government, Department of Health, 2021; NASEM, 2016; White et al., 2020).

Competency frameworks have also been developed by professional development organizations that support the pharmaceutical and medical technology industries such as the Association of Clinical Research Professionals (ACRP), the Regulatory Affairs Professionals Society (RAPS), the International Federation of Associations of Pharmaceutical Physicians (IFAPP) and the Professional Society for Health Economics and Outcomes Research (Bridges, 2019; Pizzi et al., 2020; Sonstein & Jones, 2018; Stonier et al., 2020).

SHORT COURSES

One way of addressing immediate training needs is through the delivery of short courses and micro-credentials. There is an increasing trend toward the development of micro-credentials by higher education institutions and other providers and being able to stack them to gain a qualification (McGreal & Olcott, 2022). As an example of a short course, Stanford University offers a certificate on the Principles and Practices of Gene Therapy (Stanford Online, 2022). Several organizations involved in cell and gene therapy already offer short courses and online training modules, such as the International Society of Cell and Gene Therapy (ISCT), which offers an online training module on advanced therapy manufacturing, and the US National Institute for Innovation on Manufacturing Biopharmaceuticals (NIIMBL), which offers a number of training programs in biopharmaceutical manufacturing (ISCT, 2022; NIIMBL, 2022). The UK-based Cell and Gene Therapy Catapult Program identified skills shortages and has developed online training, in-person training center and an advanced therapies apprenticeship scheme to drive education and training in this field (Cell and Gene Therapy Catapult, 2019). Informa offers training programs in cell and gene therapy manufacturing, viral vectors and regulatory challenges (Informa, 2022) and suppliers such as ThermoFisher host a cell and gene therapy learning center (ThermoFisher, 2022).

CONCLUSION

The rapid growth of cell and gene therapy products coming to market needs to be supported by a number of different types of training and education opportunities to enable an adequately trained workforce to deliver these therapies. This will require strong advocacy from the sector to ensure government policies are established to support these training and education initiatives. We have seen some innovative schemes developed, such as the UK Catapult advanced therapy apprenticeship scheme, but these need to be expanded in other countries to enable the development of trained staff. Short courses, especially recognized micro-credentials that can be stacked to obtain a university qualification, and more subjects within already established master's degree programs will also contribute to an educated workforce to deliver the expected benefits of these new therapies.

REFERENCES

Alliance for Regenerative Medicine (ARM). (2021). Regenerative medicine in 2021: A year of firsts and records. https://alliancerm.org/sector-report/h1-2021-report/#:~:text=The%20global%20cell%20%26%20gene%20therapy,in%20product%20approvals%20and%20financings

Ausbiotech. (2021). Australia's regenerative medicine manufacturing capacity and capability. https://www.ausbiotech.org/documents/item/666

Australian Government, Department of Health. (2021). National registration and accreditation scheme. https://www.health.gov.au/initiatives-and-programs/national-registration-and-accreditation-scheme

Bridges, W. (2019). The creation of a competent global regulatory workforce. *Front Pharmacol*, doi:10.3389/fphar.2019.00181

Cell and Gene Therapy Catapult. (2019). UK cell and gene therapy skills demand report 2019. https://ct.catapult.org.uk/

Chakraverty, A. (2021). Skilled labor shortages impact cell and gene therapy manufacturing. Labiotech.eu; https://www.labiotech.eu/trends-news/cell-gene-therapy-skills/

Informa. (2022). Cell and gene therapy training. https://informaconnect.com/cell-gene-therapy/training/

International Society of Cell and Gene Therapy (ISCT). (2022). Training programs. https://www.isctglobal.org/resources/training-programs

McGreal, R., and Olcott, D. (2022). A strategic reset: micro-credentials for higher education leaders. *Smart Learning Environments*, 9, 9. doi:10.1186/s40561-022-00190-1

National Academies of Sciences, Engineering, and Medicine (NASEM); Health and Medicine Division; Board on Global Health; Global Forum on Innovation in Health Professional Education. Exploring the Role of Accreditation in Enhancing Quality and Innovation in Health Professions Education: Proceedings of a Workshop. Washington (DC): National Academies Press (US); 2016 October 5. 3, Competency-Based Accreditation and Collaboration. https://www.ncbi.nlm.nih.gov/books/NBK435960/

National Institute for Innovation on Manufacturing Biopharmaceuticals (NIIMBL). (2022). Education and training resources. https://niimbl.force.com/s/education-and-training

New South Wales (NSW) government. (2020). Gene and cell therapy PhD program. https://www.medicalresearch.nsw.gov.au/cell-gene-therapy/

Pizzi, L. T., Onukwugha, E., Corey, R., Albarmawi, H., and Murray, J. (2020). Competencies for professionals in health economics and outcomes research: The ISPOR health economics and outcomes research competencies framework. *Value in Health: The Journal of the International Society for Pharmacoeconomics and Outcomes Research*, 23(9), 1120–1127. doi:10.1016/j.jval.2020.04.1834

Sonstein, S. A., and Jones, C.T. (2018). Joint taskforce for clinical trial competency and clinical research professional workforce development. *Frontiers in Pharmacology*. doi:10.3389/fphar.2018.01148

Stanford Online. (2022). Principles and practices of gene therapy. https://online.stanford.edu/courses/xgen201-principles-and-practices-gene-therapy

Stonier, P. D., Silva, H., Boyd, A., Criscuolo, D., Gabbay, F. J., Imamura, K., Kesselring, G., Kerpel-Fronius, S., Klech, H., and Klingmann, I. (2020). Evolution of the development of core competencies in pharmaceutical medicine and their potential use in education and training. *Frontiers in Pharmacology*, 11, 282. doi:10.3389/fphar.2020.00282

ThermoFisher Scientific. (2022). Cell and gene therapy learning center. https://www.thermofisher.com/au/en/home/clinical/cell-gene-therapy/cell-gene-therapy-learning-center.html

White, J. A., Gaver, D. P., Butera, R. J. et al. (2020). Core competencies for undergraduates in bioengineering and biomedical engineering: Findings, consequences, and recommendations. *Annual Review of Biomedical Engineering*, 48, 905–912. doi:10.1007/s10439-020-02468-2

Wyles, S. P., Monie, D. D., Paradise, C. R., Meyer, F. B., Hayden, R. E., and Terzic, A. (2021). Emerging workforce readiness in regenerative healthcare. *Regenerative Medicine*, 16(3), 197–206.

11 How to Distribute Cell and Gene Therapies

Andrea Zobel
World Courier GmbH, Berlin, Germany

CONTENTS

Logistics Requirements of Cell and Gene Therapies – Demand of
the Product 183
    Time Criticality 183
    Cryogenic Storage and Distribution 184
    Chain of Custody 184
    Genetically Modified Therapies 184
Regulatory Framework of Cell and Gene Therapy Logistics 185
Logistics for Time-Critical Products 186
Cryogenic Storage and Distribution 188
How to Assure Chain of Custody 192
Logistics of Genetically Modified Cell Therapies and Gene Therapies 194
Outlook 195
References 195

LOGISTICS REQUIREMENTS OF CELL AND GENE THERAPIES – DEMAND OF THE PRODUCT

In contrast to pharmaceuticals including chemical entities as active pharmaceutical ingredients and biopharmaceuticals, most of which consist of proteins, cell and gene therapies consist of cells, tissues, viruses or nucleic acids. All of these are new pharmaceutical classes with consequences for the logistics. Autologous and allogeneic therapies are also influencing logistics. For an autologous therapy, the patient is providing the raw materials for therapy manufacturing, in most therapies cells or tissues. For allogeneic therapies, the donations are coming from well-selected donors, which are the raw materials for manufacturing large batches of products for the treatment of patients. Either one batch fits all or various batches with different characteristics are manufactured that can be matched to patients according to specific criteria.

TIME CRITICALITY

Cells and tissues are only viable for a short period of time outside the tissue culture environment with 37°C and CO_2-containing atmosphere. This leads to time criticality when these therapies must be shipped to the treatment centers in fresh conditions at ambient or cooled temperatures.

CRYOGENIC STORAGE AND DISTRIBUTION

The only possibility to extend the shelf life of cells and tissues is freezing, whereas only cryogenic temperatures below −150°C extend the shelf life to the typical storage period of pharmaceuticals of at least three to five years.

CHAIN OF CUSTODY

The third new requirement is the need for documentation of the chain of custody for autologous therapies where the therapy is manufactured out of the donations of the patient. It must be assured that the patient is getting back the therapy manufactured out of their own donation.

GENETICALLY MODIFIED THERAPIES

Gene technology provides the possibility to change the genetic information of cells, viruses and nucleic acids in a way to create genetically modified therapies with the potential to replace dysfunctional genes, add missing genes or modify cells in a body leading to curing a disease or positively influencing symptoms of a disease.

The following table shows examples of cell and gene therapies and their logistics demand:

	Short Shelf Life	Frozen	Chain of Custody	Genetically Modified Organism or Cell
Autologous cell therapy, fresh	X		X	
Autologous cell therapy chimeric antigen receptor (CAR-T), fresh	X		X	X
Autologous cell therapy, frozen		<−150°C	X	
Autologous cell therapy, frozen, fresh donation	X	<−150°C	X	
Autologous cell therapy CAR-T, frozen		<−150°C	X	X
Autologous cell therapy CAR-T, frozen, fresh donation	X	X	X	X
Allogeneic cell therapy, fresh	X			
Allogeneic cell therapy CAR-T, fresh	X			X
Allogeneic cell therapy, frozen		<−150°C		

(*Continued*)

How to Distribute Cell and Gene Therapies

	Short Shelf Life	Frozen	Chain of Custody	Genetically Modified Organism or Cell
Allogeneic cell therapy, CAR-T, frozen		<−150°C		X
adeno-associated viruses (AAV) therapy,		−80°C		X
Gene therapy, DNA		−20°C		

The questions and answers that follow should be asked when the logistics of cell and gene therapy are planned. Ideally, these questions are asked in early therapy development to avoid unnecessary hurdles and assures that the right assessments and capabilities are available.

The following table summarizes key questions and answers for planning supply chain strategy:

Number	Question	Answer
1	Is the product a cell therapy or a gene therapy?	If gene therapy Ask 6 and 9 If cell therapy: ask 2
2	Is the product an allogeneic or autologous cell therapy?	If allogeneic Ask 6 If autologous: ask 3
3	Is the autologous cell therapy based on fresh or frozen donations?	If fresh donations: ask 4 and 5 If frozen donations: ask 6
4	What is the shelf life of the donation?	Enter shelf life
5	How many manufacturing sites and where are they located?	Enter sites and location
6	Which temperature, LN_2 or −80°C or −20°C?	Enter temperature
7	Is the cell therapy fresh or frozen?	If fresh: Ask 4 and 5 If frozen: Ask 6
8	Are the cells genetically modified?	If yes: ask 9
9	Is a risk assessment available showing risks for the environment? Are the criteria for a GMO (genetically modified organism) dangerous for human, animal, plant, microorganism or environment fulfilled?	If yes: Product must be shipped as dangerous goods UN3245 If no: Dangerous classification UN3245 is not applicable

REGULATORY FRAMEWORK OF CELL AND GENE THERAPY LOGISTICS

The storage and distribution conditions have the potential to affect the quality of cell and gene therapies to the point of making them ineffective. As a consequence, the regulations for these products include a lot of requirements and guidance on how to conduct storage and distribution and which documentation and quality standards are applicable.

In the European Union (EU), the regulatory framework is developing, with the European Tissues and Cells Directives giving general guidance and setting the standards. The guidelines relevant to advanced therapy medicinal products (ATMP) are published on the official website of the European Medicines Agency and extended continuously (EMA, 2022). The good manufacturing practice guidelines for ATMP (EC, 2017) and good clinical practice of advanced therapy investigational medicinal products (ATIMP) (EC, 2019) include the standards for storage, distribution and traceability of cell and gene therapies in clinical development and aftermarket authorization. In an action plan, additional guidelines are in development; for example, a guideline for good manufacturing practice of ATIMP (EMA, n.d.) For the United States, the US Food and Drug Administration (US FDA) provides regulations for cell and gene therapies on the related website (FDA, n.d. a).

In Chemistry, Manufacturing, and Control's (CMC) Information for Human Gene Therapy Investigational New Drug Applications (INDs) (FDA, n.d. b) clearly describes that the storage and distribution conditions are part of the product: "*Your IND should also include information on shipping conditions, storage conditions, expiration date/time (if applicable), and chain of custody from the manufacturer to the site of administration in the summary information of the CTD.*"

The regulatory framework for cell and gene therapies is extended continuously by the FDA to provide more guidance and standards. On the related website, these documents are published (FDA, n.d. c).

LOGISTICS FOR TIME-CRITICAL PRODUCTS

There are various reasons why cells are shipped at 2°C–8°C or ambient 15°C to 25°C as viable cells and not at frozen temperatures. Donations of cells derived from critically ill patients often have insufficient quantity and quality, which would decrease again during the freezing process. As a consequence, many donation protocols require immediate shipment of the cells to the manufacturing site to achieve a sufficient amount of cells as starting material for the manufacturing of the therapeutic product. To guarantee an efficient dosage of the therapeutic product, many manufacturing protocols of autologous and allogeneic cell therapies require shipment of viable cells at ambient or refrigerated temperatures. Another reason is to avoid freeze protection agents like dimethyl sulfoxide (DMSO), which can impact the patient's tolerability of the finished cell product. There are manufacturing and freezing protocols available without or with a very small amount of these agents, but the disadvantages of freezing can lead to the decision to accept the difficulties of fresh cell logistics.

The main difficulty is the short shelf life of cells outside the tissue culture environment. The shelf life of a cell varies per collection or manufacturing protocol from 12 up to 96 hours. Often, the short shelf life is a consequence of the limited availability of stability data, especially in early stages of development. In later

How to Distribute Cell and Gene Therapies

stages, dosage and formulation can be adapted to achieve longer shelf lives. This extremely short period of time of in hours in comparison to months and years for other medicinal products limits the availability of these therapies for patients based on their location relative to the manufacturing site. At the beginning of clinical development, a limited number of manufacturing sites is established, often just one in one region. But patients should be recruited in many countries to achieve a sufficient and representative patient population, which is the main challenge for logistics.

In this situation, a logistics feasibility study is required as a first step before treatment centers can be selected. Based on the location of the manufacturing site(s) reachable countries and cities are identified within the shelf life of the product. Options for flight connections, charter, onboard courier and road transportation for the last mile are evaluated including the required time for import processes. In the next step, treatment centers with eligible investigators and patients are identified in these reachable regions and qualified. In the third step, detailed plans for pick-up, shipment, import and last mile from the manufacturing site to these sites plus contingency plans are developed.

Very short shelf lives only allow distribution to nearby countries and regions. A few hours more or less can have a significant impact on the feasibility and country selection. For example, a shipment from a US manufacturing site to a European country is only feasible for certain lanes with a tight time plan for pick-up and delivery times, which cannot always be synchronized with the manufacturing time plan. In many situations, it is recommended to start the first clinical studies with a sufficient shelf life to restricted regions, cities and countries close to the manufacturing site. Another possibility is to establish regional manufacturing sites, bedsite manufacturing at the treatment center or to switch to frozen shipments. For autologous therapies, the shortest shelf life of either the donation or the finished product determines the logistics strategy.

The following table shows the feasibility of short shelf-life logistics in relation to the number and location of manufacturing sites.

Shelf Life	One Manufacturing Site (MA[1]) in US	One Manufacturing Site (MA) in EU	Regional Manufacturing Sites (RMA[2])	One Manufacturing Site per Country/State(CMA[3])
12 hours	US[4]/CA[5]/MEX[6] sites in large cities or close to MA	EU[7] sites in large cities or close to MA	Sites close to RMA	All sites per country/state
12 to 24 hours	US/CA/MEX sites	EU sites	Sites in region	All sites per country/state
24 to 48 hours	NA[8] + LATAM[9] large cities	EU + US/CA/EMEA[10] large cities	Sites in region	All sites per country/state
48 to 72 hours	Americas + LATAM + EU large cities	EU + EMEA + NA large cities	Sites in region	All sites per country/state

Shelf Life	One Manufacturing Site (MA[1]) in US	One Manufacturing Site (MA) in EU	Regional Manufacturing Sites (RMA[2])	One Manufacturing Site per Country/ State(CMA[3])
72 to 96 hours	Americas + EU + APAC[11] large cities	EU + EMEA + Americas large cities	Sites in region	All sites per country/state

Notes: [1]MA = Manufacturing Site; [2]RMA = Reginal Manufacturing Site; [3]CMA = Country Manufacturing Site; [4]US = United States; [5]CA = Canada; [6]MEX = Mexico; [7]EU = European Union; [8]NA = North America; [9]LATAM = Latin America; [10]EMEA = Europe, Middle East and Africa; [11]APAC = Asia Pacific

FIGURE 11.1 Logger for tracking of temperature and location of a shipment with a view on the portal (Controlant Saga-P real-time data logger).

For shipments of cells with short shelf lives, an active tracing solution is very helpful and can reduce the risks of product loss. Active tracking is provided by temperature loggers with a GPS, mobile phone or other IoT (Internet of Things) solution sending information about temperature, location and other status information to a central portal where these data are accessible in real-time or near real-time. The logistics provider, shipper and consignee can observe the course of the shipment and react when required. For example, when the shipment was not loaded on the planned plane, contingency lanes can be activated, other flights booked or switched to road transportation.

Figure 11.1 shows an example of a logger and the view during the shipment in the portal.

CRYOGENIC STORAGE AND DISTRIBUTION

Cells and tissues are stable for a long time when frozen in liquid nitrogen or at least below −140°C. This is a new temperature range for pharmaceutical logistics

How to Distribute Cell and Gene Therapies   189

and needs a new infrastructure, including suitable shippers, sensors, loggers and storage equipment.

There are three situations when cryologistics is required

- shipments of donations of cells or tissues from treatment centers to manufacturers
- shipments of cryopreserved medicinal products
- storage of cryopreserved medicinal products

Whereas the first two requirements are applicable for autologous and allogeneic therapies, storage is typically required for allogeneic cell therapies where batches of the therapies are stored at central locations or close to the treatment centers. Sometimes backup samples are stored for autologous therapies or an autologous therapy needs a short-term storage location when arriving in the region of the treatment center before quality release. For this storage time, a Good Manufacturing Practices (GMP)-qualified depot with manufacturing import authorization and an appropriate quarantine storage area is required.

Cryostorage is a new requirement in pharmaceutical logistics where ambient-controlled temperatures of 2°C–8°C, −20°C and −80°C are typical storage temperatures offered by pharmaceutical storage providers. This can lead to a lack of capabilities when new medicinal products are starting clinical trials and later for market entry. In a situation with hundreds of candidates in the pipeline but only a small number of already approved pharmaceutical storage, providers hesitate to invest in cryostorage equipment before a significant number of cryoproducts fill their hubs.

Many types of cryoshippers are available for different purposes, volumes and product types. Treatment centers at hospitals and medical centers have different requirements than the typical users of cryoshippers, for example, cell, tissue and blood banks, laboratories and research centers. The volumes are much smaller, the shipper should be easy and safe to handle, light and special packaging are required for apheresis samples and small dosages of the finished product.

Figure 11.2 shows a dry shipper with liquid nitrogen without contact with the product and an inbuilt tracking device. Depending on the usage packages for infusion, bags are available or racks and boxes for vials. The shipper is developed for the requirements of cell therapies, especially for the first CAR-T therapies where the users have given a lot of input for the specifications of the shipper. In addition, the shipper needs a long time to keep the temperature in order to allow long-distance shipments, delays during import processes and ideally some time left at the treatment center when the patient is not ready for the treatment immediately after arrival. With 10 to 20 days, the DV10 from Biolife fulfills all of these requirements.

But not all treatment centers are used to working with liquid nitrogen, some have restrictions or are not willing to accept a LN_2 shipper. For these purposes, a liquid nitrogen-free shipper was developed, shown in Figure 11.3. Another demand is the possibility for short-term storage of up to a few weeks for treatment

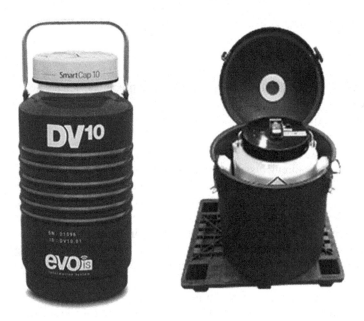
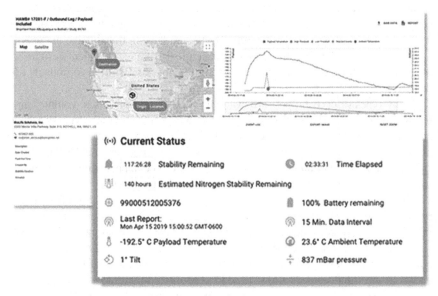

FIGURE 11.2 Two examples of dry shippers for cryogenic shipments with integrated or attached loggers including tracking device for tracking of temperature, location and shipper status with an example of tracking information below.

How to Distribute Cell and Gene Therapies 191

FIGURE 11.3 Liquid nitrogen free shipper for cryogenic shipments powered by electricity.

centers without their own storage capabilities. With the exception of treatment centers at research and stem cell centers, hospitals only in rare cases have cryostorage facilities. More and more cell therapies for wider indications are developed and therefore more hospitals need a solution for intermediate cryostorage. The LN_2 free shipper is powered by electricity using a Stirling engine, which can be used at the hospital for storage by plugging in the Stirling engine module and continuing cooling.

For long-term storage, suitable tanks are available in different sizes and levels of automation. For the protection of the stored products, an automated system allows the removal of vials without opening the tank and protects the contents from critical temperature spikes. An inventory system provides full visibility and tracking of stored products. In Figure 11.4, an automated tank with an inventory system is shown.

FIGURE 11.4 Automated storage tank for cryostorage with inventory management system (BioStore™ III Cryo −190°C "Vending" System)

HOW TO ASSURE CHAIN OF CUSTODY

In autologous therapies, the chain of custody must be provided to assure that the patient is receiving back his own therapy and to provide all data of the whole process, which is part of the product. A system to oversee and orchestrate the process is a regulatory requirement:

> A system that enables the bidirectional tracking of cells/tissues contained in ATMPs from the point of donation, through manufacturing, to the delivery of the finished product to the recipient should be created. Such system, which can be manual or electronic, should be established since the beginning of the manufacture of batches for clinical use.
>
> (EC, 2017)

How to Distribute Cell and Gene Therapies 193

This includes not only the product but also starting materials and all products coming in contact:

> The holder of a marketing authorisation for an advanced therapy medicinal product shall establish and maintain a system ensuring that the individual product and its starting and raw materials, including all substances coming into contact with the cells or tissues it may contain, can be traced through the sourcing, manufacturing, packaging, storage, transport and delivery to the hospital, institution or private practice where the product is used.
>
> (EU, 2007)

Cellular orchestration systems provide the system environment for physicians to enroll patients and to book production slots, for manufacturers to oversee and control their production and for both to plan and organize the administration. These systems are integrated with the logistics provider to book shipments and to get back information on the time and location of pick-up, the shipment method, time, location and proof of delivery and shipment data like the temperature curve.

These systems are continuously improved and adapted to new requirements. Some allogeneic therapies are derived from donors with different profiles of cell markers and other characteristics. Out of these donations, various batches of cell therapies are manufactured, which are stored in pharmaceutical storage centers. Matching modules provided by the system allow the selection of the ideal batch by the physician for each individual patient. By integrating the inventory system of the storage site with the cellular orchestration system, the right dosage of the selected batch can be ordered and taken out of stock and shipped then to the treatment center.

For traceability purposes the usage of unique identifiers is required, ideally barcodes, QR codes or eLabels, which are linked to the electronic cellular orchestration system. The requirement to comply with the labeling requirements of Regulation (EU) No 536/2014 (EC, 2016) is creating practical and technical challenges when small immediate and secondary containers at frozen conditions are used.

Next, the labeling requirements of Regulation (EU) No 536/2014, Annex VI (EC, 2016) are shown:

> A.2.2. Small immediate packaging 5. If the immediate packaging takes the form of blister packs or small units such as ampoules on which the particulars required in section A.1. cannot be displayed, the outer packaging provided shall bear a label with those particulars. The immediate packaging shall contain the following: (a) name of the main contact; (b) route of administration (may be excluded for oral solid dose forms) and, in the case of clinical trials which do not involve the blinding of the label, the name/identifier and strength/potency; (c) batch or code number identifying the contents and packaging operation; (d) a clinical trial reference code allowing identification of the trial, site, investigator and sponsor if not given elsewhere; (e) the subject identification number/treatment number and, where relevant, the visit number; and (f) period of use (expiry date or re-test date as applicable), in month and year format and in a manner that avoids any ambiguity.

For traceability, the unique identifier would be sufficient, but the regulation demands full labeling with all contents in all local languages of the EU member states where the ATIMP should be distributed. All of this text is difficult up to impossible to place and adhere to the often very small vials and tubes.

LOGISTICS OF GENETICALLY MODIFIED CELL THERAPIES AND GENE THERAPIES

For the logistics of gene therapy medicinal products, the temperature and the classification of the therapy as gene therapy should be considered.

Logistics of genetically modified cells follow the same concept as described for time-critical and cryogenic cell therapies. But in case the cells are genetically modified – for example, by a lentivirus or AAV carrying the respective modifying gene – these therapies are classified as gene therapy medicinal products (EC, 2009). For gene therapies, a risk assessment is required to assess the risk for humans, animals and the environment. As cells cannot survive outside tissue culture conditions and have no infectious potential, they are considered without any risk to the environment by the FDA (FDA, n.d. d) and recently by the European Medicine Agency (EMA) (EC, 2021), but residual viruses should be considered, and data should be provided about any remaining virus in the finished product that could provide a risk for humans, animals and the environment.

Based on the risk assessment, most of the therapies consisting of genetically modified cells are not considered dangerous goods by the International Air Transport Association (IATA) classification UN3245.

Interestingly, the Cartagena Protocol on biosafety (BCH, n.d.), which is an international agreement that aims to ensure the safe handling, transport and use of living-modified organisms (LMOs) resulting from modern biotechnology, excludes human therapies:

> Notwithstanding Article 4 and without prejudice to any right of a Party to subject all living modified organisms to risk assessment prior to the making of decisions on import, this protocol shall not apply to the transboundary movement of living modified organisms which are pharmaceuticals for humans that are addressed by other relevant international agreements or organisations.
>
> (COBD, n.d.)

According to IATA's Dangerous Goods Regulations, the protocol of Cartagena is the foundation of the UN3245 classification of genetically modified organisms and excludes authorized GMOs, which is the case for commercial and investigational cell and gene therapies: "GMMOs or GMOs are not subject to these Regulations when authorized for use by the appropriate national authorities of the States of origin, transit and destination." (IATA, 2022) As a consequence, genetically modified cells and also genetically modified viruses for human therapy are not genetically modified organisms requiring dangerous goods classification during shipment and storage within the meaning of the act.

Despite this, many genetically modified cells and viruses for human therapy are shipped as dangerous goods UN3245.

The risk of genetically modified AAV is also considered negligible by the EMA (EC, 2022). The same logic gene therapies containing genetically modified AAVs as active ingredients could be shipped without any dangerous goods classification.

Another challenge for the distribution of gene therapies is the temperature condition. Although most virus-based therapies are shipped in dry ice and stored at −80°C, some therapies are developed with other frozen temperature ranges above −80°C. Temperature ranges between −20°C and −80°C, for example −50°C or −65°C, are sometimes requested. This is a critical requirement and limits logistics feasibility because these temperatures are not standard and need qualification of equipment like shippers and freezers. This is possible but needs time for qualification and validation. As a consequence, the development can be delayed and costs are increasing. The reason for these nonstandard temperature conditions is mainly increased shelf life at this temperature and available stability data for the respective temperature at the time of the clinical study start. It should be critically analyzed if the prolonged shelf life is justifying the difficulties, time delay and costs to establish the supply chain at this temperature.

OUTLOOK

Cell and gene therapies have different requirements for distribution and storage than pharmaceuticals based on chemical entities and biopharmaceuticals, mainly cryogenic conditions, chain of custody and time criticality. With a large number of successful therapies on the market and in development, another logistics infrastructure and regulatory understanding must be developed. Correct assessment of risks, reflection of the new product characteristics in the regulations and exceptions for inappropriate provisions are required to remove hurdles for logistics, which are based on regulations not adapted to the demand of these new therapeutic products. Similar to the market entry of biopharmaceuticals with the need for 2°C–8°C cryostorage and cryoshipping, infrastructure will be available in the future with new technologies and equipment fulfilling the demand of manufacturers, physicians and patients. Shipments of time-critical therapies across continents due to a lack of local manufacturing and temperature conditions outside the standard ranges will be very likely only temporary solutions before a new normal is established that provides patients around the globe access to cell and gene therapies.

REFERENCES

European Medicines Agency (EMA). (2022). Guidelines relevant for advanced therapy medicinal products | European Medicines Agency (europa.eu) https://www.ema.europa.eu/en/human-regulatory/research-development/advanced-therapies/guidelines-relevant-advanced-therapy-medicinal-products?msclkid=e8f4f05bd10311eca59b608e74898d03

European Commision (EC). (2017). Good manufacturing practice guidelines for advanced therapy medicinal products (2017) https://ec.europa.eu/health/system/files/2017-11/2017_11_22_guidelines_gmp_for_atmps_0.pdf

European Commision (EC). (2019). Good clinical practice of advanced therapy investigational medicinal products ATIMP's https://ec.europa.eu/health/system/files/2019-10/atmp_guidelines_en_0.pdf

EMA. (n.d.) Good manufacturing practice of advanced therapy investigational medicinal products (n.d.) https://www.ema.europa.eu/en/documents/other/european-commission-dg-health-food-safety-european-medicines-agency-action-plan-advanced-therapy_en-0.pdf

FDA. (n.d. a) https://www.fda.gov/vaccines-blood-biologics/biologics-guidances/cellular-gene-therapy-guidances

FDA. (n.d. b) Chemistry, Manufacturing, and Control (CMC) Information for human gene therapy Investigational New Drug Applications (INDs) https://www.fda.gov/media/113760/download

FDA. (n.d. c) https://www.fda.gov/vaccines-blood-biologics/cellular-gene-therapy-products?msclkid=5da91d67d10b11ecae6ba052aca17c63

European Commision (EC). (2017). Paragraph 6.35 Guidelines on good manufacturing practice specific to advanced therapy medicinal products https://ec.europa.eu/health/system/files/2017-11/2017_11_22_guidelines_gmp_for_atmps_0.pdf

European Union (EU). (2007). Article 15 of REGULATION (EC) No 1394/2007 https://eur-lex.europa.eu/legal-content/EN/TXT/PDF/?uri=CELEX:32007R1394&from=EN

European Commision (EC). (2016). Regulation (EU) No 536/2014 (2016) (https://ec.europa.eu/health/system/files/2016-11/reg_2014_536_en_0.pdf

European Commision (EC). (2009). Commission Directive 2009/120/EC of 14 September 2009 amending Directive 2001/83/EC of the European Parliament and of the Council on the Community code relating to medicinal products for human use as regards advanced therapy medicinal productsText with EEA relevance (europa.eu) https://eur-lex.europa.eu/legal-content/EN/TXT/PDF/?uri=CELEX:32009L0120

FDA. (n.d. d) Determining the need for and content of environmental assessments for gene therapies, vectored vaccines, and related recombinant viral or microbial products (n.d.) https://www.fda.gov/regulatory-information/search-fda-guidance-documents/determining-need-and-content-environmental-assessments-gene-therapies-vectored-vaccines-and-related

European Commision (EC). (2021). Good Practice on the assessment of GMO-related aspects in the context of clinical trials with human cells genetically modified (2021) https://ec.europa.eu/health/system/files/2021-11/gmcells_gp_en_0.pdf

Biosafety Celaring-House (BCH). (n.d.) (https://bch.cbd.int/protocol/

Convention on Biological Diversity. (n.d.) Article 5 of the Cartagena Protocol on biosafety (n.d.) https://bch.cbd.int/protocol/text/

International Air Trasport Association (IATA) (2022). 3.9.2.5.3 IATA's Dangerous Goods Regulations. IATA – Dangerous Goods Regulations (DGR).

European Commision (EC). (2022). Good Practice on the assessment of Genetically Modified Organism (GMO) related aspects in the context of clinical trials with Adeno-associated viral (AAV) clinical vectors https://ec.europa.eu/health/system/files/2022-01/aavs_gp_en.pdf

12 Regulatory Compliance and Approval

Siegfried Schmitt
Parexel, Braintree, UK

CONTENTS

General Considerations 197
Importance of cGMP in Product Manufacture 197
Structuring and Streamlining Your Regulatory Submission 202
Summary and Outlook 206
References 206

Many cell and gene therapy (CGT) products offer novel ways of treating diseases, often offering cures rather than mere treatment of symptoms and often for rare diseases. For that reason, regulators provide a number of different expedited review and submission pathways. The implications of shorter time lines for establishing the Chemistry, Manufacturing and Control (CMC) section are discussed.

GENERAL CONSIDERATIONS

The term "cell and gene therapy products" is used throughout this chapter to collectively refer to cell therapies, gene-modified cell therapies, gene therapies, and tissue-engineered products, also known as regenerative medicines or, in the European Union (EU), advanced therapy medicinal products (ATMPs).

For detailed regulations overviews and discussions of the regulatory pathways, please refer to the other chapters in this book. This chapter covers general considerations for regulatory compliance and regulatory submissions for cell and gene products.

IMPORTANCE OF cGMP IN PRODUCT MANUFACTURE

Regulatory agencies everywhere require the manufacture of medicines to follow the principles of Good Manufacturing Practice (GMP). To be more precise, regulators expect industry to abide by the laws, adhere to regulatory guidance and apply state of the art science. This is referred to as current good manufacturing practices (cGMP).

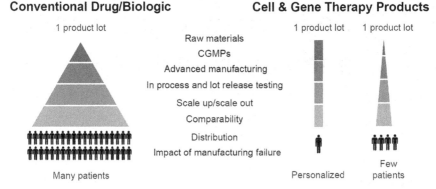

FIGURE 12.1 CGT Product Manufacturing – A New Paradigm (Steven S. Oh, 2019).

GMP as a concept has been in place for half a century, but, of course, it was developed first for chemical products and then for biotechnological products. CGT products are different to either of these, as is visualised in Figure 12.1.

CGT products are products of very high complexity, making their description very challenging. Uniformity and quality of the product must therefore be assured through process controls and process knowledge. All of this necessitated adaptation (amendments or revisions) to the existing GMP regulations. While much of the reworking is still ongoing, basic GMP principles always apply.

Along the product life cycle from development to commercialisation to removal from the market, companies gain increasing knowledge about product and process. The GMP regulations reflect this, which is referred to as phase-appropriate cGMPs. There are certain prerequisites that must be complied with.

These include a Pharmaceutical Quality System (PQS) according to the International Conference On Harmonisation Of Technical Requirements For Registration Of Pharmaceuticals For Human Use (International Council for Harmonization (ICH) Q10) (ICH). It is irrelevant how big or small a company is, even a one-person company (yes, they do exist) has to implement such a system. As a substantial number of companies active in the field of cell and gene therapies come from academia as start-up companies, it is important to mention this here. Depending on the activities, e.g., performing clinical trials or manufacturing, the system has to cover all activities relevant to the product and therewith patient safety. In its simplest form, such a PQS will consist of a dozen or so Standard Operating Procedures (SOPs) or thousands of procedures at the other end of the scale.

Having a PQS is not enough; companies also need to have the resources, the organisation and the staffing levels that can support such a PQS. If, for example, a small company wishes to run clinical trials in several countries, they are likely to rely on third parties to manufacture the cell or gene therapy product, or to run the clinical trials or to interact with regulatory agencies, to name but some of the most obvious areas of outsourcing. While the company can outsource as much as they like, the primary regulatory responsibility remains with them. This means they must have staff who can review documentation and approve decisions by third parties.

The development, testing and manufacture of CGT products requires facilities, which depending on the product may have to be as close to the location of the patient as possible, sometimes within a hospital (point-of-care manufacture). These facilities may not normally be operated by staff familiar with GMP or in a manner conducive to GMP. Thus, location selection, also bearing in mind that some regulators or governments wish to have domestic manufacture, can be crucial to success. Commercial considerations typically call for multipurpose, rather than single-product, facilities, which are perfectly manageable but require designs that prevent cross-contamination and allow for easy cleaning and disinfection. Examples of multi-product facilities are those with different starting materials (e.g., stem cells and T cells), autologous treatments and those with similar starting materials, but different processes or different final products (see e.g., A. Rietveld, 2019). Because of the concern of cross-contamination, concurrent production of two different ATMPs or batches in the same area is not acceptable.

For CGT facilities, all regular GMP requirements, such as qualification, maintenance, calibration, apply (for a concise overview of GMP requirements for cell and gene therapies, see, e.g., Kasia Averall, 2019).

The US Food and Drug Administration (US FDA) has also provided information regarding product, process and facility design in conference presentations (Y. Koo Lily, 2019):

- Facility design and layout should be appropriate for the intended operations (closed versus open operations, aseptic processing requirements, multi-product and manufacturing capacity considerations)
- There is no one facility design that is best – the key is process-appropriate control and containment

Manufacture and distribution rely on third-party organisations, such as raw material suppliers, contract development and manufacturing organisations (CDMOs), logistics partners for supply chain and distribution and service companies (e.g., pest control, equipment maintenance, metrology). Finding the right contract organisations, e.g., warehouses, freezing centres and customs clearance sites that are capable of handling and are familiar with CGT products, can be challenging. Volumes are typically a lot smaller than for typical pharmaceutical products, and many of the CGT products require very delicate handling. There might be the temptation to use whatever supplier is available, but if this results in violations of GMP or Good Distribution Practice (GDP) regulations, the outcome may be a refusal of the marketing authorisation.

Talking about drug applications/regulatory submissions, these need to showcase a company's process and product understanding (see, e.g., ARM and NIIMBL, 2021). Collating, analysing and interpreting the data from in-house and external sources is crucial to being able to put together a convincing dossier for the regulatory authorities.

TABLE 12.1
Unique Challenges in the Manufacture of CGT Products
(S. Oh Steven, 2019)

Prior knowledge	Companies have limited product manufacturing experience prior to licensure (incomplete knowledge of Critical Process Parameters (CPP), limited number of batches or lots made).
Product quality	Critical Quality Attributes (CQAs) are not entirely understood due to limited characterisation of drug product, drug substance and in-process materials (e.g. cells).
Repeat product quality	Product variability arising from source materials (often as individual as a person).
Supply chain	Increased demand for qualified reagents and materials. Many materials are from specialist sources, limiting supplies.
Analytical methodology	Assays not fully developed and qualified. Many are highly complex assays.
Analytical testing	Limited time for testing due to limited material or short shelf life. Instead of weeks and months for testing, often only days.
Product stability	Limited product stability data (Note: CGT products are out of scope for ICH Q5C (Stability Testing of Biotechnological/Biological Products)). Shelf life of sometimes only days.
Repeat product quality	Reproducibility of replacement cell banks.
Process reproducibility	Complicated planning for advanced manufacturing, process automation, scale-up/scale-out. For example, each individual lot may have to be tested instead of a batch containing many lots.
Repeat product quality	Comparability studies in the absence of reliable reference standards and validated assays.
Patient impact	Definite direct impact of manufacturing failure on patient.

Key to linking clinical studies with the CMC section of the dossier are comparability studies; they demonstrate consistent manufacture of the product (Tables 12.1 and 12.2).

Changes to materials (e.g., source) and process (e.g., equipment) during the development up to commercialisation are almost inevitable. Companies need to have a robust change management procedure, coupled with quality risk management, in order to be able to assess changes for their impact on the product and therewith on the patient. The more representative CQAs are of clinical safety and efficacy, the easier it is to evaluate the consequences of a manufacturing change.

Full understanding of CQAs and CPPs facilitates post-approval change management. The interlinkage is shown in Figure 12.2.

In the CGT space, the FDA and the European Medicine Agency (EMA) have published several documents that offer guidance about early stage filings; however, these have provided limited reference to process validation and product commercialisation. The expectation – just as it is with biologics – is that the process is the product. However, CGT products have unique features that must be considered during process validation.

Regulatory Compliance and Approval

TABLE 12.2
Describing CGT Product and Process Understanding

Product and Process Understanding Attributes	Description
The mechanism of action	This describes how the product works in the patient.
The Quality Target Product Profile (QTPP)	This is a scientific description of the product.
The CQA	For CGT products, these are identity, purity and potency. CQAs are linked to product efficacy and safety.
The CPP	These manufacturing parameters (e.g., temperature, time, concentration) are critical to product quality. CPPs have to have defined operation values or ranges and must be controlled and monitored.
The material attributes	Raw materials and ingredients are critical to product quality and process performance. Materials have to have defined specifications.

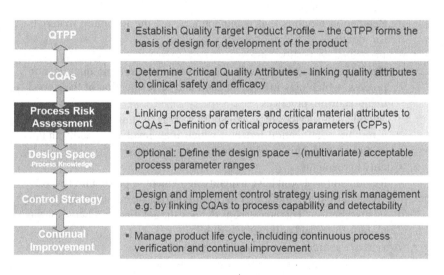

FIGURE 12.2 Quality and Process Understanding (Kimberley Buytaert-Hoefen, 2020).

The complexity and the novelty of CGT products are significant. CGT products can be composed of proteins, nucleic acids and membranes. Each of these is a highly complex, biologically active component, and they are each impactful to the safety and efficacy of those products (Kyle Zingaro, 2020).

For a discussion on CMC challenges in the manufacture and regulation of cell, gene, and tissue therapies, see, e.g., Anjali Apte (2020).

For a discussion on manufacturing controls in the manufacture and regulation of cell, gene, and tissue therapies, see, e.g., Heidaran M (2019a).

For a discussion on the difficulties of manufacturing cell and gene therapies at scale, see, e.g., Allan Bream and Brita Salzmann (2021).

Because of the complexity and often uniqueness of the supply chain for CGT products, regulatory agencies regularly request information on (Kimberley Buytaert-Hoefen, 2020):

- Raw materials, ancillary materials, equipment, containers
  - Source material collection equipment
  - Media and growth factors, others
  - Manufacturing tools (flask, bags, etc.)
  - Container closures (bags or vials)
  - Quality Control (QC) test platforms
  - Grade of materials (cGMP) alone is not sufficient to ensure quality
  - Shipping, and storage conditions
- Material qualification
  - Verification of the safety, identity, purity and potency
  - Only having a certificate of analysis is not necessarily sufficient
  - Quality (fit for purpose) and reliability
  - Regulation requires that manufacturers test the incoming materials for identity
- Vendor/supplier qualification
  - Vendors of critical materials should undergo a routine qualification process, which may involve an audit and/or verification of their good manufacturing practices
  - Suppliers of services (e.g., transport, storage, testing) should undergo a routine qualification process, which may involve an audit and/or verification of their good manufacturing practices
- Quality agreements
  - Manufacturer should have a quality agreement in place with key vendors, particularly those that perform contract manufacturing, testing or distribution. This agreement defines the relationship between the manufacturer and contract acceptors
- Alternative sources (supply chain uncertainty)
  - Determine long-term sustainability of the supply
  - Determine potential alternative sources

Don't forget that full compliance with cGMPs is a requirement at the time of licensure application, which may be verified at the time of the pre-license inspection.

STRUCTURING AND STREAMLINING YOUR REGULATORY SUBMISSION

The regulatory submission process is interlinked with the CGT product life cycle as visualised in Figure 12.3.

Regulatory Compliance and Approval

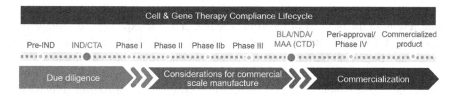

FIGURE 12.3 CGT Compliance Life Cycle.

As CGT products have often new and unrivalled patient benefits, regulatory agencies are very supportive of expedited submission and approval time lines compared with traditional biotechnology products. You can read more about the actual cycle times (for standard process, rolling review process, etc.) in the relevant chapters of this book. Please remember though that aligning CMC product and process knowledge with clinical development is crucial for approval success.

Accelerated approval pathways exist in most countries, and as a representative example, here are some accelerated approval pathways in emerging markets (Cristina Mota, 2018):

- Expedited review:
  - Regulatory authorities speed up the review of certain products to enable faster approval: Brazil, China, Egypt, Saudi Arabia, Singapore, Indonesia, South Korea and Israel.
- Expedited submission (rolling submissions):
  - Information and data packages can be submitted and reviewed as they become available: South Korea.
- Accelerated development:
  - Earlier submission and approval with a data set, which may be less complete than from a standard development programme (e.g., surrogate endpoints, clinical phase II data only): Brazil, South Korea and Taiwan.

For more detailed information on expedited reviews and submissions, please refer to the relevant chapters in this book. Where companies have the most leverage is over accelerated development, which is keenly supported by management, as it will reduce cost. Yet, this comes with its own challenges, not least with regard to CMC (Alexandra Beumer Sassi, 2019).

As mentioned earlier, companies need to develop sound processes and product understanding, which they gain during the development phase. Overall, accelerated pathways condense CMC development, occurring usually during clinical phase III. This results not only in limited manufacturing experience at the time of submission but also often means that the experience is gained in facilities designed for clinical trial products, rather than commercial operations. The consequences of such a set-up must not be underestimated, as there will be limited room for optimisations of product and process over development. A quite simple example may be buffer production. At scale-up, this can easily interfere with the product production schedule and, consequently, the company may buy in buffer, rather than manufacture it in-house. Even though the in-house and bought-in buffers meet the same specifications, they can interact differently with the product or raw materials. Such seemingly small differences can have major consequences later. Furthermore, commercial operations typically have a higher grade of automation, allowing for increased data collation, which can provide much greater insight into process behaviour. Without such information, the process may well be a suboptimal process at the time of dossier submission.

It is not only the manufacturing process where accelerated development will lead to fewer data and less knowledge, but it is also affecting the development and testing of analytical methods, leaving less time for characterisation, and this can lead to difficulties in establishing meaningful specifications. And lastly, long-term stability data may not be as complete as under normal circumstances.

Potency testing is optional for clinical phase I but is required for clinical phase III and is a product-specific attribute. Even though a potency assay is not a requirement for clinical phase I, it is highly recommended to develop multiple potential assays early on. Typically, biological assays must be developed for each product that can demonstrate its bioactivity, including cell-based assays to measure infectivity and product-specific potency (C. Challener, 2019).

It is challenging to establish in vitro assays that serve as reliable indicators of product activity/potency in vivo. Ultimately, assays must be validated. The question is at what point is it sufficient to qualify an assay, and when does it need to be validated? Assay qualification demonstrates that an assay is suitable for measuring the analyte, and assay validation demonstrates that an assay is suitable for the intended use under the worst conditions of use. Table 12.3 compares assay qualification and assay validation.

TABLE 12.3
Assay Qualification and Assay Validation Requirements

Assay Qualification	Assay Validation
Sensitive	Sensitive
Accurate	Accurate
Reproducible (identical samples measured at two different times in the same lab give similar results)	Precision
Requirement for early phase for all critical analytical methods	Robustness
If compendial methods are used, you must follow the sampling requirements	Ruggedness
	Assay gives similar results when identical samples are tested in different laboratories by different operators

Regulatory Compliance and Approval

Manufacturers who choose to use a qualified, not a validated, assay to collect critical information during clinical phase III or during pivotal studies potentially collect data sets, which are not fully representative of their product quality. This could potentially impact the usefulness of assay results that are relied upon to define meaningful specification or acceptance criteria to assess and verify product quality (Mo Heidaran, 2019b).

Yet, moving faster does not mean that regulatory agencies will accept less in terms of CMC (Alexandra Beumer Sassi, 2019):

- CMC cannot compromise product supply and patient safety
- Anticipate characterisation effort earlier in development
- Focus on core process and product data needed to ensure reproducible and high-quality product manufacturing, even if formal validation is postponed
- Plan for completion of the CMC exercise over post-approval activities and a commitment to meet conditional approval requirements as detailed, e.g., in ICH Q12 Lifecycle Management (ICH)

When planning for submission, consult the relevant regulatory agency websites to find the list of CMC information required. As an example, this is the list by US FDA for an Investigational New Drug Application (IND):

- Description of product
- Manufacturing process
- Mechanism of action
- Analytical methods qualification
- Ancillary materials (human versus animal derived)
- Master cell bank (MCB) and master virus seed (MVS) qualification
- Donor screening and testing
- In-process release specifications
- Drug substance release specifications
- Drug product release specification
  - Sterility, identity, purity (endotoxin level), viability, mycoplasma if cultured and potency
  - Other list of release tests (product specific)
- Action plan for sterility failure when results are not available prior to administration
- Container label
- Stability studies
- Shipping qualifications
- Categorical exclusion (environmental)

It is not at all trivial to pull all this information and data together. Remember though that CMC readiness alone may not be sufficient for receiving expedited programme designation from the respective agencies.

The question is what to do when not all CMC information is available. As defined in ICH Q9 Quality Risk Management (ICH), apply a risk-based approach. Please note, ICH Q9 is currently undergoing a revision. Such an approach needs to marry business and compliance needs; thus it is important to have defined priorities. In the formal and documented risk assessment process, you can include applicable prior knowledge, which can help when classifying the potential impact of the risk and defining appropriate mitigation measures.

We can and we should always learn from mistakes others have made. Therefore, hereafter a list of common license application review issues (Kimberley Buytaert-Hoefen, 2020) in no particular order:

- Lack of adequate process validation/plan
- Lack of sufficient stability
- Lack of proper validation for analytical procedures
- Insufficient information extractables and leachables
- Inadequate SOPs
- Lack of retention samples
- Lack of shipping validation
- Inadequate donor eligibility determination
- Insufficient information about CQA/CPP
- Lack of container closure integrity test
- Lack of validation of starting biological material collection
- Lack of detailed information about formulation
- Lack of validation of dilution or thaw and wash step at clinical site
- Inadequate instructions for use
- Deficient content of label

SUMMARY AND OUTLOOK

CGT products are a fascinating field for regulatory affairs professionals to work in, from a regulatory process and a product perspective. Being involved in bringing such valuable new products to market and helping patients get access to such modern treatments is something to be truly proud of. It poses, of course, also many challenges as the harmonisation of regulations (but not necessarily the associated terminology) is still ongoing. And as much as regulators are catching up on scientific advances, so too is the regulatory professional.

REFERENCES

Rietveld. (2019, June). Design considerations for ATMP premises from the EU point of view, 2nd Cell & Gene Therapy Products Symposium, Bethesda, MD. https://www.casss.org/home/papers-and-presentations

Alexandra Beumer Sassi. (2019, June). Breaking the traditional cmc development pathway, 2nd Cell & Gene Therapy Products Symposium Bethesda. https://www.casss.org/home/papers-and-presentations

Allan Bream and Brita Salzmann. (2021, April). The difficulties of manufacturing cell and gene therapies at scale. *BioProcess International* 19(4). https://bioprocessintl.com/manufacturing/cell-therapies/the-difficulties-of-manufacturing-cell-and-gene-therapies-at-scale-moving-to-flexible-facilities-and-closed-cell-processing-systems/?hsCtaTracking=47d4e21b-83cd-4e58-a745-41fc80ded89e%7C8a39e88a-f180-4c8a-9691-0c19

Anjali Apte, Adeyemi Afuwape, Zaklina Buljovcic, and Zeb Younes. (2020, November–December). Manufacture and regulation of cell, gene, and tissue therapies part 1: Chemistry, manufacturing, and control challenges, *BioProcess International* 18, 11–12. https://bioprocessintl.com/manufacturing/cell-therapies/manufacture-and-regulation-of-cell-gene-and-tissue-therapies-part-1-chemistry-manufacturing-and-control-challenges-for-atmps/

ARM and NIIMBL. (2021, June). Release project a-gene to bring quality by design principles to gene therapy manufacturing. https://alliancerm.org/manufacturing/a-gene-2021

Challener. (2019). Cell and gene therapies necessitate new lot release Test methods. *BioPharm International* 32(11). https://www.biopharminternational.com/cell-and-gene-therapies-necessitate-new-lot-release-test-methods

Cristina Mota, Jarbas Barbosa, and Lawrence Liberti. (2018). Denise Bonamici and Alberto Grignolo, Expedited Regulatory Pathways in Established and Emerging Markets, Global Forum. https://globalforum.diaglobal.org/issue/september-2018/expedited-regulatory-pathways-in-established-and-emerging-markets/

Heidaran, M. (2019a, April). Manufacturing controls at commercial scale: a major hurdle for the cell and gene therapy industry, regulatory focus. regulatory affairs professionals society. https://www.raps.org/news-and-articles/news-articles/2019/4/establishing-manufacturing-controls-a-hurdle-for

ICH, https://www.ich.org/page/quality-guidelines

Kasia Averall. (2019, November–December). Cell and gene therapies and their GMP requirements. *Pharmaceutical Engineering*. https://ispe.org/pharmaceutical-engineering/cell-gene-therapies-their-gmp-requirements

Kimberley Buytaert-Hoefen. (2020, July). GMPs for Gene and Cellular Therapies, PDA Mountain States Chapter Webinar: cGMPs for Gene and Cellular Therapies. https://www.pda.org/docs/default-source/website-document-library/chapters/presentations/mountain-states/gmps-for-gene-and-cellular-therapy-7-16-2020.pdf?sfvrsn=a2b06281_5

Kyle Zingaro. (2020, September). Addressing the complexity of process validation for cell and gene therapy products. *Pharmaceutical* Online. https://www.pharmaceuticalonline.com/doc/addressing-the-complexity-of-process-validation-for-cell-and-gene-therapy-products-0001?vm_tId=2260638&vm_nId=62871&user=c8f4559b-ef6d-4ec9-adc9-d7d0d2dfadfb%E2%80%A6

Y. Koo Lily. (2019, June). FDA perspective on commercial facility design for cell and gene therapy products, 2nd Cell & Gene Therapy Products Symposium Bethesda, MD. https://ipq.org/wp-content/uploads/2019/10/Koo_CASSS.June2019.CGT_.pdf

Mo Heidaran. (2019b, March). Qualification as a step toward assay validation for cber regulated cell and gene therapy products. Parexel internal communications

S. Oh Steven. (2019, June). Facilitating expedited development of advanced therapy products, 2nd Cell & Gene Therapy Products Symposium Bethesda, MD. https://www.casss.org/meetings-and-events/multi-day-symposiums/cell-gene-therapy-products

13a Regulatory Landscape in US, EU and Canada

Kirsten Messmer

Agency IQ, Garner, USA

CONTENTS

Introductory Note 209
    Regulation in the EU 209
        Definitions 209
        EU Regulatory Framework 211
        Regulatory Oversight 212
        Regulatory Tools 213
        Approval Pathways 215
    The Regulatory Framework in the US 217
        Definitions 217
        Regulatory Framework 217
        Regulatory Oversight 218
        Regulatory Tools 219
        Approval Pathways 222
References 224

INTRODUCTORY NOTE

Advanced therapies (see the following section for statutory definitions) are regulated as biological medicines in both the European Union (EU) and the United States (US). Therefore, all biologics laws, regulations and guidelines also apply to advanced therapies. This chapter will note some of the basic applicable regulations for biologics but will not provide an in-depth discussion. It will focus on regulations, designations and guidelines that are particularly applicable or even specific to advanced therapies. This chapter will refer to advanced therapies or cell and gene therapies even though terminology may vary.

REGULATION IN THE EU

Definitions

Medicines, both chemical and biological, are called "medicinal products." In the EU, cell and gene therapies fall under Advanced Therapy Medicinal Products (ATMPs). The product types included in this medicinal product category are defined in legal texts such as directives and regulations.

The regulatory framework consists of three types of documents:

"Regulations" – binding legislative acts applied across the entire EU.

"Directives" – legislative acts providing goals for the EU Member States (i.e., countries that are members of the European Union) to achieve. How these goals are achieved is determined by laws that the individual EU countries develop and implement.

"Scientific guidelines" – provide recommendations and aim to harmonize the approach to apply the legislative requirements.

A dedicated regulation, the Advanced Therapy Medicinal Product Regulation (ATMP Regulation, Regulation (EC) N 1394/2007; EC, 2007), addresses specific regulatory requirements for advanced therapies. Article 2 of this regulation defines advanced therapies as any of the following products:

- a gene therapy medicinal product as defined in Part IV of Annex I to Directive 2001/83/EC (EC, 2001),
- a somatic cell therapy medicinal product as defined in Part IV of Annex I to Directive 2001/83/EC (EC, 2001),
- a tissue engineered product as defined in point (b).

According to Directive 2001/83/CE (EC, 2001), gene therapy medicinal products are defined as *biological medicinal product which has the following characteristics*:

a) *it contains an active substance which contains or consists of a recombinant nucleic acid used in or administered to human beings with a view to regulating, repairing, replacing, adding or deleting a genetic sequence*;
b) *its therapeutic, prophylactic or diagnostic effect relates directly to the recombinant nucleic acid sequence it contains, or to the product of genetic expression of this sequence. Gene therapy medicinal products shall not include vaccines against infectious diseases.*

The same directive defines a somatic cell therapy medicinal product as *biological medicinal product which has the following characteristics*:

a) *contains or consists of cells or tissues that have been subject to substantial manipulation so that biological characteristics, physiological functions or structural properties relevant for the intended clinical use have been altered, or of cells or tissues that are not intended to be used for the same essential function(s) in the recipient and the donor*;
b) *is presented as having properties for, or is used in or administered to human beings with a view to treating, preventing or diagnosing a disease through the pharmacological, immunological or metabolic action of its cells or tissues.*

Article 2 of Regulation (EC) N 1394/2007 (EC, 2007) Point 2 specifies that the pharmacological, immunological or metabolic action of any viable cells or tissues in an advanced therapy product will be regarded as the primary mode of action.

Cell therapy products that are only minimally manipulated (i.e., have only undergone treatments listed in Annex I of the ATMP Regulation) are not advanced therapies (and outside the scope of this book). They fall under different regulatory requirements. Minimal manipulation types include cutting, grinding, shaping, filtering and cell separation.

According to Article 2 of Regulation (EC) N 1394/2007 (EC, 2007) Point 5, any product that falls within both the definition of "somatic cell therapy medicinal product" (or a tissue-engineered product) and the definition of "gene therapy medicinal product" will be regulated as a gene therapy.

EU Regulatory Framework

Although advanced therapies are regarded as a specific group of medicine within human biological medicines, the standard human medicines directives and regulations apply. The ATMP Regulation provides specific requirements for advanced therapies. The applicable human medicines regulatory framework comprises the following:

Directive 2001/83/EC (EC, 2001) covers medicinal products for human use as the overarching directive applicable to all *"medicinal products for human use intended to be placed on the market in a Member State and either prepared industrially or manufactured by a method involving an industrial process."* This directive governs the "centralized authorization," manufacture and distribution of all medicinal products within EU Member States. It is important to note that advanced therapies must always be approved through a centralized procedure. This procedure involves a scientific review by the European Medicines Agency (EMA) followed by the evaluation and decision on authorization by the European Commission (EC). This approval is valid across all EU Member States. While it is possible to receive approval for medicines by other processes for only some specific countries (i.e., decentralized or mutual-recognition procedures), these approval pathways do not apply to advanced therapies.

The directive applies to products manufactured using an industrial process and does not apply to certain products such as compounded drugs.

Directive 2009/120/EC replaces Part IV of Annex I in the then-valid Directive 2001/83/EC. The 2009 directive accounts for scientific progress and includes specific requirements for advanced therapies, including gene and somatic cell therapies as well as tissue-engineered products.

Regulation (EC) No 726/2004 (EC, 2004) provides procedures for marketing authorization and oversight of medicines in the EU through the centralized procedure, which is applicable to advanced therapies. Individual Member States retain sovereignty in setting prices for medicines. Additionally, the decision to include various medicines in the national health-care or social security systems rests with the EU Member States. This is particularly crucial to consider for advanced therapies since their cost per patient will likely be much higher than more conventional medicines. Sponsors should expect extensive negotiation with each Member State's pricing and reimbursement bodies to bring the cost down. For some countries, the cost of treatment per patient may be prohibitive.

Regulation EC (No) 1394/2007 (EC, 2007) discusses specific requirements for advanced therapies, including labeling and post-authorization activities.

Other EU directives and regulations apply as applicable and some may apply depending on product type. For example, if the cell or gene therapy product is combined with a device or it is used in conjunction with a companion diagnostic, the medical device and *in vitro* diagnostic regulations would also apply. Regulatory requirements for these types of combination products are beyond the scope of this chapter.

Regulatory Oversight

At the EU level, the EMA facilitates and oversees the development of new medicines including drugs and biologics. Each EU Member State has its own regulatory authority and ethics committees to oversee clinical trials and authorize medicines that seek approval only in one or a few EU countries through decentralized or mutual-recognition procedures. However, as mentioned earlier, advanced therapies can only be approved through a centralized procedure. Therefore, the EMA will be the responsible agency for all oversight while EU Member States oversee clinical trials in their country. National requirements for clinical trials are outside the scope of this chapter.

Two of EMA's scientific committees are mainly responsible for overseeing advanced therapy development and evaluation: the Committee for Advanced Therapies (CAT) and the Committee for Medicinal Products for Human Use (CHMP). Other committees may be involved as necessary throughout the development program. For example, the Committee for Orphan Medicinal Products will be involved in the assessment and development of orphan drugs, and the Pediatric Committee will support pediatric drug development. For the purposes of this chapter, we will focus on the two main committees.

The CAT follows scientific developments for advanced therapies closely, is involved in providing advice to sponsors during development and evaluates marketing authorization applications and post-authorization procedures.

Some other responsibilities include:

- Supporting scientific recommendations on the classification of advanced therapies. The classification procedure is a consultation with the EMA to confirm whether a medicinal product based on a cell or gene therapy is indeed an advanced therapy.
- Providing scientific advice together with the Scientific Advice Working Party to address development questions from sponsors. Scientific Advice or Protocol Assistance for orphan drugs aims to support the optimization of the development program to ensure evidence supporting a marketing authorization can be achieved.
- CAT also contributes to the post-authorization oversight through advice on efficacy follow-up, pharmacovigilance or risk-management systems. Due to the rapid development in the field, the long-term effects are less known for advanced therapies, and post-authorization safety and efficacy

monitoring needs to be more stringent. For example, the currently valid guideline mentions "years" will be needed for efficacy follow-up but is not specific. The draft revision notes that for gene therapies, commonly 15-year follow-ups are expected.
- The committee also supports the CHMP with advanced therapy-related expertise for any product and/or procedure that may require specialist input.
- The CAT also provides its scientific expertise for development of specific regulations, the CHMP work program, any EU initiatives that aim to foster innovation in medicines and any other actions that require specialist input on the topic of advanced therapy development and regulation.

Regulatory Tools

The EMA has developed a series of scientific guidelines that can serve as a starting point for determining the regulatory requirement for advanced therapies. Table 13a.1 provides an overview of the guidelines available at the time of writing of this chapter that are specific and suggested reading for advanced therapies. Guidelines for biologics and general medicinal product development from the EMA and International Council for Harmonization (ICH) will also apply. For a more in-depth listing of all applicable guidelines, readers should visit the EMA website provided in the table footer.

The first three guidelines are the overarching guidelines that provide a broad overview of the regulatory framework and the development requirements. They address key considerations at each of the development stages through to the marketing authorization application. The overarching guidelines are supported by additional guidelines that focus on specific aspects of product development and/or treatment applications.

Advanced therapy products, similar to other products, must conform with Good Manufacturing Practice (GMPs) requirements. However, some bridging of requirements early in development has been made possible since many cell and gene therapy (CGT) products are initially developed by small biotechnology companies or academic institutions that may not have access to full GMP facilities.

The EC has developed a GMP guideline specifically addressing early stage development of cell and gene therapies to allow for environments that may not be fully GMP certified to address the gaps in capabilities. The guideline addresses specific aspects of these therapies, including novel and complex manufacturing scenarios. A risk-based approach to advanced therapy manufacturing and testing is taken. Additionally, guidelines specifically tailored to advanced therapies address Good Laboratory Practices and Good Clinical Practices.

The EMA also offers the opportunity to meet with agency staff to discuss the product and/or product development plan:

- Innovation Task Force Briefing Meetings: An informal exchange forum to discuss regulatory, technical and scientific issues of innovative medicines at a very early stage.

TABLE 13A.1
EMA Guidelines for Advanced Therapy Development

	Overarching guidelines
GT	Guideline on the quality, nonclinical, and clinical aspects of gene therapy medicinal products (EMA/CAT/80183/2014)
CT	Guideline on human cell-based medicinal products (EMA/CHMP/410869/2006)
GT + CT	Draft guideline on quality, nonclinical and clinical requirements for investigational ATMPs in clinical trials (EMA/CAT/852602/2018)
	Guidelines addressing specific aspects of development
GT + CT	Quality, nonclinical, and clinical aspects of medicinal products containing genetically modified cells (CAT/CHMP/GTWP/671639/2008)
GT + CT	Questions and answers on comparability considerations for ATMPs (EMA/CAT/499821/2019)
GT	Question and Answer on gene therapy (EMA/CAT/80183/2014)
GT	Reflection paper on quality, nonclinical and clinical issues relating specifically to recombinant adeno-associated viral vectors (CHMP/GTWP/587488/07)
GT	Reflection paper on design modifications of gene therapy medicinal products during development (EMA/CAT/GTWP/44236/06)
GT	Reflection paper on management of clinical risks derived from insertional mutagenesis (CAT/190186/2012)
GT	Guideline on safety and efficacy follow-up and risk management of ATMPs (EMEA/149995/2008)
	Development and manufacture of lentiviral vectors
GT	Guideline on the nonclinical studies required before first clinical use of gene therapy medicinal products (EMA/CHMP/GTWP/125459/2006)
GT	Guideline on nonclinical testing for inadvertent germline transmission of gene transfer vectors (EMEA/273974/2005)
	Risk-based approach according to Annex I, part IV of Directive 2001/83/EC applied to ATMPs
GT	Guideline on follow-up of patients administered with gene therapy medicinal products (EMA/CHMP/GTWP/60436/2007)
GT	Scientific requirements for the environmental risk assessment of gene therapy medicinal products (CHMP/GTWP/125491/06)
CT	Guideline on xenogeneic cell-based medicinal products (EMEA/CHMP/CPWP/83508/2009)
CT	Reflection paper on stem cell-based medicinal products (EMA/CAT/571134/2009)
	Therapeutic area/indication-specific guidelines
CT	Guideline on potency testing of cell-based immunotherapy medicinal products for the treatment of cancer ((CHMP/BWP/271475/06)
CT	Reflection paper on in vitro cultured chondrocyte-containing products for cartilage repair of the knee (EMA/CAT/CPWP/568181/2009)
	Applicable ICH guidelines
	ICH S12 – Nonclinical biodistribution considerations for gene therapy products
	ICH Considerations: Oncolytic Viruses (EMEA/CHMP/ICH/607698/2008)

Abbreviations: CT – cell therapy, GT – gene therapy; Source: https://www.ema.europa.eu/en/human-regulatory/research-development/advanced-therapies/guidelines-relevant-advanced-therapy-medicinal-productsUp to date as of July 2022

- Scientific advice (or Protocol Assistance for Orphan Drugs): A formal procedure that obtains advice from the EMA "on the most appropriate way to generate robust evidence of a medicine's benefits and risks." Sponsors can seek scientific advice at any stage of product development and multiple times.

The development of advanced therapies can also be supported through various designations that offer advantages during development and/or after approval:

- Advanced Therapy Medicinal Product Classification – consultation with the EMA to determine whether a product under development is classified as advanced therapy.
- Orphan Drug Designation – available for products addressing life-threatening or chronically debilitating disease, no available satisfactory treatment and prevalence is not more than 5 in 10,000. The designation offers for example reduced fees for regulatory activities including protocol assistance and ten-year market exclusivity.
- PRIME – innovative products that address an unmet medical need providing advantages such as a clinically meaningful improvement compared to available medicines. Benefits include enhanced interactions with the EMA, dedicated contact point, potential eligibility for accelerated assessment.
- Accelerated Assessment – reduces statutory review time from 210 to 150 days for medicines of major public health interest (not including clock stops, see p. 000). Although accelerated assessment may be granted for a marketing authorization application, this designation may be reverted to standard assessment times if the EMA has too many questions during the evaluation of the application that would significantly extend the assessment time.

Approval Pathways

Marketing authorization applications for advanced therapies are initially evaluated by the CAT. The committee develops a draft opinion on whether a product can be approved or not based on the quality, safety and efficacy of the product. This opinion is then sent to the CHMP for evaluation.

The CHMP conducts its review of the CAT opinion and assessment of the application. Based on the findings, the committee provides a final recommendation on the approvability of the product to the EC. The EMA publishes the outcome of the assessment in the CHMP meeting minutes. CHMP's other roles include but are not limited to:

- The CHMP and CAT assess modifications and extensions (also called "variations") of existing authorizations and evaluate any safety considerations provided by the Pharmacovigilance Risk Assessment Committee

(PRAC) recommendations. Depending on the issue at hand, recommendations for additional safety measures, suspension or revocation of the authorization may be necessary. Changes to the original marketing authorization would then be notified to the EC for evaluation and approval.
- Supporting development of medicines and regulation through providing scientific advice, preparing scientific guidelines and international collaboration.

The statutory time for the assessment of a marketing authorization application is 210 days. However, there may be questions from the review team to the sponsor, which is particularly likely for advanced therapies. The EU legislation allows for the clock counting those days during the review to be halted twice: at 120 days and at 180 days. During the first stop, which could last three to six months, the sponsor will need to answer a list of questions provided by the review team. The responses will be taken into account during further review. Any outstanding issues will be addressed during the potential second "clock stop" at 180 days. This review break is usually shorter, lasting one to three months. Therefore, the overall time for a product review typically lasts about one year. The final CHMP opinion, drafts for the *Summary of Product Characteristics* (i.e., information about the medicinal product that allows efficient use by healthcare professionals), package leaflet (i.e., information for the patient) and proposed labeling are sent to the EC within 15 days.

The EC review and approval process should usually not take more than 67 days. However, in practice, it can take substantially longer. This process has multiple steps, including the review of translations of information for all EU languages by national authorities. Sponsors will be notified of the assessment outcome by the EC. The final decision will be published in the *Union Register*. The approval publication in the *Union Register* signifies the authorization to market the product in all EU Member States. EU (standard) marketing authorizations are valid for five years and renewable through application by the marketing authorization holder.

There are three types of marketing authorization in the EU that may apply to advanced therapies:

- Marketing authorization – full approval based on a complete quality, safety and efficacy data package (five-year validity, renewable)
- Conditional marketing authorization – can rely on a less comprehensive data package if the benefit of immediate availability of the drug outweighs the risks; generally used for orphan drugs or to treat seriously debilitating and/or life-threatening diseases or in emergency situations (post-marketing conditions apply, generally additional studies)
- Authorization under exceptional circumstances – generally this type of authorization is intended for rare diseases, and a full data package for safety and efficacy cannot be provided due to ethical reasons

After marketing authorization through the EC, it is the individual EU Member States' responsibility to determine pricing, reimbursement and inclusion in their national health care and social security systems in negotiations with the sponsor.

EMA will publish various documents for the approved product, which include the European Public Assessment Report or EPAR, the *Summary of Product Characteristics* and will update the website with any post-approval procedures. Generally, information on ongoing assessments can be found in the applicable committee meeting minutes.

THE REGULATORY FRAMEWORK IN THE US

Definitions

There is no statutory definition for gene therapy, although one has been provided for somatic cell therapies. However, definitions may be included in applicable guidance documents. Perhaps the most detailed definition of gene therapy was provided by the Department of Health and Human Services in October 1993 as follows:

> *Gene therapy is a medical intervention based on modification of the genetic material of living cells. Cells may be modified ex vivo for subsequent administration or may be altered in vivo by gene therapy products given directly to the subject. When the genetic manipulation is performed ex vivo on cells that are then administered to the patient, this is also a type of somatic cell therapy. The genetic manipulation may be intended to prevent, treat, cure, diagnose, or mitigate disease or injuries in humans.*

The 1993 *Federal Register* notice on the application of statutory authorities to human somatic cell therapy products defines this term as

> *autologous (i.e., self), allogeneic (i.e., intra-species), or xenogeneic (i.e., inter-species) cells that have been propagated, expanded, selected, pharmacologically treated, or otherwise altered in biological characteristics ex vivo to be administered to humans and applicable to the prevention, treatment, cure, diagnosis, or mitigation of disease or injuries.*

Regulatory Framework

The US regulatory framework is composed of acts and rules, while the US Food and Drug Administration (US FDA) provides guidance documents:

- "Acts" are laws addressing regulatory needs. The responsible authority is generally named, in the case of advanced therapies it is the FDA, and required to implement regulations providing the requirements set by the law. The US Code (USC) contains the federal statutes.
- "Rules" are made by federal authorities to achieve the goals set out in laws and are legally binding. The *Code of Federal Regulations* (CFR) contains the regulations (i.e., interpretation).

- "Guidance" is the recommendations provided by the FDA detailing the agency's current thinking on the application of regulatory requirements. They are not legally binding.

Advanced therapies are regulated as biological medicinal products under Section 351 of the Public Health Service (PHS) Act in the US (codified 42 USC §262). However, if an advanced therapy is a combination product that includes a medical device and the device's function is the primary mode of action, then these products would be regulated as medical devices. Additionally, the sections of the Food, Drug and Cosmetics Act (21 USC Chapter 9) also apply to biologics including advanced therapies. A sample of applicable regulations include (FDA, n.d.):

- 21 CFR 312 sets out the requirements for investigational new drugs (INDs). This section includes requirements for application, when and what administrative actions such as clinical holds can be taken and formal meetings between FDA staff and the sponsor.
- 21 CFR 600 – 680 covers definitions, good manufacturing practice, establishment registration requirements and applicable standards for biologics and some other products. 21 CFR 601 covers the requirements for Biologics License Applications to request approval for the marketing of a new biological product.
- 21 CFR 1271 addresses Human Cells, Tissues, and Cellular and Tissue-Based Products (HCTPs) and covers for example donor testing to avoid transmission of infectious diseases. These requirements apply to all products involving cells extracted from human donors (e.g., CAR-T cell therapy development).

In the US, "advanced therapies" generally referred to CGT products. Human tissues and some cell therapy products are regulated as a separate product category known as HCTPs, which is outside the scope of this chapter. Regulation of cell therapies outside of advanced therapies depends on various conditions (e.g., amount of manipulation, donor-recipient relationship and mode of action determine regulatory requirements).

Regulatory Oversight

The Food and Drug Administration oversees almost all of the drug development process and approval for medicinal products, which include drugs, biologics and medical devices. The FDA also regulates food, radiation-emitting products, veterinary medicines, cosmetics and tobacco products.

Within the FDA, the Center for Biologics Evaluation and Research (CBER) is responsible for regulating the development and approval of biological products. Within CBER the Office of Tissues and Advanced Therapies (OTAT) oversees all products involving tissues, cellular and gene therapy products. Within OTAT, the Division of Cellular and Gene Therapies (DCGT) is responsible for overseeing products involving cell and gene therapies.

Clinical trials are managed through the FDA IND process for regulatory aspects in cooperation with institutional review and ethics boards for ethical considerations.

Regulatory Tools

The FDA has issued specific guidance documents that can serve as a starting point for determining the regulatory requirement for advanced therapies. Table 13a.2 provides an overview of the guidance documents available at the time of writing of this chapter that are specific to gene and/or cell therapies. Other guidance that addresses biologics and general medicinal product development developed by the FDA and/or ICH also apply. For a more in-depth listing of guidelines, readers should visit the FDA website provided.

The first few guidance documents grouped as "overarching guidance" documents provide considerations for a broader aspect of the advanced therapy development program (e.g., donor eligibility determination and chemistry, manufacturing, control requirements for cell and gene therapies). The next group of guidance documents looks at specific aspects (e.g., shedding studies) or technologies (e.g., genome editing). The FDA has also released several guidance documents for specific indications (e.g., pancreatic cancer, retinal disorders).

The FDA has also released a guidance specifically addressing gene and cell therapy considerations for products as it applies to COVID-19-related products. This short guidance discusses considerations regarding donor assessment, cellular or tissue source material, manufacturing and material testing, risk assessment and mitigation during product manufacture.

Since many gene therapy products address rare diseases, the guideline on interpreting sameness with respect to orphan drugs is also of particular interest. One of the benefits of the orphan drug designations is seven years of market exclusivity during which the "same" product by a different developer cannot be approved or enter the market. In this guidance, the FDA provides specific examples that would make gene therapies the same or different with respect to orphan drugs. The EMA has not released a guideline addressing this topic. Of note is that the EMA also requires differences in structure to have an impact on biological characteristics and/or activity relating to the therapeutic effect.

Similar as mentioned for the EU, following GMP requirements is very important for advanced therapies. The FDA has released a GMP guideline for broader investigational products in phase I development that is applicable to all products, including advanced therapies (Current Good Manufacturing Practice for Phase 1 Investigational Drugs). However, it is an important guidance since many issues arising during the approval evaluation relate back to manufacturing and comparability issues.

The FDA also offers opportunities to meet with agency staff to discuss the product and/or development plan through informal and formal meetings. The INTERACT meetings have recently been particularly highlighted by FDA speakers at various conferences. There are several meeting types, including four formal meeting types, that fulfill various purposes:

TABLE 13A.2
FDA Guidelines for Advanced Therapy Development

	Overarching guidelines
CT + GT	Guidance for Human Somatic Cell Therapy and Gene Therapy
CT	Eligibility Determination for Donors of Human Cells, Tissues, and Cellular and Tissue-Based Products
CT + GT	Studying Multiple Versions of a Cellular or Gene Therapy Product in an Early-Phase Clinical Trial
CT + GT	Potency Tests for Cellular and Gene Therapy Products
CT	Content and Review of Chemistry, Manufacturing, and Control (CMC) Information for Human Somatic Cell Therapy Investigational New Drug Applications (INDs)
GT	Chemistry, Manufacturing, and Control (CMC) Information for Human Gene Therapy Investigational New Drug Applications (INDs)
CT + GT	Considerations for the Design of Early-Phase Clinical Trials of Cellular and Gene Therapy Products
	Guidelines addressing specific aspects of development
GT	Determining the Need for and Content of Environmental Assessments for Gene Therapies, Vectored Vaccines, and Related Recombinant Viral or Microbial Products
GT	Testing of Retroviral Vector-Based Human Gene Therapy Products for Replication Competent Retrovirus During Product Manufacture and Patient Follow-up
GT	Design and Analysis of Shedding Studies for Virus or Bacteria-Based Gene Therapy and Oncolytic Products
CT + GT	Expedited Programs for Regenerative Medicine Therapies for Serious Conditions
GT	Long-Term Follow-Up After Administration of Human Gene Therapy Products
GT	Human Gene Therapy Products Incorporating Human Genome Editing
CT + GT	Considerations for the Development of Chimeric Antigen Receptor (CAR) T Cell Products
	Therapeutic area/indication-specific guidelines
CT	Considerations for Allogeneic Pancreatic Islet Cell Products
CT	Cellular Therapy for Cardiac Disease
GT	Human Gene Therapy for Retinal Disorders
GT	Human Gene Therapy for Rare Diseases
GT	Human Gene Therapy for Hemophilia
GT	Human Gene Therapy for Neurodegenerative Diseases
CT + GT	Manufacturing Considerations for Licensed and Investigational Cellular and Gene Therapy Products During COVID-19 Public Health Emergency
GT	Interpreting Sameness of Gene Therapy Products Under the Orphan Drug Regulations
	Applicable ICH guidelines
	ICH S12 – Non-clinical Biodistribution Considerations for Gene Therapy Products
	ICH Considerations: Oncolytic Viruses (EMEA/CHMP/ICH/607698/2008)

Abbreviations: CT – cell therapy, GT – gene therapy; Source: https://www.fda.gov/vaccines-blood-biologics/biologics-guidances/cellular-gene-therapy-guidancesUp to date as of July 2022

Regulatory Landscape in US, EU and Canada

- INTERACT meetings – INitial Targeted Engagement for Regulatory Advice on CBER producTs, also sometimes called pre-pre-IND meetings, are informal nonbinding consultations at an early stage of product development.
- Type A meetings – Formal meetings to address safety issues associated with a stalled development program.
- Type B meetings – Formal meetings usually held at a development program milestone such as prior to submission of an IND or new biologics license application. Additionally, the overall development program can be discussed or specific aspects such as risk evaluation and mitigation strategies.
- Type B (EOP) meetings – meetings held at "End of Phase" (EOP) generally for EOP Phase 2 and 3. EOP meetings for Phase 1 are only applicable to specific product types.
- Type C meetings – These meetings are for any consultations that fall outside the other three meeting types. Typical examples for requesting a Type C meeting would be to discuss the use of a biomarker or new surrogate endpoint.

The development of advanced therapies can also be supported through various designations that offer advantages during development and/or approval:

- Orphan Designation – The prevalence of the disease must be less than 200,000 people in the US in order for a disease to be classified as rare. Obtaining orphan designation provides several benefits, including tax credits for qualified clinical trials, fee exemptions and potentially seven years of market exclusivity.
- Fast-Track Designation – The designation can be granted for treatments addressing serious conditions and nonclinical and/or clinical data indicate that the product addresses an unmet need. It may also be applicable to products designated as qualified infectious disease products. Benefits of the program aim to expedite the development and review and the product may be eligible for rolling review.
- Breakthrough Therapy Designation – The designation can be granted for treatments addressing serious conditions and preliminary clinical data support that the product may provide substantial improvement on a clinically significant end point(s) compared to available treatments. Benefits aim to expedite the program through intensive guidance to support an efficient development program, FDA organizational commitment, other actions and eligibility for rolling review.
- Regenerative Medicine Advanced Therapy (RMAT) Designation – The 21st Century Cures Act added Section (g) to 21 U.S.C. 356 providing an expedited pathway for regenerative medicines. The designation was implemented for cells, therapeutic tissue engineering products, human cell and tissue products intended to "treat, modify, reverse or cure a serious or life-threatening disease or condition." According to guidance

issued by the FDA, *"human gene therapies, including genetically modified cells, that lead to a sustained effect on cells or tissues, may meet the definition of a regenerative medicine therapy."* Preliminary evidence is needed to show that the product has the potential to address an unmet medical need. The benefits of Fast-Track and Breakthrough Therapy Designations are applied to this designation, and sponsors benefit from early interactions with the FDA.
- Priority Review Designation – The designation can be granted for treatments addressing serious conditions and would provide a significant safety or effectiveness benefit over already approved products. Review time is reduced to six months from the standard ten-month approval.

Approval Pathways

An FDA review team will evaluate biologics license applications for cell and gene therapies. The standard review timeline of 10 months is set by the Prescription Drug User Fee Act (PDUFA) and applies to cell and gene therapies. The review team may send questions and request additional information during the ongoing application review. There is no clock stop like in the EU. The ten-month review time includes these consultations with the sponsors as well as any need to consult an advisory committee.

The FDA may request a meeting of the Cellular, Tissue, and Gene Therapies Advisory Committee to address specific scientific, technical and/or policy questions that remain open from the application review. The need for convening an advisory committee is technology and product dependent, with more innovative products being more likely to require discussion by the committee. The advisory committee is composed of FDA-independent experts in the areas to be discussed. They must declare any financial interest in relation to the product to be discussed. During the meeting, which usually lasts several hours, the advisors hear information provided by the sponsor for the product to be discussed and the FDA's view on data provided to support an approval. After thorough discussion of a specific question, the committee will provide recommendations on the question. The meeting agency may also include a vote on whether the product should be approved or not.

However, it is the FDA's responsibility to decide whether a product is approved for marketing within the US. While the vote and recommendations from the advisory committee are taken into consideration, the agency does not have to agree with the committee.

There are several approval mechanisms that may apply to CGT products:

- Biologics License Application – Generally, a Biologics License Application would be the mechanism to obtain approval to introduce a cell and/or gene therapy into the interstate market.
- Accelerated Approval - The designation can be granted for treatments addressing serious conditions, provides a meaningful advantage over available products and shows an effect on a surrogate endpoint that is "reasonably likely to predict clinical benefit" or on an endpoint

TABLE 13A.3
Regulatory Overview for Advanced Therapies in the EU and the US (EC, 2001, 2004, 2007; FDA, n.d.)

	EU	US
Legal Framework	Directive 2001/83/EC (overarching human medicines)	Public Health Service Act – Section 351 (biologics license)
	Directive 2009/120/EC (advanced therapies)	Food, Drug and Cosmetics Act
		21 CFR 600 – 680 (biologics)
	Regulation 726/2004/EC (procedures)	21 CFR 312 (INDs)
		21 CFR 1271 (infectious diseases)
	Regulation 1394/2007/EC (advanced therapies)	21 U.S.C. 356(g) (RMAT designation)
Product Types	Gene Therapy	Gene Therapy
	Somatic Cell Therapy	Cellular Therapy
	Tissue Engineered Product	Combination product (combination with drug and/or device)
	Combined ATMP (combination with device)	Excludes HCT/Ps regulated solely under PHS Act section 361
General Oversight	EMA	FDA
Guidance source		
Options for sponsors to meet with regulators	Innovation Task Force Meetings (early stage, innovative products)	INTERACT meetings
		Type A meetings (stalled programs)
		Type B meetings (milestones)
	Scientific Advice (at any development stage)	Type B (EOP) meetings (at end of Phase 2 and/or 3)
		Type C meetings (all else)
Expedited pathways (these programs differ significantly between EU and US)	Orphan Drug Designation	Orphan Drug Designation
	Advanced Therapy Product Classification	Fast-Track Designation
		Breakthrough Therapy Designation
	PRIME	RMAT Designation
	Accelerated Assessment	Priority Review Designation
Product Evaluation	CAT – draft opinion	FDA – CBER/OTAT
	CHMP – product opinion	Advisory Committee (as needed)
Review Timeline (standard)	210 days (plus clock stops)	Up to 10 months
Approval Types	Conditional Marketing Authorization	Accelerated Approval
	(standard) Marketing Authorization	(standard) Approval
	Authorized under Exceptional Circumstances	
Product Approval	EC (67 for EC decision from CHMP opinion receipt)	FDA
Approval validity	Reapply five years in	Continuous

measurable earlier than irreversible morbidity or mortality. The approval will typically be based on the surrogate or 'other' endpoint and clinical benefit must be demonstrated post-marketing. The conduct of post-authorization studies is generally a condition of approval.

The FDA will announce any new approval of an advanced therapy and the product will be listed under 'Approved Cellular and Gene Therapy Products'. Approval documentation including product reviews are available on the FDA's website and provide further insight into the FDA's decision-making process.

Table 13a.3 provides an overview of important regulatory aspects for the EU and US side-by-side.

REFERENCES

European Commission. (2001). Part IV of Annex I. Directive 2001/83/EC.

European Commission. (2004). Regulation (EC) No. 726. Community Procedures for the Authorization and Supervision of Medicinal Products for Human and Veterinary Use and Establishing a European Medicines Agency.

European Commission. (2007). Advanced Therapy, Medicinal Product Regulation. N 1394/2007.

FDA. (n.d.). Code of Federal Regulations (CFR). 21CFR 312: Investigational New Drugs (IND). 21CFR 600-680: Biologics License Applications. 21CFR 1271: Human Cells, Tissues, and Cell ad Tissues Based Products.

13b Regulatory Landscape in South America

Heloisa Mizrahy
HMC Consultoria Ltda. – Independent Consultant, Rio de Janeiro, Brazil

CONTENTS

Introduction: South America Pharmaceutical Medicines Market 225
    Argentina's Regulatory Background 225
    Brazil Regulatory Background 226
Argentina Regulatory Framework on ATMP 226
Brazil's Regulatory Framework on ATMP 227
References 230

INTRODUCTION: SOUTH AMERICA PHARMACEUTICAL MEDICINES MARKET

Up to now, Argentina and Brazil are the countries that have established a more robust regulatory framework related to advanced therapy medicinal products (ATMP), and the chapter will be dedicated to these two countries.

South American regulatory agencies are being requested to adapt themselves by issuing new regulatory requirements, most of which are aligned with international standards and guidelines. This movement will allow the population access to innovative medicines, such as ATMP.

The amount of ATMPs being registered across the globe is increasing day by day, and clinical trials are being requested to be performed at a fast speed; therefore, regulatory agencies need to establish a framework to accommodate these products.

Most countries in this region are still developing their regulatory frameworks, such as Chile, Colombia and Mexico. Latin American countries are conducting meetings to engage governments, academia and industries to discuss the next steps on ATMP regulations.

ARGENTINA'S REGULATORY BACKGROUND

Argentina's medicine market is mainly supplied by domestic pharmaceutical companies, according to the Argentinian Pharmaceutical Laboratories Industrial

Chamber (Cámara Industrial de Laboratorios Farmacéuticos Argentinos, CILFA). Unlike the rest of the countries in the region, more than half of the drug market is supplied by local companies, manufacturing around 690 million units per year. Biological and biotechnological medicine products are still mainly imported (CILFA).

Argentina´s regulatory agency, ANMAT (Administración Nacional de Medicamentos, Alimentos y Tecnología Médica), was created in 1992, after issues related to adulteration of medicines, in some cases leading to death. ANMAT's mission is the prevention and protection of the health care of the population through the control and supervision of the quality and health of products, substances, elements and materials that are consumed or used in medicine, food and cosmetics and the control of the activities, processes and technologies that mediate or are included in these matters (Administración Nacional de Medicamentos, Alimentos y Tecnología Médica (ANMAT) 2017). Depending on the type of ATMP, ANMAT may regulate it or not.

Brazil Regulatory Background

Brazil has the sixth biggest world population with more than 210 million inhabitants, and according to IQVIA 2020 data, Brazil is in the seventh position in the rank of global pharmaceutical market sales (2.6%) and is projected to be in the fifth position by 2023 (Sindusfarma, 2020).

Brazilian domestic pharmaceutical companies represent around 76% of the sales. While local companies are dedicated to generic medicines mainly, multinational companies offer innovative products, and the advanced therapy products market is not different, as will be explained later in this chapter. The Brazilian government is planning to invest 407 million reais (something around $90 million) in 2021/2022 in advanced therapy research, including preclinical and clinical trials (Serviços e Informações do Brasil).

Brazil's regulatory agency, ANVISA (Agência Nacional de Vigilância Sanitária), was created in 1999 by law no. 9.782 (Lei 9782, 1999). According to article 8 of this law, products and goods that may bring risks to public health are subject to sanitary vigilance, including the ones obtained by genetic engineering or another related procedure. It is in this context that cell and gene therapies fit when associated with clinical trials or human therapy.

ARGENTINA REGULATORY FRAMEWORK ON ATMP

There are two governmental entities in Argentina in charge of the regulatory requirements for ATMPs: ANMAT and INCUCAI (Instituto Nacional Central Único Coordinador de Ablación e Implante) – National Implant Institute. INCUCAI is the competent body to regulate the processes linked to the use of cells of human origin for implantation in human beings. To this end, INCUCAI has the power to establish quality and safety standards for the donation, procurement, evaluation, processing, traceability, preservation, storage and distribution of such human cells and tissues for clinical application within the scope of its competence. It is the responsibility of INCUCAI to dictate rules for obtaining cells and tissues to be used as starting material or at some stage of the production process of advanced therapy drugs.

In the regulation of the production of AMTPs, both organisms are competent in some of the stages involved in the production of these products when the production process requires the use of cells or tissues obtained from human beings.

Argentina's government created a commission on research and medicines for advanced therapies by Joint Resolution 1-E/2017 (2017) within the scope of the Ministry of Health and the Ministry of Science, Technology and Productive Innovation in order to update the existing regulatory framework on advanced therapies and develop a proposal in alignment with international standards.

According to this resolution, it was agreed that when a product does not involve substantial manipulation, it will remain exclusively under the competence of INCUCAI. When an ATMP does not involve the donation, procurement or control of cells or tissues of human origin, it shall be exclusively under the competence of ANMAT. When an ATMP has been developed for an individual patient (autologous) within the same facility and used within that same facility, the stages of donation, obtaining and controlling are under the regulations of INCUCAI. For its subsequent processing and preparation of the medicine, INCUCAI may request the intervention of ANMAT. Any product subject to studies in clinical pharmacology regulated by ANMAT must be produced in establishments that meet the requirements of Good Manufacturing Practices of ANMAT. The requirements for the donation, obtaining and control of the starting material will be those established by INCUCAI.

In 2018, ANMAT established a regulatory milestone by launching requirement no. 179 (ANMAT Disposición 179/2018), considering ATMPs as biological products. According to this document, ATMPs are classified as gene therapy products, cell therapy products and tissue engineering products. The document also establishes the requirements for these products' approval.

Since 2008, ANMAT has been a member of PIC/s (Pharmaceutical Inspection Co-operation Scheme), and their Good Manufacturing Practices (GMP) guidelines are aligned. GMPs for ATMPs are described in Disposición 3602/2018 (ANMAT Disposición 3602/2018) Annex 17 – Biological Products Manufacturing.

Although ANMAT is a member of PIC/s, guidelines need to be approved by the agency before being internalized. Having said that, a memo was issued in 2019 called Circular 12/2019, which listed all guidelines that are to be referred to in the production, control, registration and surveillance of advanced therapy medicines.

BRAZIL'S REGULATORY FRAMEWORK ON ATMP

According to the Brazilian Federal Constitution from 1988, article 199, paragraph 4°, organs, tissues and human substances aimed at transplants, research and treatments cannot be marketed, bringing a lot of juridical insecurity to companies that wanted to invest in cell and gene therapy in Brazil. Before issuing specific resolutions related to advanced therapy medicines, ANVISA requested the Federal Attorney's Office, PGF/AGU, in 2016 to issue a legal opinion related to the registration and marketing of products based on advanced therapy. It reached the opinion that advanced therapy products could be registered and marketed provided that the starting human materials (cells and tissues) are donated (Garcia et al., 2018). The opinion also highlighted the need to create a regulatory framework to be applied to advanced therapy products.

In 2003, ANVISA launched a resolution defining the minimum handling requirements for umbilical cord and placental blood banks, RDC N° 190/2003, replaced by RDC 34/2014.

Brazilian law no. 11.105 (Lei 11.105, 2005) was approved in 2005, allowing the use of embryonic stem cells for research and therapy, and decree no. 5.591 (Decreto 5591/2005) demanded ANVISA create procedures for collecting, processing, laboratory testing, storing, transporting, quality control testing and using human embryonic stem cells. Moreover, the same law established biosafety standards and mechanisms for monitoring activities involving genetically modified organisms (GMOs) and their derivatives. The National Biosafety Council – CNBS, was created and then restructured as the National Technical Commission on Biosafety – CTNBio.

CTNBio is a multidisciplinary collegiate body created by law no. 11,105 of March 24, 2005, whose purpose is to provide technical advisory support and advice to the federal government in the formulation, updating and implementation of the National Biosafety Policy on GMOs, as well as in the establishment of technical safety standards and technical opinions regarding the protection of human health, living organisms and the environment for activities involving the construction, experimentation, cultivation, manipulation, transportation, marketing, consumption, storage, release and disposal of GMOs and derivatives.

In 2012, ANVISA approved the creation of the first Cell Therapy Technical Chamber (Portaria No. 1701/2012), similar to the European Chamber in Advanced Therapy (CAT) (EMA — Committee of Advanced Therapy), constituted by independent experts to support ANVISA in the creation of resolutions, recommendations on clinical trials, technical advice on safety and efficacy of ATMP, as well as manufacturing processes. In 2016, ANVISA established this committee (Portaria 1.731/2016), appointing members to join it.

At ANVISA, ATMP dossiers are evaluated by a team of experts, who can at any time request evaluation of external ad hoc experts enrolled in the Network of Specialists in Advanced Therapy – RENETA – or members of the Technical Advisory CAT of ANVISA, depending on the product and its clinical application.

In Brazil, three public institutions can act with complementary functions supervision of clinical trials in ATMP. The approval of ethical and social aspects of studies involving human beings is the responsibility of the National Ethics Commission (CONEP). In parallel, ANVISA assesses the aspects of quality and safety of the investigational product through a careful review of the manufacturing process (and its controls), from characterization tests (identity, purity, potency, others) and sterility finished products and process intermediaries, as well as the scientific reliability of the proposed protocol, i.e., the design capacity in proving the efficacy and safety of the product, to enable a risk and benefit assessment. In the phase of clinical development of therapy products, biosafety assessment of components identified as GMOs is carried out by the National Technical Commission for Biosafety – CTNBio.

As part of an international harmonization strategy, in 2016, the Brazilian Health Regulatory Agency (ANVISA) became a member of the International Council for Harmonization of Technical Requirements for Pharmaceuticals for Human Use (ICH), and since January 2021, it has been a member of PIC/s.

Brazil laid the regulatory path for advanced therapies with the publication of the regulatory framework by ANVISA. The regulations had different objectives, including fostering clinical development and, consequently, speeding market approval of gene and cell products by optimizing the risk-based authorization process. This is also thought to prevent the use of unproven treatments in the country.

Brazil has split ATMPs into two classes:

- Class I – Advanced therapy medicine product submitted to minimal manipulation that performs in the recipient a function distinct from that performed in the donor, and
- Class II – Advanced therapy medicine product subjected to extensive handling, tissue engineering product and gene therapy product.

A comprehensive comparison with Brazilian, US, Europe and Japanese ATMP regulatory framework, denomination, requirements and other points have been written by Gomes et al. (2022). According to the article, both class I and class II products have the same requirements for clinical and preclinical report content, but quality dossier differs, as explained next:

- Class I: Technical report with a summary of the general characteristics of the product and production information (minimum handling) to highlight critical product quality parameters (e.g., equipment, starting material, raw materials, laboratory controls)
- Class II: Description of starting materials, raw materials and excipients used; materials necessary for the production of vectors and genetic manipulation of cells; analysis of genetic sequence, attenuation of virulence and description of tropisms; information on selection and collection of human material; information on active component and final product; analytical methodologies employed; impurities, descriptions and flows of production steps; validation reports of critical steps; analytical method validation reports; traceability mechanisms; transport validation; stability studies; and storage precautions

ANVISA's ATMP regulatory framework is a combination of US Food and Drug Administration (US FDA), European Medicines Agency (EMA) and Pharmaceuticals and Medicinal Devices Agency (PMDA, Japan) standards. Currently, ANVISA has three main resolutions related to advanced therapy products:

- RDC 505/2021, which replaced RDC 338/2020, defines minimum requirements for the registration of advanced therapy products with a view toward proving efficacy, safety and quality for use and commercialization in Brazil. For the purposes of this resolution, the products of advanced therapies that can be registered are advanced cellular therapy products, gene therapy products and tissue engineering products.
- RDC 506/2021, which replaced RDC 260/2018, defines the procedures and regulatory requirements for conducting clinical trials on investigational therapy products in Brazil. Any research with advanced therapy products involving humans must be previously analyzed and approved by ANVISA.
- RDC 508/2021, replacing RDC 214/2018, establishes Good Practices in Human Cells for Therapeutic Use and Clinical Research. Advanced therapy products may only be made available for clinical research after the approval of the clinical research project by the CONEP system and ANVISA and can only be made available for therapy by submitting the product to ANVISA.

Only ATMPs that are approved by ANVISA can be manufactured, marketed or administered to a patient. The approval of an ATMP in Brazil follows the same approach used for any other drug. The company must demonstrate (i) quality aspects (characterization, production process and control, risk management and compliance with GMP); (ii) safety profile (essential pharmacological/toxicological characteristics) and proof of concept of the product in in vitro and in vivo models; and (iii) human safety profile and efficacy results for the established indication, dosage and target population (Gomes et al., 2022).

REFERENCES

Administración Nacional de Medicamentos, Alimentos y Tecnología Médica (ANMAT). (2017). https://salud.gob.ar/dels/printpdf/134

ANMAT Disposición 179/2018. https://www.argentina.gob.ar/normativa/nacional/disposici%C3%B3n-179-2018-314546/texto

ANMAT Disposición 3602/2018. http://www.anmat.gov.ar/boletin_anmat/BO/Disposicion_3602-2018.pdf

Cámara Industrial de Laboratorios Farmacéuticos Argentinos (CILFA). (n.d.). https://cilfa.org.ar/wp1/la-industria-farmaceutica-argentina-su-caracter-estrategico-y-perspectivas/

Circular ANMAT 12/2019. https://www.boletinoficial.gob.ar/detalleAviso/primera/223014/20191205

Decreto 5591//2005. http://www.planalto.gov.br/ccivil_03/_Ato2004-2006/2005/Decreto/D5591.htm

European Medicines Agency. Committee for Advanced Therapies – CAT. London: European Medicines Agency. https://www.ema.europa.eu/en/committees/committee-advanced-therapies-cat

Foro terapias avanzadas. (2021). https://terapiasavanzadas.org/images/memorias/memorias_terapias_avanzadas_2021.pdf

Garcia, L. A. O., Takao, M. R. M., Parca, R. M., & Silva Junior, J. B. da. (2018). Legal possibility of marketing authorization and commercialization of advanced therapy medicinal products in Brazil. *Health Surveillance under Debate: Society, Science & Technology*, 6 (1), 6–14. https://doi.org/10.22239/2317-269x.01075

Gomes, K.L.G. et al. (2022). Comparison of new Brazilian legislation for the approval of advanced therapy medicinal products with existing systems in the USA, European Union and Japan. *Cytotherapy*, 24 (5), 1–10.

Joint Resolution - E 1/2017. (2017). https://www.argentina.gob.ar/normativa/nacional/resoluci%C3%B3n-1-2017-273954/texto

Lei 11.105 de/2005. http://www.planalto.gov.br/ccivil_03/_ato2004-2006/2005/lei/l11105.htm

Lei 9782 de. (1999). http://www.planalto.gov.br/ccivil_03/leis/l9782.htm

Portaria 1.731/2016. https://www.in.gov.br/materia/-/asset_publisher/Kujrw0TZC2Mb/content/id/23531679

Portaria No. 1701/2012. https://bvsms.saude.gov.br/bvs/saudelegis/anvisa/2012/prt1701_12_12_2012.html

RDC 34/2014. http://antigo.anvisa.gov.br/documents/10181/2867975/%282%29RDC_34_2014_COMP.pdf/140dc780-ac2e-4829-8e2a-6fbc680677dc

RDC 505/2021. https://www.in.gov.br/en/web/dou/-/resolucao-rdc-n-505-de-27-de-maio-de-2021-323002775

RDC 506/2021. https://www.in.gov.br/en/web/dou/-/resolucao-rdc-n-506-de-27-de-maio-de-2021-323008725

RDC 508/2021. https://www.in.gov.br/en/web/dou/-/resolucao-rdc-n-508-de-27-de-maio-de-2021-323013606

Rede Nacional de Especialistas em Terapias Avancadas (RENETA). https://www.reneta.org.br/

Serviços e Informações do Brasil. (2021). https://www.gov.br/pt-br/noticias/educacao-e-pesquisa/2021/06/mais-de-r-407-milhoes-serao-destinados-a-bolsas-de-estudo

Sindusfarma. (2020). Relatorio Anual de Atividades 2020. https://sindusfarma.org.br/publicacoes/exibir/15189-relatorio-anual-de-atividades-2020

13c Regulatory Landscape in Australia and New Zealand

Orin Chisholm
Principal Consultant, Pharma Med, UNSW Sydney, Australia

CONTENTS

Australia 233
    Research Sector 234
    Regulation of Therapeutic Goods 235
        Registration of Prescription Medicines 235
        Registration of Biologicals 237
    Regulation of Genetically Modified Organisms 239
New Zealand 240
References 241

AUSTRALIA

Cell and gene therapies are a subset of regenerative medicines. Regenerative medicines seek to harness the use of stem cells, biomaterials and molecules to repair, regenerate or replace diseased cells, tissues, genes and organs. However, only a subset of regenerative medicines are considered to be therapeutic goods in Australia; other forms of regenerative medicine come under the definition of medical practice and are thus not regulated by the national regulatory authority in Australia.

Anyone wishing to manufacture or import cell and gene therapy products into Australia will need to understand two separate regulatory systems: the regulatory system for therapeutic goods and the regulatory system for genetically modified organisms (O'Sullivan et al., 2019). They will also need to understand the complex health technology systems in Australia in order to obtain reimbursement from the government for their products. Furthermore, an understanding of the responsibilities of Commonwealth versus State government responsibilities within the healthcare system and the roles of public versus private support for funding healthcare is vital to navigating the complex scenarios involved in the development and delivery of cell and gene therapy products in Australia.

The first Chimeric Antigen Receptor (CAR)-T cell therapy product was approved in Australia in 2018 and has been funded under a risk-sharing agreement between the sponsor and the Australian government since 2019. The first gene therapy product was approved in Australia in 2020 and was recommended for reimbursement in 2021; however, the sponsor is still under negotiation with the government regarding funding arrangements.

Research Sector

Australia has a robust research sector involved in the discovery and development of cell and gene therapies. In addition to basic and early translational research, which is primarily funded by government grant schemes, there are over 60 companies developing regenerative medicine products and over 130 clinical trials in progress in Australia (Regenerative Medicine Catalyst Project, 2021; Australia's Regenerative Medicine Clinical Trials Database – https://www.ausbiotech.org/programs/regenerative-medicines-catalyst-project). The number of clinical trials for regenerative medicines in Australia has been growing at a rate of approximately 15% per year for the past several years, with the rate of growth of industry-sponsored trials outstripping non-industry-sponsored trials. The major areas of application are in oncology, followed by the central nervous system, infectious diseases, rare genetic disorders and haematological disorders (Regenerative Medicine Catalyst Project, 2021; Australia's Regenerative Medicine Clinical Trials Database – https://www.ausbiotech.org/programs/regenerative-medicines-catalyst-project). Clinical trials underway in Australia are included in the Australia New Zealand Clinical Trials Registry (ANZCTR; https://www.anzctr.org.au). The regulatory requirements for establishing clinical trials will be discussed further in this chapter.

The National Health and Medical Research Council (NHMRC), the Australian Research Council and Universities Australia have developed the National Statement on Ethical Conduct in Human Research (updated in 2018; https://www.nhmrc.gov.au/about-us/publications/national-statement-ethical-conduct-human-research-2007-updated-2018), and compliance with this statement is a prerequisite for receipt of NHMRC clinical research funding. The purpose of the statement is to promote ethical conduct in human clinical research, and it clarifies the responsibilities of institutions and researchers involved in human research, as well as the human research ethics committees (HREC). The statement requires that researchers show compliance with the International Council for Harmonisation requirements for Good Clinical Practice, the International Standards Organisation requirements for clinical investigation of medical devices, the World Health Organisation International Clinical trials Registry Platform and the Therapeutic Goods Administration (TGA) requirements for clinical trials involving therapeutic products.

The majority of clinical trials performed in Australia are done so under a Clinical Trials Notification (CTN) scheme by which companies submit payment and notification to the TGA of their intention to sponsor a clinical trial involving an unapproved therapeutic good prior to commencement of the trial. Under this scheme, the TGA does not evaluate any data pertaining to the trial. The HREC is responsible for the review and monitoring of the trial. TGA target time for processing CTNs is five to seven working days. The sponsor should notify the TGA of any variations to the CTN during the trial. Sponsors are required to comply with relevant adverse event reporting requirements, as outlined in the NHMRC document "Safety Monitoring and Reporting in Clinical Trials involving Therapeutic Goods." Once the trial has been completed, sponsors should again notify the TGA that the CTN is no longer required.

Some clinical trials may be conducted under a Clinical Trials Approval (CTA) scheme where the clinical trial is approved by the TGA prior to its commencement. The choice of route lies with the sponsor and HREC; however, if the sponsor is considering using the CTA scheme, they are strongly encouraged to discuss the trial with the TGA before submitting their trial for approval. The CTA route is designed for high-risk or novel therapeutic goods where there is limited or no safety knowledge available. Under the TGA regulatory framework for biologicals, Class 4 biologicals should proceed via the CTA scheme unless use of the biological in a trial is supported by evidence from previous clinical use or the therapeutic good has been approved for an equivalent indication from a comparable national regulatory authority.

REGULATION OF THERAPEUTIC GOODS

The TGA was established under the *Therapeutic Goods Act 1989 (Cth)*. This legislation is available on the Federal Register of Legislation (https://www.legislation.gov.au/). The Act defines a therapeutic good and provides for regulatory schemes for the regulation of medicines, biologicals, medical devices and other therapeutic goods. Therapeutic goods must be listed on the Australian Register of Therapeutic Goods (ARTG) before they can be supplied, imported into or exported from Australia, unless they are exempt or otherwise authorised by the TGA. The sponsor of the therapeutic good must be a legal entity in Australia and is responsible for applying for and maintaining the registration of the therapeutic good.

Gene therapy products delivered *in vivo* (often via a viral vector) are considered to be prescription medicines in Australia. Gene therapy products that are delivered *ex vivo* (such as CAR-T cells) are considered to be biologicals, as are other cell therapies.

Registration of Prescription Medicines

Prescription medicines are those medicines that require a prescription from a healthcare professional to purchase from a pharmacist. The process for registering prescription medicines requires the sponsor to submit a package of data that supports the quality, safety and efficacy of the product and pay the relevant fees for evaluation. The data package required is described in the Australian Regulatory Guidelines for Prescription Medicines (ARGPM; https://www.tga.gov.au/

publication/australian-regulatory-guidelines-prescription-medicines-argpm) and should be supplied in eCTD format, with Module 1 containing Australian-specific information and forms. The standard evaluation pathway for prescription medicines is 255 working days and proceeds through a number of set milestones, including pre-submission, submission, first-round assessment, consolidated section 31 request response (30- or 60-day response timeframe), second-round assessment, expert advisory panel review, TGA Delegate's decision and post-decision finalisation of documents such as the Product Information and Australian Public Assessment Report (AusPAR).

The sponsor may choose to engage in a pre-submission meeting with the TGA, and sponsors of gene therapy medicines are highly encouraged to do so. If the gene therapy medicine is considered to be an orphan drug (the disease affects less than 5 in 10,000 individuals in Australia), the sponsor should apply for orphan drug designation prior to submitting their dossier for evaluation. If granted, the designation will result in the waiver of fees for the prescription medicine registration application and evaluation. Once the designation is granted, the sponsor has a six-month period in which to submit their dossier of information for evaluation by the TGA. Once the product is registered and listed on the ARTG, the sponsor may commence supply of the product in Australia. The approval letter from the TGA will include post-registration commitments such as compliance with the risk management plan, any testing requirements for batches and maintenance of Good Manufacturing Practice (GMP) certification.

Over the past several years, the TGA has introduced a number of facilitated registration pathways for prescription medicines, and sponsors should be aware of the requirements for these pathways, as they may take advantage of them to gain more rapid approval of their products.

If the gene therapy prescription medicine is to treat a life-threatening or seriously debilitating condition where there are no other products available, or the product provides significant improvement on existing therapies and the sponsor has a full data package available, they may apply for priority determination and have their dossier evaluated in a target timeframe of 150 working days (https://www.tga.gov.au/publication/priority-determination). The priority registration process has greater flexibility including the use of rolling questions to accelerate the review process (https://www.tga.gov.au/publication/priority-registration-process).

If the gene therapy prescription medicine is to treat a life-threatening or seriously debilitating condition where there are no other products available, or the product provides significant improvement on existing therapies but the sponsor only has preliminary clinical evidence (Phase 2 clinical data), but that evidence suggests the product is a major therapeutic advance, then the sponsor may apply for a provisional determination. The sponsor has a six-month period in which to submit their dossier following the granting of a provisional determination (https://www.tga.gov.au/publication/provisional-determination). Evaluation of the data package proceeds with the standard evaluation timeframe, although the TGA aims to shorten the period slightly to 220 working days. If provisional approval is granted once the dossier has been evaluated, the sponsor will have a period of six years to provide the full data package to convert the registration to a full registration, otherwise the registration lapses.

Other pathways available include through the Project Orbis initiative with the USA, Canada, Switzerland, Singapore, Brazil and Israel participating in this FDA-initiated scheme to facilitate review of oncology products, through the ACCESS Consortium, where a shared dossier review process exists between the TGA, Health Canada, SwissMedic, Health Sciences Agency, Singapore and the Medicine and Healthcare Product Regulatory Agency in the UK. A strategic review and early discussion with participating regulatory authorities may provide access to these alternative pathways to product approval in Australia.

Gene therapy prescription medicines must meet GMP requirements for supply on the Australian market. Manufacturers based in Australia must hold a manufacturing license issued by the TGA, and the TGA will conduct inspections of the site before issuing a license. If the manufacturer is based overseas, the TGA will require GMP certification of the manufacturing site. The sponsor is responsible for applying for a GMP certification from the TGA. The TGA will inspect the manufacturing premises before issuing the GMP certification, or they may work with comparable national regulatory authorities under a mutual recognition agreement or compliance verification to issue a GMP clearance for the manufacturer. The TGA has adopted the *PIC/S Guide to GMP for Medical Products*, version PE009-14, excluding Annexes 4, 5 and 14, for medicines and active pharmaceutical ingredients.

Gene therapy prescription medicines currently registered in Australia include voretigene neparvovec (Luxturna) and onasemnogene abeparvovec (Zolgensma). Both products were granted orphan drug designation and were evaluated under the standard prescription medicine evaluation pathway.

Registration of Biologicals

The majority of products classified as regenerative medicines are considered to be biologicals under the *Therapeutic Goods Act 1989 (Cth)*. Biologicals are things made from or containing human cells or tissues that are used to treat or prevent a disease, ailment, defect of injury; are used to diagnose such conditions; alter the physiological processes; test for susceptibility to disease; replace or modify parts of the body; fecal microbiota transplant products or things that comprise or contain live animal cells, tissues or organs, unless they are specifically excluded or regulated under another category.

The TGA applies a risk-based classification system for biologicals ranging from Class 1 biologicals, which are low risk and have appropriate levels of external governance and clinical oversight; through Class 2, or low-risk biologicals; Class 3, or medium-risk biologicals; and Class 4, which are high-risk biologicals. Human cells that have been altered by genetic modification are classified as Class 4 biologicals. The following information will focus on this class of biologicals.

The process for including biologicals on the ARTG is contained in the Australian Regulatory Guidelines for Biologicals (ARGB; https://www.tga.gov.au/publication/australian-regulatory-guidelines-biologicals-argb). Pre-submission meetings with the TGA are highly recommended for sponsors submitting applications to include Class 4 biologicals. Biologicals that comprise or contain or are derived from human cells and tissues must comply with the *Australian Code of GMP for Human Blood and Blood Components, Human Tissues and Human Cellular Therapy* (https://www.tga.gov.au/publication/australian-code-good-manufacturing-practice-human-blood-and-blood-components-human-tissues-and-human-cellular-therapy-products). This is different to the Pharmaceutical Inspection Co-operation Scheme (PIC/S) GMP requirements for prescription medicines. Evidence of the sponsor's GMP license application must be provided in the dossier. The sponsor may submit the dossier in either eCTD format or the TGA Biologicals Dossier Structure (https://www.tga.gov.au/book-page/dossier-structure#tga). Once the dossier has been submitted, there is a 30-day legislated timeframe for the TGA to respond with a notification of outcome of preliminary assessment. If the dossier is accepted for evaluation, there is a 255 working-day timeframe for evaluation. Dossiers for biologicals may undergo three rounds of evaluation followed by consultation with an expert Advisory Committee on Biologicals before the final TGA decision is made. Post-decision requirements include finalisation of the product information, risk management plans and ARTG entries. Risk management plans are required for biologicals, as part of the biovigilance responsibilities of sponsors (https://www.tga.gov.au/publication/biovigilance-responsibilities-sponsors-biologicals). Unlike the processes for prescription medicines, there are no accelerated pathways currently available for biologicals, nor are Australian Public Assessment Reports publicly available for biologicals. However, the TGA has released a consultation document regarding the introduction of a priority approval pathway and AusPARs for priority biologicals(https://consultations.tga.gov.au/medicines-regulation-division/priority-review-pathway-for-biologicals/).

There are several specific biological standards that apply to human cell therapy products, as outlined in Table 13c.1.

Cell therapy biologicals currently registered in Australia include tisagenlecleucel (Kymriah), axicabtagene ciloleucel (Yescarta) and brexucabtagene autoleucel (Tecartus). Only tisagenlecelucel and axicabtagene ciloleucel are reimbursed, while brexucabtagene autoleucel has been recommended for funding.

TABLE 13C.1
Biological Standards Relevant to Cell Therapy Products

Standard	Link
Therapeutic Goods (Standard for biologicals—labelling requirements) (TGO 107) Order 2021	https://www.legislation.gov.au/Details/F2021L01325
Therapeutic Goods (Standard for human cell and tissue products—donor screening requirements) (TGO 108) Order 2021	https://www.legislation.gov.au/Details/F2021L01326
Therapeutic Goods (Standards for biologicals—general and specific requirements) (TGO 109) Order 2021	https://www.legislation.gov.au/Details/F2021L01332

REGULATION OF GENETICALLY MODIFIED ORGANISMS

In Australia, genetically modified organisms (GMOs) are regulated under the *Gene Technology Act, 2000 (Cth)*. The Act establishes a statutory officer, the gene technology regulator, and the Office of the Gene Technology Regulator (OGTR) to oversee the operation of the Act and make decisions regarding the licensing of dealings with GMOs in Australia. It defines what is included as a GMO under the Act and what 'dealings' are regulated under the Act. The Act is supported by the *Gene Technology Regulations 2001 (Cth)* and various other documents available on the OGTR website (https://www.ogtr.gov.au/). Due to limitations under the Australian Constitution, the gene technology regulatory scheme is complex with the federal Act mirrored in various State and Territory legislation to give it full effect. The aim of the Act is to identify risks posed by or as a result of gene technology and managing those risks appropriately. The Act is primarily concerned with the risks to people handling the GMO or to the environment.

All dealings with GMOs require approval, which is a three-step process. The first step is obtaining accreditation for the organisation that will be dealing with the GMO. In order to obtain accreditation, the organisation must have access to an Institutional Biosafety Committee (IBC). They may establish their own IBC or enter into an agreement to utilise an existing IBC. Facilities where the GMO will be handled may also require certification by the OGTR.

There are different levels of approvals for various types of dealing with GMOs, but once you have obtained your approval, you may then proceed with your dealing with the GMO under the conditions granted by the gene technology regulator. The organisation will also need to comply with any enforcement or compliance requests by OGTR and may be subject to inspections by the OGTR.

Dealings that require a license from the gene technology regulator are:

- Dealings not involving intentional release (DNIR)
- Dealings involving intentional release (DIR)
- Inadvertent dealings

Some GMOs may be exempt under the Act. Other dealings, such as in controlled laboratory research settings, are considered notifiable low-risk dealings and are notified to the OGTR once assessed by an IBC. Some dealings may no longer require a license but are entered on the GMO Register if they are deemed sufficiently safe to be undertaken by anyone without the need for oversight. Emergency dealing determinations may be granted by the regulator for a six-month period in an emergency situation.

Most dealings for commercial release or clinical trial use of GMO therapeutics will require a DNIR or DIR license. In either case, applicants are strongly urged to discuss their application with the OGTR before deciding on the type of license required.

For a DNIR license, the applicant should submit their application containing the information required by the OGTR to their IBC, who will review the application and provide their assessment to the OGTR, along with the application. Once received by the OGTR, the office will review the application and prepare a risk assessment and risk management plan for the application. OGTR may seek further advice from their Advisory Committee before making their final decision and notifying the applicant. The decision timeframe is 90 working days from receipt of the application by OGTR. The record of the decision is listed on the public GMO Record (https://www.ogtr.gov.au/what-weve-approved).

For a DIR license, the steps are similar. However, once the risk assessment and risk management plan (RARMP) has been prepared, the regulatory opens it for public consultation. The Advisory Committee, prescribed agencies, other regulatory authorities and the public may comment on the RARMP. Once feedback has been reviewed, the regulator will make their final decision and notify the applicant. The timeframes for DIR licenses range from 150 working days for limited and controlled release applications to 255 working days where release is not limited or controlled. Again, the decision is listed on the public GMO Record. There are no accelerated pathways available under the Gene Technology Act.

Gene therapies that have received licenses under this system include voretigene neparvovec (Luxturna) and onasemnogene abeparvovec (Zolgensma). In both cases, since the therapy is contained, the sponsor has been granted DNIR licenses. Human cell therapies containing genetically modified human cells are exempted under this Act.

NEW ZEALAND

New Zealand (NZ) has a tax-funded public health system which is supplemented by private health insurance. The Ministry of Health oversees the health and disability system. The primary regulatory agency for human therapeutics is the NZ Medicines and Medical Devices Safety Authority (Medsafe; https://www.medsafe.govt.nz/). The regulation of GMOs in NZ is by the Environmental Protection Agency (EPA). NZ has active research in cell and gene therapies and the Genetic Technology Advisory Committee reviews clinical trial applications involving the introduction of nucleic acids or GMOs into humans (https://www.hrc.govt.nz/resources/genetic-technology-advisory-committee-approval).

Medsafe operates under the *Medicines Act, 1981* and the *Medicines Regulations, 1984*, which are available from the NZ legislation website (https://www.legislation.govt.nz/). The *Current Guidelines on the Regulation of Therapeutic Products in New Zealand* should be followed by sponsors wishing to register cell and gene therapy products in NZ (https://www.medsafe.govt.nz/regulatory/current-guidelines.asp). Cell and gene therapy products are classified as high-risk products, with a standard review timeline of 200 calendar days. Data may be submitted in eCTD format. If full unredacted evaluation reports are available from recognised overseas regulatory authorities, then the product may undergo an abbreviated approval process.

GMO therapeutics are captured by the *Hazardous Substances and New Organisms Act, 1996*, which is administered by the EPA. GMOs reviewed under this Act include a vaccine for yellow fever, which has an unconditional release for use and experimental CAR-T cells for a clinical trial (conditional release). Sponsors should factor in this additional pathway if considering the introduction of GMO therapeutics in NZ. While the actual application process under the HSNO Act is relatively short, the pre-application phase to determine the appropriate pathway for the application can take several months. This is done in consultation with the EPA.

REFERENCES

Australia New Zealand Clinical Trials Registry (ANZCTR) https://www.anzctr.org.au

Australia's regenerative medicine clinical trials database – https://www.ausbiotech.org/programs/regenerative-medicines-catalyst-project

Australian code of GMP for human blood and blood components, human tissues and human cellular therapy (https://www.tga.gov.au/publication/australian-code-good-manufacturing-practice-human-blood-and-blood-components-human-tissues-and-human-cellular-therapy-products

Australian Regulatory Guidelines for Biologicals (ARGB) https://www.tga.gov.au/publication/australian-regulatory-guidelines-biologicals-argb

Australian Regulatory Guidelines for Prescription Medicines (ARGPM) https://www.tga.gov.au/publication/australian-regulatory-guidelines-prescription-medicines-argpm

Federal register of legislation (https://www.legislation.gov.au/)

Genetic technology advisory committee approval forms https://www.hrc.govt.nz/resources/genetic-technology-advisory-committee-approval

GMO Record https://www.ogtr.gov.au/what-weve-approved

Medsafe https://www.medsafe.govt.nz/

Medsafe current regulatory guidelines https://www.medsafe.govt.nz/regulatory/current-guidelines.asp

National statement on ethical conduct in human research (updated in 2018) https://www.nhmrc.gov.au/about-us/publications/national-statement-ethical-conduct-human-research-2007-updated-2018

NZ legislation website (https://www.legislation.govt.nz/

OGTR website https://www.ogtr.gov.au/

O'Sullivan, G. M., et al. (2019). Cell and gene therapy manufacturing capabilities in Australia and New Zealand. *Cytotherapy* 2019;21(12), 1258–1273. https://doi.org/10.1016/j.jcyt.2019.10.010.

Priority review pathway for biologicals consultation. https://consultations.tga.gov.au/medicines-regulation-division/priority-review-pathway-for-biologicals/

TGA biologicals dossier structure (https://www.tga.gov.au/book-page/dossier-structure#tga

TGA biovigilance responsibilities of sponsors https://www.tga.gov.au/publication/biovigilance-responsibilities-sponsors-biologicals

TGA priority determination process https://www.tga.gov.au/publication/priority-determination

TGA priority registration process https://www.tga.gov.au/publication/priority-registration-process

TGA provisional process determinationhttps://www.tga.gov.au/publication/provisional-determination

13d1 Regulatory Landscape in Singapore

Stefanie Fasshauer
PharmaLex, Peng Chau, Hong Kong

CONTENTS

Introduction	244
CTGTPs in Singapore	244
Definition	244
Classification	245
Legislation	245
Regulatory Review	245
Product Consultation	245
Chemistry, Manufacturing and Controls	246
(Non-)clinical Requirements	246
Regulatory Pathway	246
Guidance Documents	246
Regulatory Registration Requirements and Process	247
Applicant	247
Application Route	247
Evaluation Route	248
Documentation	248
Application Process	248
Life Cycle Management – Variation Application (of a Registered CTGTP)	249
Fees and Turnaround Time	250
Clinical Trials	251
GMP Inspection of Manufacturers	252
Import, Manufacture and Wholesale of CTGTP	252
Dealer's Licensing or Notification and Certification	252
Advertisement and Sales Promotion	253
Adverse Events Reporting	253
Product Defect Reporting	253
References	253

DOI: 10.1201/9781003285069-16

INTRODUCTION

On 1 March 2021, Singapore introduced a new regulation for cell and tissue-based therapeutics and gene therapy products. Regulated as biological products in the past, Singapore introduced in the first schedule to the *Health Products Act* (HPA) a new and stand-alone health product category, so-called Cell, Tissue and Gene Therapy Products (CTGTPs). The *HPA* (Cap. 122D) and its regulations regulate all health products and provide the legislative basis for interested parties to manufacture, import, supply or advertise health products in Singapore. In recognition of the newly implemented category, Singapore issued the Health Products (Cell, Tissue and Gene Therapy Products) Regulations 2021. Followed by several guidance documents available on the Health Sciences Authority (HSA) website to assist CTGTP manufacturers, importers, suppliers and registrants to meet local standards before placing these products on the market.

This chapter will introduce the newly established regulations for CTGTPs and implications for product registration in Singapore.

CTGTPs IN SINGAPORE

DEFINITION

According to the first schedule of the *HPA*, a CTGTP means any substance that is intended for use by and in humans for a therapeutic, preventive, palliative or diagnostic purpose. CTGTPs may have any component or combination of viable or non-viable human cells or tissues, viable animal cells or tissues or recombinant nucleic acids. A CTGTP achieves its primary intended action by pharmacological, immunological, physiological, metabolic or physical means leading to its intended use (HPA 2007, 2020).

CTGTPs are not any of the following products:

- A recombinant vaccine for a preventive purpose (considered therapeutic products)
- An in-vitro diagnostic product
- Bone marrow, peripheral blood or umbilical or placental cord blood from a human that is minimally manipulated and intended for homologous use
- Cells and tissues obtained from a patient that are minimally manipulated and re-implanted for homologous use into the same patient during the same surgical procedure
- Organs and tissues that are minimally manipulated and intended for transplant
- Reproductive cells (sperm, eggs) and embryos intended for assisted reproduction
- Whole blood or any blood component that is minimally manipulated and intended for treating blood loss or blood disorders (HPA 2007, 2020)

Classification

According to the Health Products (Cell, Tissue and Gene Therapy Products) Regulations 2021, CTGTPs are risk-stratified into two classes:

1. Class I (lower risk) means a CTGT product, which must satisfy ALL of the following
   Class I CTGTP criteria:
   - the result of only minimal manipulation of human cell or tissue
   - intended for homologous use
   - not combined or used with a product categorised as therapeutic products or medical devices
   - a product assigned by the authority as Class I CTGTP due to a lower health risk to a user
2. Class II (higher risk) means any CTGTPs that do not meet the criteria for Class I CTGTPs (HPA) (Chapter 122D).

Legislation

In Singapore, all health products are regulated under the *HPA* and its regulations, including the Health Products (Cell, Tissue and Gene Therapy Products) Regulations 2021. Interested stakeholders can find and are advised to check on the HSA website the various guidance documents with relevant forms and templates to assist with:

- Product registration and post-approval variation
- Dealer's licensing and certification
- Safety monitoring and product recall
- Clinical trials

Reference to well-known and relevant international guidelines and standards is usually accepted. HSA has implemented guidelines and standards of, for example, the International Conference on Harmonisation (ICH), Pharmaceutical Inspection Convention and Pharmaceutical Inspection Co-operation Scheme (PIC/S) or specific guidelines by HSA reference authorities.

REGULATORY REVIEW

PRODUCT CONSULTATION

HSA established an Innovation Office (as a pilot) allowing stakeholders to seek early scientific and regulatory advice on CTGTP development. This support or advice includes (but is not limited to) (non-)clinical and quality development, Good Manufacturing Practice (GMP) consultation and regulatory environment. Interested parties can email the Innovation Office with queries or a request to meet, stating a description of the company, product, scientific questions and any background information (Health Science Authority (HAS), 2021a).

CHEMISTRY, MANUFACTURING AND CONTROLS

The Module 3 (ICH) or Part II (ASEAN CTD) – Chemistry, Manufacturing and Controls (CMC) dossier part – details all information on CTGTPs manufacturing and quality controls, including the origin of the starting materials, quality of reagents, manufacturing process and process controls, analytical methods and stability studies. The *Guidance on Cell, Tissue and Gene Therapy Products Registration in Singapore – Appendix 8 Chemistry, Manufacturing and Controls Requirements for Cell, Tissue or Gene Therapy Products for Clinical Trials and Product Registration* provides HSA expectation and clarity on the CMC requirements to be included in the respective sections in the quality dossier section, relevant for CTGTP clinical trials and product registration applications. The guidance document lists the requirements from upstream cells/tissue procurement, testing to downstream product manufacturing, analytical methods and release to the clinical site to assure that product safety, quality and efficacy are maintained throughout the complete product life cycle. The CMC information should reflect all involved manufacturers and testing laboratories for clinical trials and product registration. The ultimate goal is to reflect the ability of each involved manufacturer and/or testing site to consistently manufacture, analyse and store the finished product as stated and approved in the CTGTP dossier (Health Science Authority (HAS), 2021b).

(NON-)CLINICAL REQUIREMENTS

In Module 4 of the ICH CTD or Part III of the Association of Southeast Asian Nations Common Technical Dossier (ASEAN CTD or ACTD), all related CTGTP non-clinical information should be presented. Relevant stakeholders should consult the ICH CTD Guidelines M4S (Safety) technical guidelines or the ACTD Part III: Nonclinical Guidelines for further guidance. The clinical documents relate to Module 5 of the ICH CTD or Part IV of the ACTD; all relevant clinical information should be presented. Further guidance is provided in the ICH CTD Guideline M4E (Efficacy) technical guidelines, specifically the ICH E3 guidance document on *Structure and Contents of Clinical Study Reports*, or the ACTD Part IV: Clinical Guidelines. The new drug application should present the clinical studies conducted with the CTGT product as submitted and the patient population for the indication(s) and/or dosing regimen(s) as requested in the application (Health Science Authority (HAS), 2021c). A risk management plan (RMPs) in support of all new CTGTP applications is mandatory. Appendix 9 Guideline on the submission of RMPs documents provides further information (Health Science Authority (HAS), 2021d).

REGULATORY PATHWAY

GUIDANCE DOCUMENTS

HSA published various guidance documents, templates, forms and checklists on its website to assist manufacturers, importers, suppliers or applicants to meet regulatory requirements and local standards before placing CTGTP on the market. Guidance documents include, but are not limited to:

- Main guidance document and appendices for product registration and variation application: Guidance on Cell Tissue and Gene Therapy Products Registration in Singapore
- Good Distribution Practice (GDP) and GMP
- Post-marketing Vigilance Requirements and Reporting product defects and recall for CTGTP
- Guidance documents for clinical trials assisting clinical conducting and submission of clinical trials, Good Clinical Practice (GCP), investigational product and other guidelines

REGULATORY REGISTRATION REQUIREMENTS AND PROCESS

Applicant

In Singapore, marketing authorisation applications for all health products must be submitted by a locally registered company (Health Science Authority (HAS), 2021c).

APPLICATION ROUTE

A Class I CTGTP requires a notification and written acceptance of the notification by HSA before placing the product on the market. In contrast, a Class II CTGTP must be registered and approved with/by HSA. The applicant may choose from one of the three new drug application (NDA) routes:

• NDA 1	The product has not been registered in Singapore before.
• NDA 2	First strength of a CTGTP as a new combination of approved CTGTPs or approved CTGTP in a new dosage form, new presentation, formulation, new administration route or products that do not fall in category NDA 1 or NDA 2.
• NDA 3	Subsequent strengths of approved or submitted CTGTP via NDA-1 or NDA-2.

The application route determines the required application documentation, turn-around time and fees (Health Science Authority (HAS), 2021c).

HSA allows in specific cases to import or supply Class II CTGTP via the Special Access Routes (SAR). This exemption includes the import and supply

- of an unregistered Class II CTGTP at the request of a qualified practitioner for the use on its patient, and
- of a registered Class II CTGTP on consignment basis by companies that are neither registrants nor authorised by the registrant.

Both exemptions need prior approval by HAS (Special Access Routes (SAR), 2022).

Evaluation Route

An NDA application for a new Class II CTGTP may be submitted via the full evaluation route, unless the product has been approved by at least one comparable overseas regulatory authority, such as Australia's Therapeutic Goods Administration (TGA), Health Canada, US Food and Drug Administration (US FDA), European Medicines Agency (EMA) or the UK's Medicines and Healthcare Product Regulatory Agency (MHRA). In this event, the application might be eligible for an abridged evaluation route (Health Science Authority (HAS), 2021c).

Documentation

A company seeking to supply a Class I CTGTP in Singapore must submit an electronic notification via dedicated HSA notification route, including the following documents (in English) as coloured softcopy:

- Administrative documents
  - Table of content, cover letter and FormSG application form
  - List of Class I CTGTP
  - Valid certificate of accreditation (certified true copy)
  - Evidence demonstrating that the establishment is registered with local regulatory agency
- Product specific documents
- Release specification or certificate of analysis
- Package insert and product label
- Product shelf life and container closure information (Health Science Authority (HAS), 2021e)

The application dossier for a Class II CTGTP requires the submission of Modules 1–5 (in English) in either ICH CTD or ASEAN CTD format, as illustrated in Table 13d1.1. The choice of format determines the future product life cycle submission format (Health Science Authority (HAS). 2021c). CTGTP product labels, package insert (PI) and/or patient information leaflet (PIL) must be submitted in English to HAS (Health Science Authority (HAS), 2021f).

Application Process

As illustrated in Figure 13d1.1, after submission, HSA screens the application dossier for completeness and evaluates submitted Modules 1–5 for a Class II CTGTP. HSA notifies the applicant in the event of deficiencies and the stop clock starts. Any deficiencies or requests for additional information need to be resolved within process-specific defined working days. The stop clock ends upon receipt of a satisfactory response. The application evaluation is completed when a regulatory decision has been concluded and issued to the applicant. A positive outcome results in a listing of the product in the list of Class I and Class II CTGTPs, respectively (Health Science Authority (HAS), 2021c).

TABLE 13D1.1
Class II CTGTP, Modules 1–5 ICH CTD versus Part I–IV ASEAN CTD

Documents	Located in the CTD		Sections required for	
	ICH CTD	ACTD	Full Evaluation	Abridged Evaluation
Administrative Documents	Module 1	Part I	Yes	Yes
Overview and Summaries	Module 2	Incorporated in Parts II, III and IV	Yes	Yes
Quality	Module 3	Part II	Yes, complete quality documents for active substance and finished product	Yes, complete quality documents for active substance and finished product
Non-clinical	Module 4	Part III	Yes, complete pharmaco-toxicological or non-clinical documents	Non-clinical overview
Clinical	Module 5	Part IV	Yes, complete clinical documents, i.e., all study reports from phases I to III, including tables and appendices	Clinical overview Study report(s) of pivotal studies and synopses of all studies (phases I–IV) relevant to requested indication, dosing and/or patient group

Source: Health Sciences Authority, Singapore

LIFE CYCLE MANAGEMENT – VARIATION APPLICATION (OF A REGISTERED CTGTP)

Throughout the life cycle of a product, it may be required to apply for changes to an approved product. The change application shall be submitted electronically in the same format as used for the original application.

For a Class I CTGTP, a product variation application may include the accreditation status of the tissue establishments, product label and shelf life. The applicant will be informed by HSA of the outcome within 14 working days.

For a Class II CTGTP, a product variation application can be classified into one of three types of applications:

1. MAV-1: Major Variation-1
   - Change(s) to the approved indication, dosing regimen or patient group(s)

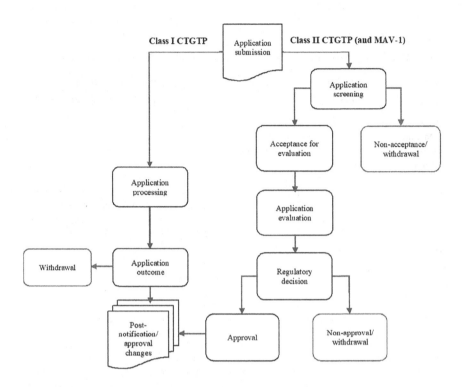

FIGURE 13D1.1 Application process of Class I/II CTGTP (and MAV-1).
Source: Health Sciences Authority, Singapore.

- Inclusion of clinical information extending the usage of the product
- Prior approval is required before implementation
2. MIV-1: Minor Variation-1
    - Any variation that is specified under Part A of Appendix 10 Guideline on Minor Variation (MIV) Application for Class II cell, tissue or gene therapy products
    - Prior approval is required before implementation
3. MIV-2: Minor Variation-2
    - Any variation that is specified under Part B of Appendix 10 Guideline on Minor Variation (MIV) Application for Class II cell, tissue or gene therapy products
    - Change(s) may be implemented within 40 days upon application submission (Health Science Authority (HAS), 2021c)

FEES AND TURNAROUND TIME

HSA published an overview of fees and turnaround time on its website. All concerned stakeholders are advised to check the fees overview, as the authority reserves the right to change the fees from time to time.

TABLE 13D1.2
Turnaround Time Class II CTGTP NDA and Variation Applications in Working Days

Turnaround Time		Abridged Evaluation	Full Evaluation
NDA/MAV-1	Screening	50	50
NDA/MAV-1	Evaluation	180	270
MIV-1		120	
MIV-2		40	

Source: Health Sciences Authority, Singapore.

The target turnaround time for a Class I CTGTP notification is 14 working days. Table 13d1.2 lists further information on target turnaround times for a Class II CTGTP NDA and the different variation application procedures (Health Science Authority (HAS), 2021g).

CLINICAL TRIALS

All clinical trials of a Class II CTGTP are regulated by has, except for observational clinical trials. Class I CTGTP trials are not regulated by HSA but must comply with the requirements of the *Human Biomedical Research Act*. Table 13d1.3 illustrates the key regulations and submission routes of a Class II CTGTP.

Clinical trials must comply with the protocol, *Clinical Trial Regulations*, GCP and applicable Standard Operating Procedures, including adverse events reporting.

With a view towards starting a clinical trial, the local sponsor has to apply either for a Clinical Trial Authorisation (CTA) or submit a Clinical Trial

TABLE 13D1.3
Key Regulations and Submission Routes of a Class II CTGTP

Requirements	Submission Route		Regulation
	CTA	CTN	*HPA*,
Conditions	• One or more locally unregistered CTGTPs, or • Unapproved use of a locally registered CTGTP	• Locally registered CTGTPs used according to local approved labels	*Health Products (Clinical Trials) Regulations*
Turnaround time	60 working days	5 working days	
Outcome	Authorisation	Acceptance of Notification	

Source: Health Sciences Authority, Singapore.

Notification (CTN), respectively. The application should be endorsed by relevant stakeholders before online submission through the Pharmaceutical Regulatory Information System (PRISM). CTN is submitted after Institutional Review Board (IRB) approval, while CTA may be initiated in parallel with the IRB. No fees are involved for clinical trial applications.

The local sponsors may be required to submit online via PRISM substantial amendments, updates or reports to HSA after obtaining a CTA or CTN. No fees are involved for such submissions (Health Science Authority (HAS), 2021h).

Clinical trial sponsor must report unexpected serious adverse drug reactions (USADR) not fatal or life-threatening no later than 15 calendar days; all others should be reported immediately but no later than 7 days after first knowledge by the sponsor to the respective authority (Health Science Authority (HAS), 2021i).

GMP INSPECTION OF MANUFACTURERS

Overseas CTGTP manufacturers may be subject to a GMP conformity assessment by HSA, following one of the two assessment routes:

4. Manufacturers audited by at least one PIC/S member authority may submit a valid GMP certificate, for evaluation only. An on-site GMP audit may be skipped if the evidence is found to be acceptable.
5. On-site GMP audit in case the overseas manufacturer does not fulfil the criteria will be subject to a GMP audit to assess compliance according to the *HSA Guidelines on Good Manufacturing Practice for CTGTP* (Health Science Authority (HAS), 2022a).

IMPORT, MANUFACTURE AND WHOLESALE OF CTGTP

Dealer's Licensing or Notification and Certification

Every manufacturer, importer and wholesaler of CTGTP is subject to certain regulatory control depending on the degree of manipulation of the CTGTP they are dealing with.

Companies dealing with minimally manipulated CTGTP do not require licenses but must notify HSA about their intentions prior to the start of any activity. Companies must ensure compliance with their duties and obligations according to the CTGTP regulations.

Companies dealing with CTGTPs not minimally manipulated must obtain the appropriate licenses depending on the use of the CTGTPs, which could range from manufacture, import and wholesale for marketing; clinical use or research; scientific education; or research and development. The license holder must appoint one or more responsible persons to fulfil all requirements and duties as imposed by the terms of the respective license (Health Science Authority (HAS), 2019).

Advertisement and Sales Promotion

A company marketing a CTGTP in Singapore does not require prior approval by HSA for its advertisements. Advertisements have to comply with the principles and requirements as stipulated in the *HPA* and the *Health Products (Advertisement of Specified Health Products) Regulations* (Health Science Authority (HAS), 2022b).

Adverse Events Reporting

Importers, manufacturers, suppliers or registrants of CTGTPs are obliged to report all serious adverse events (AEs), unless AEs are not suspected of being product-related by the healthcare professional and the company has reasons to suspect a causal association. AEs are submitted electronically via the Council for International Organisations of Medical Sciences (CIOMS) I form as stipulated in the legislation as soon as possible but not later than 15 calendar days (Health Science Authority (HAS), 2021j).

Product Defect Reporting

Any product defect that may have an impact on the safety, quality and efficacy of a CTGTP must be reported to HSA. Hence, companies dealing with CTGTP should have systems and processes in place to investigate, review and report any defects to HSA to ensure the safety, quality and efficacy of its products and recall affected products, if deemed necessary.

Product defects are reported electronically via the Product Defect Reporting Form within 48 hours (excluding Sundays and public holidays) for critical defects or within 15 calendar days for non-critical defects. Depending on the potential risk, HSA may require additional risk control measures, such as (but not limited to) a product recall or Dear Healthcare Professional Letter (Health Science Authority (HAS), 2021k).

REFERENCES

Health Products Act (HPA) 2007, 2020 revised edition. Last updated March 2022. Singapore Statutes Online. https://sso.agc.gov.sg/Act/HPA2007?WholeDoc=1

Health Products Act (Chapter 122D), Health products (cell, tissue and gene therapy products regulations 2021). Last updated March 2022. Singapore Statutes Online. https://sso.agc.gov.sg/SL/HPA2007-S104-2021

Health Science Authority (HAS). (2021a). Cell, tissue and gene therapy products. Last updated 2021. HSA website. https://www.hsa.gov.sg/ctgtp

Health Science Authority (HAS). (2021b). Appendix 8. Chemistry, manufacturing and controls requirements for cell, tissue or gene therapy products for clinical trials and product registration. March 2021. Final. Last updated March 2021. HSA website. https://www.hsa.gov.sg/docs/default-source/hprg-atpb/appendices/appendix-8-chemistry-manufacturing-and-controls-requirements-for-cell-tissue-or-gene-therapy-product-for-clinical-trials-and-product-registration.pdf

Health Science Authority (HAS). (2021c). Guidance on cell, tissue and gene therapy products registration in Singapore. March 2021. Final. Last updated March 2021. HSA website. https://www.hsa.gov.sg/docs/default-source/hprg-atpb/guidance-documents/guidance-on-cell-tissue-and-gene-therapy-products-registration-in-singapore.pdf

Health Science Authority (HAS). (2021d). Appendix 9 Guideline on the submission of risk management plan documents. March 2021. Final. Last updated March 2021. HSA Website. https://www.hsa.gov.sg/docs/default-source/hprg-atpb/appendices/appendix-9-guideline-on-the-submission-of-risk-managemnet-plan-documents.pdf

Health Science Authority (HAS). (2021e). Appendix 1 Guideline on notification process for class i cell, tissue or gene therapy product. March 2021. Final. Last updated March 2021. HSA website. https://www.hsa.gov.sg/docs/default-source/hprg-atpb/appendices/appendix-1-guideline-on-notification-process-for-class-1-cell-tissue-or-gene-therapy-product.pdf

Health Science Authority (HAS). (2021f). Appendix 7 Points to consider for class II CTGTP labelling. March 2021. Final. Last updated March 2021. HSA website. https://www.hsa.gov.sg/docs/default-source/hprg-atpb/appendices/appendix-7-points-to-consider-for-class-2-ctgtp-labelling.pdf

Health Science Authority (HAS). (2021g). Fees and turnaround time for CTGTP. Last updated 2021. HSA website. https://www.hsa.gov.sg/ctgtp/fees-and-turnaround-time

Health Science Authority (HAS). (2021h). Clinical trials guidance. Regulatory requirements for new applications and subsequent submissions. March 2021. Final. Last updated March 2021. HSA Website. https://www.hsa.gov.sg/docs/default-source/hprg-io-ctb/hsa_gn-ioctb-04_new_and_subsequent_appl_28apr2021.pdf

Health Science Authority (HAS). (2021i). Clinical trials guidance. Expedited safety reporting requirements for clinical trials. March 2021. Final. Last updated May 2017. HSA website. https://www.hsa.gov.sg/docs/default-source/hprg-io-ctb/hsa_gn-ioctb-10_safety_reporting_1mar2021.pdf

Health Science Authority (HAS). (2022a). Good manufacturing practice (GMP) conformity assessment of overseas manufacturers of CTGTP. Last updated 2022. HSA website. https://www.hsa.gov.sg/ctgtp/registration/gmp-comformity-assessment

Health Science Authority (HAS). (2019). Dealer's licensing and certification. Last updated 2019. HSA website. https://www.hsa.gov.sg/ctgtp/dealers-licence

Health Science Authority (HAS). (2022b). Advertisements and promotions of cell, tissue or gene therapy products (CTGTP). Last updated 2022. HSA website. https://www.hsa.gov.sg/ctgtp/advertisements

Health Science Authority (HAS). (2021j). Adverse event reporting of cell, tissue or gene therapy products (CTGTP). Last updated 2021. HSA website. https://www.hsa.gov.sg/ctgtp/adverse-events

Health Science Authority (HAS). (2021k). Product defect reporting and recall procedures for cell, tissue and gene therapy products (CTGTP). Last updated 2021. HSA website. https://www.hsa.gov.sg/ctgtp/defect-and-recalls. All URLs were accessed on 6th March 2022.

Special Access Routes (SAR). Last updated 2022. HSA website. https://www.hsa.gov.sg/ctgtp/registration/sar

13d2 Regulatory Landscape in Malaysia

Stefanie Fasshauer
PharmaLex, Peng Chau, Hong Kong

CONTENTS

Introduction	256
Regulatory Pathway for CTGTPs	256
Definition	256
Classification	257
Legislation	257
Regulatory Review	258
Pre-submission Meeting (PSM)	258
Chemistry, Manufacturing and Controls	258
(Non-)clinical Requirements	259
Regulatory Pathway	259
Guidance Documents	259
Regulatory Registration Requirements and Process	259
Applicant	259
Application Route	260
Evaluation Route	260
Documentation	261
Application Process	262
Lifecycle Management – Variation Application (of a Registered Class II CGTP)	264
Fees and Turnaround Time	264
Clinical Trials	264
GMP Inspection of Manufacturers	266
Import, Manufacture and Wholesale of CGTP	266
Dealer's Licencing or Notification and Certification	266
Advertisement and Sales Promotion	266
Adverse Events Reporting	267
Product Defect Reporting and Recalls	267
References	267

DOI: 10.1201/9781003285069-17

INTRODUCTION

In Malaysia, pharmaceutical products are governed by the National Pharmaceutical Regulatory Agency (NPRA), respectively its Drug Control Authority (DCA). Key legislations for drugs include, among others, the Sale of Drugs Act 1952: Control of Drugs and Cosmetics Regulations (CDCR) 1984 [P.U.(A) 223/84]. Cell and Gene Therapy Products (CGTPs) are classified as biologics; even so, CGTPs are regulated under a separate framework. In 2016, Malaysia published the Guidance Document and Guidelines for Registration of Cell and Gene Therapy Products, which became compulsory on 1 January 2021. The guidance document assists CGTP applicants, manufacturers and importers to comply with governing acts and regulations, including the regulatory framework and requirements. The governing body for CGTP is the NPRA.

This chapter will provide an overview of established regulations for CGTPs and implications for product registration in Malaysia.

REGULATORY PATHWAY FOR CTGTPs

Definition

CGTPs are classified as medicinal products with the purpose to treat or prevent diseases in humans, or used in or administered to humans to restore, correct or modify physiological functions by exerting principally pharmacological, immunological or metabolic action.

The CGTP Guidance Document provides direction for somatic cell therapy, tissue engineering and gene therapy products with the following characteristics:

- Contains viable human cells of allogeneic or autologous origin undergoing a manufacturing process (ESCs, iPSCs, HSCs, PSCs)
- May be combined with non-cellular components
- Cells may be genetically modified

The regulatory CGTP framework is broadly divided into three parts: cell therapy, xenotransplantation and gene therapy (NPRA, 2022a).

The framework includes:

- Human stem cells, tissue and cellular therapy products
- Genetically modified cellular products
- Cell-based cancer vaccines and cell-based immunotherapies
- Dendritic cells, lymphocyte-based therapies, cell-based therapies for cancer, peptides, proteins (NPRA, 2022a)

The framework does not include:

- Fresh viable or parts of human organs, for direct donor-to-host transplantation
- Fresh viable human haematopoietic stem/progenitor cells for direct donor-to-host transplantation for the purpose of haematopoietic reconstitution

Regulatory Landscape in Malaysia

- Labile (fresh) blood and blood components
- Unprocessed reproductive tissues
- Secreted or extracted human products
- Samples of human cells or tissues that are solely for diagnostic purposes in the same individual
- *In vitro* diagnostic devices (NPRA, 2022a)

The inclusion and exclusion list are not absolute but may be amended as required. Applicable national standards or guidelines may have been implemented for products not falling within the scope.

CLASSIFICATION

Malaysia adopted the risk-based approach, meaning products with a greater risk of adverse clinical outcomes require more and better control, hence a more stringent framework and oversight. Two classes of CTG products have been adopted:

1. Class 1: Lower-risk cell therapy products
   - products are not subject to pre-market approval but must be listed at the practitioner's premises
   - products must meet all four criteria:
     - minimally manipulated
     - intended for homologous use only
     - it is not a combination with another drug/device/article
     - does not have a systemic effect, and for its primary function, it is not dependent upon the metabolic activity of living cells
2. Class 2: Higher-risk cell therapy products
   - does not meet all four criteria of Class I
   - product is "highly processed," used for other than normal function, combined with non-tissue components or used for metabolic purposes
   - regulated as biologic product (NPRA, 2022a)

LEGISLATION

In Malaysia, pharmaceutical products are governed by the NPRA and regulated under the Sale of Drugs Act 1952: Control of Drugs and Cosmetic Regulations 1984 and associated legislations. NPRA published on its website several guidance documents to assist interested stakeholders along the complete registration process, including (but not limited to) registration activities and requirements, quality control or licensing. One of the main guidance documents is the Drug Registration Guidance Document (DRGD). Stakeholders interested in developing or marketing a CGTP should consult applicable guidance documents on biologics, specifically the Guidance Document and Guidelines for Registration of Cell and Gene Therapy Products. The CGTP guidance documents were developed and modelled based on established international regulatory frameworks and benchmark authorities, including International Conference on Harmonisation (ICH), World Health Organization (WHO), European Medicines Agency (EMA), United States Food and Drug Administration (US FDA), Therapeutic Goods Administration (TGA), etc. (NPRA, 2022b)

REGULATORY REVIEW

PRE-SUBMISSION MEETING (PSM)

The NPRA offers all applicants a PSM to provide regulatory advice, including quality, safety and efficacy aspects prior to submission of an application. The PSM entails a formal meeting between the applicant and NPRA to request feedback during the developmental stages of a product. The PSM is not mandatory and can be used if technical or scientific issues or questions cannot be addressed easily by other means (NPRA, 2020a).

CHEMISTRY, MANUFACTURING AND CONTROLS

As described in *Annex 2. Manufacture of Biological Medicinal Substances and Products for Human Use, PIC/S, Current Version*, it is crucial to develop a suitable quality management system that covers the entire process from collection to transplant (PIC/S, 2019). Once a CTGP, is developed appropriate controlled processes for manufacturing and testing must be in place. The quality of a CTGP depends heavily on its manufacturing process. CTGP manufacturers should follow the principles of current Good Manufacturing Practises (cGMP) and Good Tissue Practices (cGTP).

Since CGTPs contain genetic and other materials of biological origin most of the available quality guidelines for biotechnological products apply. The CMC section of a CGTP product dossier should include the following information:

- **Starting and raw materials** – clear definition of the source, origin and suitability of biological starting and raw materials as appropriate to its stage of manufacture
- **Cell banking system** – sufficient information of the properties of the original cell source to ensure product quality
- **Characterisation** - all components as present in the finished product, including matrices, scaffolds and devices
- **Description of the entire manufacturing process and its validation**
- **Quality control** and specific tests for in-process and product lot release
- **Stability data** – where possible ICH Q5C: stability testing of biotechnological/ biological products should be consulted
- **Description of container closure system**

Regulatory Landscape in Malaysia 259

- **Product traceability** – implementation of a system allowing complete traceability of the patient, including product and starting materials is essential
- **Summary of CMC data requirements** (NPRA, 2022a)

(Non-)Clinical Requirements

Parts III and IV of the ASEAN Common Technical Dossier (ACTD) should present all related CGTP non-clinical and clinical information. Relevant stakeholders should consult the ACTD Part III: Nonclinical Guidelines and the ACTD Part IV: Clinical Guidelines for further guidance. In the drug application, clinical studies should be presented with the CGT product as submitted and the patient population for the indication(s) and/or dosing regimen(s) as requested in the application. A risk management plan (RMPs) and Periodic Safety Update Reports (PSURs) should be submitted in support of all new CGTP applications and updated as relevant cases occur (NPRA, 2022a).

REGULATORY PATHWAY

Guidance Documents

NPRA has published several guidance documents and forms on its website to assist manufacturers, importers and applicants to meet local requirements and standards before placing a CGT product on the market. Guidance documents include, but are not limited to:

- DRGD, particularly Appendix 4 Guideline on Registration of Biologics
- Guidance Document and Guidelines for Registration of Cell and Gene Therapy Products
- Guidelines for Good Clinical Practice, Good Tissue Practice Guideline and Good Manufacturing Practices
- Malaysian Variation Guideline for Pharmaceutical Products
- ACTD for the Registration of Pharmaceuticals for Human Use

The CGTP guidance documents should be read in conjunction with other relevant/applicable well-established, international guidelines.

Regulatory Registration Requirements and Process

Applicant

In Malaysia, the applicant for a product registration, also known as the Product Registration Holder (PRH), must be a local registered company. The PRH is responsible for all quality, safety and efficacy information submitted for a medicinal product to the competent authority (NPRA, 2020a).

Application Route

CGTPs are subject to the same product registration process and requirements as any other biologics.

Class I CGTPs are not subject to pre-market review requirements or approval. These products must be listed at a practitioner's premise. This product class is further regulated by:

- Site/facility licensure and listing by the Medical Practice Division under the purview of the Private Healthcare Facilities and Services Act 1998 (Act 586)
- Donor screening and testing
- Good tissue practices
- Labelling
- Adverse event reporting
- Inspection and enforcement (NPRA, 2022a)

NPRA oversees the product registration of Class II CGTPs, which are categorised and regulated as biologic products. Every biologic product is regulated as a new product, considered 'high risk' and must be evaluated on its own merits. Despite the fact that CGTPs are regulated under a separate framework, applicants are advised to read *Appendix 4: Guideline on Registration of Biologics*. If an applicant is unclear on the product category, a Classification Form can be submitted to NPRA for verification. The risk-based approach should be adopted throughout all stages of a product's life cycle (NPRA, 2020a).

Malaysia established special pathways for CGTPs, such as compassionate use to allow the administration of an unregistered product or to obtain orphan product status. Malaysia published on its website further information for applicants to apply for these special pathways (NPRA, 2022a).

Evaluation Route

Class II CGTP applications shall be submitted as new biological applications via the full evaluation route unless the product qualifies and fulfils the conditions for the priority review. Priority review may be granted in special cases such as for products intended for unmet medical needs or life-saving treatment or emergency supplies.

There are three types of full evaluation routes:

- Standard pathway
- Conditional Registration
  - product is approved by at least one DCA reference authority
  - registration is valid for two years and may be renewed two times
- Abbreviated and Verification Review
  - Abbreviated Review: product is approved by one DCA reference authority
  - Verification Review: product is approved by two DCA reference authority
  - Reference Authorities: EMA and US FDA

Regulatory Landscape in Malaysia

The applicant can apply for the abbreviated or verification route if the product meets applicable requirements, such as, but not limited to

- Application has to be submitted within two years from the date of approval by the primary reference authority
- Manufacturing site holds a valid Good Manufacturing Practise (GMP)/PICs certificate
- All aspects of the drug products quality are identical, an exception is, for example, the container closure system to meet stability requirements
- Proposed indication, dosage regime, patient group or direction for use should be the most stringent among those approved by the reference drug regulatory authority (NPRA, 2022c).

Documentation

Class I CGTP are exempted from the marketing authorisation process. Class II CGTP application dossiers should be organised according to the ACTD format, as illustrated in Table 13d2.1. Due to the nature of CGTPs, manufacturers must understand the unique science behind their products and associated regulations to communicate with authorities. Data and information included in the dossier can be tailored to a specific product and should be backed up with sound data-driven scientific justifications (NPRA, 2022d).

The guidance document on CGTP points out some special considerations, such as

- the potential risk of genetically modified organisms (GMOs) should be evaluated, and
- an environmental risk assessment (ERA) must be provided (NPRA, 2022a).

Considering using the risk-based approach the extent of the quality, non-clinical and clinical data to be included in the application dossier have to be determined.

TABLE 13D2.1

Application Dossier of a Class II CGTP (Full Evaluation) Based on ACTD/ACTR

Dossier Section	Part I Administrative Information	Part II Quality Document	Part III Non-clinical Document	Part IV Clinical Document
Biologics	Yes	Complete CMC data set	Complete pre-clinical data set	Complete clinical data set

Source: DRGD, Malaysia

The analysis usually covers the complete development process. Risk factors to be considered are (but not limited to):

- origin of the cells and their ability to proliferate, differentiate or initiate an immune response
- cell manipulation level
- a combination of cells with bioactive molecules or structural materials
- level of integration of nucleic acid sequences or genes into the genome
- long-term functionality
- risk of oncogenicity
- mode of administration or use
- nature of gene therapy product
- extent of replication competence of viruses or microorganisms used *in vivo* (NPRA, 2022a)

The product label shall comply with the labelling requirements as defined in DRGD *Appendix 19 General Labelling Requirements, Appendix 20 Specific Labelling Requirements* and labelling requirements for immediate and outer packaging as described in the *Guidance Document and Guidelines for Registration of Cell and Gene Therapy Products*.

Application Process

A Class II CGT product registration application shall be submitted to NPRA, including the complete CMC, pre-clinical and clinical data set. The product application dossier is to be submitted via web-based online submissions, namely the QUEST+3 system. All data and information for product registration shall be submitted in English or *Bahasa Malaysia*.

As illustrated in Figure 13d2.1, after online submission, the dossier undergoes a screening and evaluation process. The applicant will be informed via the system of any deficiency and request for additional information, which has to be provided in process-specific timelines.

Hard copies may be provided, for example, for the complete dossier on CD or documents under Part I. Regulatory decision is made based on the submitted dossier. Upon approval, the applicant is granted a product registration number and certificate, and the products get to be listed in the respective CGTP list. Product registrations are usually five years valid or as specified in the authority database (NPRA, 2022a).

In December 2020, Malaysia NPRA issued Directive 19/2020 on Registration and Enforcement of Cell and Gene Therapy Products in Stages. The scope of the directive addresses all CGTPs already on the market starting since 1 January 2020 and before but does not meet regulatory requirements of the full dossier registration.

Malaysia provides some flexibility for CGTPs that meet the scope and pass the two stages in the transition period:

Regulatory Landscape in Malaysia

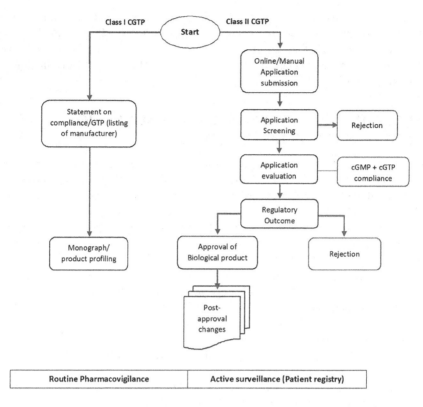

FIGURE 13D2.1 Application process for a class I/II CGTP new drug application

Source: Guidance Document and Guidelines for Registration of Cell and Gene Therapy Products (CGTPs), Malaysia.

1st Stage – Screening: All stakeholders can apply via manual submission if the documents are complete. If the application passes the screening process, the product will be included in the list and the applicant is allowed to market the product till 1 July 2022. At this stage, documents such as the screening form, invoice/sales records, evidence of safety and efficacy, Manufacturer GMP status, label and commitment of safety reporting have to be submitted. The processing timeline at NPRA is 30 working days (no correspondence).

2nd Stage – Screening: requires a manual application submission before 1st July 2022 and allows the applicant to remain with its product on the list and the product can be marketed until 31st December 2024. At this stage, documents such as the respective form, GMP certificate, Certificate of Analysis and release specifications and clinical trial information/data must be submitted. The processing timeline at NPRA is 60 working days (no correspondence).

Consequently, starting from 1 January 2025, Class II CGTPs can only be marketed after the product has been registered with the DCA and has been granted a valid product registration number, MAL.

All CGTPs that already meet full dossier requirements can apply online via Quest3+ for approval. The processing timeline at NPRA is 245 working days (no correspondence). Further information can be found in Directive 19/2020 on Registration and Enforcement of Cell and Gene Therapy Products in Stages (NPRA, 2020b).

LIFECYCLE MANAGEMENT – VARIATION APPLICATION (OF A REGISTERED CLASS II CGTP)

A biologic product will undergo various changes throughout its lifetime. In Malaysia, a variation application for biologics shall be prepared and submitted via the QUEST 3+ system according to the Malaysian Variation Guidelines for Biologics (MVGB).

Changes are categorised as per the Malaysian Variation Guideline for Pharmaceutical Products (MVG) as follows:

- Major Variation (MaVB):
  - change may affect significantly and/or directly the aspects of quality, efficacy and safety
  - change requires approval before implementation
- Minor Variation (MiVB-PA):
  - minimal changes with no or minimal impact on the aspects of efficacy, quality and safety
  - change requires approval before implementation
- Minor Variation (MiVB-N, "Do and Tell"):
  - administrative data and or/ changes with minimal/ no significant impact on the aspects of efficacy, quality and safety
  - PRH must notify NPRA, which shall acknowledge receipt of a notification within 30 working days MVGB (2017)

FEES AND TURNAROUND TIME

NPRA published an overview of fees and turnaround time in the applicable guidance documents. NPRA reserves the right to change its fee structure, thus, applicants are advised to check the respective guidance documents from time to time.

Table 13d2.2 lists further information on target turnaround times for a Class II CGTP full application process and the different variation application procedures (NPRA, 2020a).

CLINICAL TRIALS

A requisite for most CGTPs before submission of an application for marketing authorisation is to conduct clinical trials. Exceptions may include, for example, lower-risk haematological products that classify as minimally manipulated,

TABLE 13D2.2
Turnaround Time Full Evaluation of a Class II CGTP and Variation Application

Category	Type	Evaluation Time*	Maximum Review Period***
Biologics	Full evaluation	245 working days	
	Priority Review/Orphan drug	120 working days**	
MiVB-N	Minor Variation/Do&Tell		30 Working Days
MiVB-PA	Minor Variation		60 Working Days
MaVB	Major Variation		90 Working Days

* Upon processing and analysis fee for product registration payment confirmation
** Starting from the date of approval of priority review
*** Review period starts once all documents have been received
Source: DRGD, Malaysia

perform the same basic function in the donor as the recipient, are not combined with other agents and have no systemic effect. These kinds of products are regulated clinically under the Medical Practice/Development Divisions, Ministry of Health Malaysia.

Prior to investigating the use of a CGT product in a local clinical trial, an application, including data from pre-clinical studies must be filed to NPRA. An unregistered CGTP can be administered to human trial subjects only if the applicant obtained approval for a Clinical Trial Import Licence (CTIL) or a Clinical Trial Exemption (CTX).

Sufficient data demonstrating that the CGT product is safe and effective in humans are required, following cGTP and cGMP requirements. Ahead of the clinical development phase, manufacturing description and pre-clinical pharmacology/toxicology data must sufficiently characterise product quality and safety (NPRA, 2022a).

CGTPs follow the same principles and requirements as other biologic products in Malaysia, while the necessity and extent of studies are considered on a product-specific case-by-case basis. The design of the clinical development plan should follow existing general and specific guidance documents. Applicants are advised to conduct a careful risk-benefit analysis in the context of the specific clinical indication investigated in the study.

Deviations of Phases I to III clinical trial progression has to be justified with the CGTP particularity, pre-clinical studies and previous clinical experience, as well as treated pathology. In the event that only limited guidance exists for new therapeutic applications, applicants are encouraged to seek consultation with the ethics committee and the regulatory authority on the clinical development plan, including the confirmatory studies (NPRA, 2022a).

Malaysia adopted the EMA's Guideline on human cell-based medicinal products and US FDA Preclinical Assessment of Investigational Cellular and Gene Therapy Products.

For pharmacology and toxicology, the recommendations of ICH S6(R1): Preclinical Safety Evaluation of Biotechnology-Derived Pharmaceuticals (Revision 1) (June 2011) should be taken into consideration. Further information on CTIL/CTX requirements can be found in the Malaysian Guideline for Application of CTIL/CTX, current version.

GMP INSPECTION OF MANUFACTURERS

The manufacture of a CTG product should follow the principles of current Good Manufacturing Practices (cGMP). The Pharmaceutical Inspection Convention/ Scheme (PIC/S) cGMP regulations include relevant basic requirements, as well as detailed requirements for specific products. Of relevance to CGTPs are:

- Annex 1: Manufacture of Sterile Medicinal Products
- Annex 2: Manufacture of Biological Medicinal Products for Human Use
- Annex 11: Computerised System
- Annex 13: Manufacture of Investigational Medicinal Products
- Annex 14: Human Blood and Plasma Products
- Annex 15: Qualification and Validation

Inspections of manufacturing facilities are only mandatory for Class II CGTPs, but not for Class I CGTPs. The GMP compliance status is valid for three years (NPRA, 2021). In addition, the principles of current Good Tissue Practices (cGTP) should be applied in the quality assurance of a CGT product (NPRA, 2015).

IMPORT, MANUFACTURE AND WHOLESALE OF CGTP

Dealer's Licencing or Notification and Certification

Licensing of manufacturing facilities, importers and wholesalers of registered products following the principles of GMP and Good Distribution Practise (GDP) are important elements of drug control and pre-requisite for the application for respective licenses. Regulation 12, CDCR 1984 further outlines the licensing process and requirements any company has to fulfil that wants to manufacture, import or wholesale any registered products (NPRA, 2020a).

Advertisement and Sales Promotion

The *Medicines (Advertisement & Sale) Act 1956* regulates the advertisement relating to medical matters and the sale of substances recommended as a medicine. A company marketing a CGT product in Malaysia should follow its principles and requirements and associated guidance documents (Act 290 Medicines (Advertisement and Sales) Act 1956).

ADVERSE EVENTS REPORTING

According to Regulation 28, CDCR 1984, any stakeholder who possesses a registered product shall inform the director of pharmaceutical services of any adverse reaction immediately about any adverse reaction arising from the use of a registered CGT product. This approach is in line with *Regulation 28: Reporting Adverse Reaction under Control of Drugs and Cosmetics Regulations 1984, Sale of Drugs Act 1952 (Amendment 2006)* and *Malaysian Guidelines for the Reporting & Monitoring*. Any local or international safety issues should be reported to the NPRA and comply with safety-related directives by the DCA. The WHO encourages reporting of all Adverse Drug Reactions (ADR) and Adverse Events Following Immunisation (AEFI).

Safety signals observed through observational studies can be tracked via a designed non-randomised observatory trial, an example of such a trial is a patient registry for Class II CGTPs. The design and conduct of a registry follow the *ICH Topic E2E: Pharmacovigilance Planning. Note for Guidance on Planning Pharmacovigilance Activities (ICH, June 2005)* and *Good Pharmacovigilance Practices and Pharmacoepidemiologic Assessment. US FDA, March 2005* may be referred to and the NPRA shall be consulted (NPRA, 2020a).

PRODUCT DEFECT REPORTING AND RECALLS

NPRA should be informed by the licence holder of any product-related quality problems. It is also the responsibility of other stakeholders, such as pharmacists or health professionals to report any product complaints to NPRA using the applicable form. All complaints will be investigated by NPRA and the respective license holder/manufacturing site. The manufacturing site is responsible for taking appropriate corrective and preventive actions. Recall of a product shall only take place after consultation with and receiving information from the authority. The concerned license holder may refer to Chapters 6 and 7 of the *Guidelines on Good Distribution Practice, Third Edition, 1 January 2018* (NPRA, 2020a).

REFERENCES

Act 290 Medicines (Advertisement and Sales) Act. (1956). Last update 1990. https://www.pharmacy.gov.my/v2/en/documents/medicines-advertisement-sale-act-1956-and-regulations.html. All URLs were accessed on 6th March 2022

Malaysian Variation Guideline for Biologics (MVGB). (2017). Last updated. 2017. https://www.npra.gov.my/images/Guidelines_Central/Guidelines_on_Regulatory/2017/MVGB_final_post_DCA_with_editorial_changes.pdf

National Pharmaceutical Regulatory Agency (NPRA). (2015). Good tissue practice guideline. Last updated 2015. https://www.npra.gov.my/index.php/en/guideline-bio/1532-good-tissue-practice-guideline.html

National Pharmaceutical Regulatory Agency (NPRA). (2020a). Makluman Berkaitan Pelaksanaan: Guidance document for pre-submission meeting. Last updated 2020. https://npra.gov.my/easyarticles/images/users/1048/gambar/Pekelililing-PSM_guidance.pdf

National Pharmaceutical Regulatory Agency (NPRA). (2020b). Direktif berkenaan pelaksanaan pendaftaran produk dan penguatkuasaan secara berperingkat bagi produk terapi sel dan gen (CGTPs) serta tambahan senarai produk di luar skop kawalan CGTPs oleh PBKD. Last updated 2020. https://npra.gov.my/easyarticles/images/users/1048/gambar/DIREKTIF.pdf

National Pharmaceutical Regulatory Agency (NPRA). (2021). Cell and gene therapy products (CGTPS) manufacturing facility in Malaysia. Last updated 2021. https://www.npra.gov.my/easyarticles/images/users/1133/Guidance-Note-for-CGTP-Manufacturing-Facility-in-Malaysia_first-edition_July-2021.pdf

National Pharmaceutical Regulatory Agency (NPRA). (2022a). Guidance document and guidelines for registration of cell and gene therapy products (CGTPs) in Malaysia, December 2015. Last updated 2022a. NPRA website. https://www.npra.gov.my/index.php/en/guideline-bio/1527314-guidance-document-and-guidelines-for-registration-of-cell-and-gene-therapy-products-cgtps-in-malaysia-2.html

National Pharmaceutical Regulatory Agency (NPRA). (2022b). Drug registration guidance document (DRGD). 3rd Ed. 2nd Rev. Last updated 2022. NPRA website. https://www.npra.gov.my/index.php/en/component/sppagebuilder/925-drug-registration-guidance-document.html

National Pharmaceutical Regulatory Agency (NPRA). (2022c). Drug registration guidance document (DRGD). Appendix 14 Evaluation routes. Last updated 2022. https://www.npra.gov.my/easyarticles/images/users/1047/drgd/APPENDIX-14--Evaluation-Routes.pdf

National Pharmaceutical Regulatory Agency (NPRA). (2022d). Drug registration guidance document (DRGD). Appendix 15 Requirements for full evaluation and abridged evaluation. Last updated 2022. https://www.npra.gov.my/easyarticles/images/users/1047/drgd/APPENDIX-15--Requirements-for-Full-Evaluation-and-Abridged-Evaluation.pdf

PIC/S. (2019). Annex 2 Manufacture of biological medicinal substances and products for human use, PIC/S, 2019. Last updated 2019. https://picscheme.org/docview/2230

13e Regulatory Landscape in China

Kai Zhang
Fresenius Kabi, Beijing, China

CONTENTS

Background 269
Administration of CGT Products in China 270
    Relevant Authorities in China 270
    Relevant Key Regulations in China 271
Registration Pathways of CGT Products in China 271
    Communication Application for Consultation 271
    Clinical Trial Application 272
    Market Authorization Application 275
    Application Document Requirements 277
    Accelerated Procedures 277
    Registration Fees 277
Post-market Requirements 279
    Conditions Shown in the License 279
    Adverse Events Reporting 279
    PSUR 280
    Post-market Changes 280
    Annual Report 280
    Drug Recall 281
Others 281
    Advertisement and Sales Promotion 281
    Special Considerations for Overseeing CGT Products in China 282
    Clinical Trial Management for Stem Cell Products 282
References 283

BACKGROUND

China, as one of the forerunners in advancing the development of innovative technologies, began research on CGTs at an early stage, conducting the world's first clinical trial for gene therapy in 1991 and approving the world's first gene therapy (Gendicine) in 2003. However, due to the loose and ambiguous regulation of CGTs, the development of CGTs was once in chaos in China, with unapproved

CGTs being widely used for disease prevention and treatment in private clinics or even large hospitals.

In 2016, the event of "Wei Zexi" alarmed the whole nation; a college student died from receiving unapproved DC-CIK cell therapy for synovial sarcoma. This tragedy resulted in an immediate government "shutdown" of almost all research on CGTs in China, which severely impeded research and development of CGTs. The application of unproven CGTs in patients has also perpetuated a global health problem leading to multiple patient injuries and deaths, threatening legitimate research efforts and undermining regulatory authority to safeguard public health. All stakeholders on national and international levels must collaborate to efficiently address this complicated and multifactorial problem.

The rapid advancements of CGTs worldwide coupled with their therapeutic potential in fulfilling unmet clinical needs of patients in China have prompted the Chinese government to establish effective regulatory frameworks for CGTs. Over the past few years, a series of technical guidelines related to CGTs have been released by the Chinese government to provide incentives and encourage investments in the field. Most recently, thanks to government support, substantial progress in the field of CGTs has been achieved in China. Currently, China is ranked second in terms of the number of patent applications and registered clinical trials for CGTs worldwide, following the United States. On June 23, 2021, the first chimeric antigen receptor T (CAR-T) cell therapy, Yescarta, was approved in China for the treatment of relapsed or refractory large B cell lymphoma after two or more lines of systemic therapy in adult patients, which is a remarkable milestone for cell therapy in China.

ADMINISTRATION OF CGT PRODUCTS IN CHINA

There is no specific definition available in China for CGT products. CGT products are considered drugs, falling into the category of therapeutic biological products. This can be confirmed via the definition of therapeutic biological products in the regulation for classification of biological products: "Biological products for therapeutic are biological products which are used for treat human diseases, such as proteins, polypeptides and their derivatives prepared from engineering cells (such as bacteria, yeast and insect, plant and mammalian cells) with different expression systems; single-component endogenous proteins extracted from human or animal tissues; cell and gene therapy products, allergen products, microecological products, biologically active blood products and multicomponent products extracted from human or animal tissues or body fluids or prepared by fermentation."

Thus the development and registration of CGT products in China shall follow the drug administration law, drug registration regulation and other relevant drug regulations.

Relevant Authorities in China

The Chinese health and regulatory organization is made up of a national authority and local levels of authority. In 2018, the State Council, the highest executive body

Regulatory Landscape in China

in China, was restructured, and some commissions, organizations and administrations were dissolved, reorganized or newly established.

The National Health Commission (NHC) was established after the National Health and Family Planning Commission (NHFPC) was dissolved, and its functions were integrated with the former Ministry of Health (MoH) in 2018. NHC reports the State Council is responsible for high-level national health policies, laws and regulations.

Established in 2018, the National Healthcare Security Administration (NHSA) reports to the State Council and is mainly responsible for regulations related to national insurance, price setting and reimbursement, payment methods, bids and procurement.

National Medical Products Administration (NMPA) was reorganized and renamed (formerly China Food and Drug Administration) in 2018. NMPA reports to the State Administration for Market Regulation (SAMR). SAMR, one of the direct affiliates of the State Council, was established in 2018. NMPA is the highest executive body for all regulatory registration matters related to biological products, chemical drugs and traditional Chinese medicines.

Center for Drug Evaluation (CDE) is a direct affiliate of NMPA and is responsible for the technical review of all registration matters, scientific advice and guidance documents. The technical review of the clinical trial applications, new drug applications and communication meeting applications for consultation for CGT products, etc., is conducted by CDE (Figure 13e.1).

Relevant Key Regulations in China

Table 13e.1 summarizes the relevant key regulations in China, listing the type of regulation (e.g., law or regulation), year of implementation, title and the web link where the information can be found.

REGISTRATION PATHWAYS OF CGT PRODUCTS IN CHINA

Communication Application for Consultation

In China, it is also recommended to apply for communication with CDE for drugs under development. The communication can be applied in any stage of drug development, such as preclinical trial application, end of phase II, pre-phase III clinical trial, pre-market authorization application. The forms of communication include face-to-face meetings, video conferences, teleconferences and written correspondence.

- Mock timeline from request preparation to meeting minutes: two to three months (type 1 meeting), four to six months (type 2 meeting), six to nine months (type 3 meeting).
- Document requirements:
  - Application form
  - List of topics

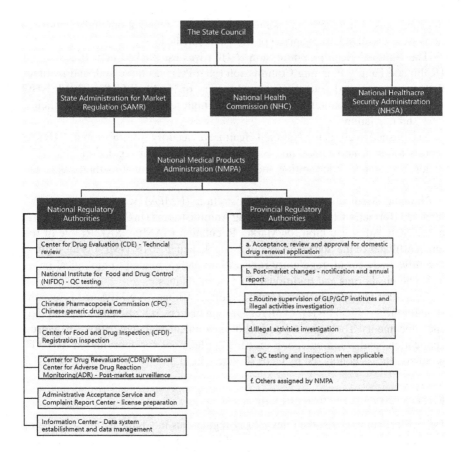

FIGURE 13E.1 Relevant authorities in China.

- Summary of supportive data (briefing package)
- Detailed study reports/draft clinical protocol which is relevant to the discussion topics with Chinese translations
- Presentation slides (PPT) for the discussion meeting

Although it is possible to apply for face-to-face (F2F) meeting or teleconference/video conference (TC/VC) with CDE, it is subject to the decision of CDE.

CLINICAL TRIAL APPLICATION

In China, before the conduction of registration clinical trials that are intended to support drug registration, clinical trial approval issued by CDE shall be obtained. However, any Investigator-Initiated Trials (IIT) studies, which are not managed as strictly as the registration clinical trial, CDE approval could not be mandatory.

Figure 13e.2 contains the flowchart for the clinical trial application in China.

TABLE 13E.1
Relevant Key Regulation in China

Type	Year	Title of Regulation	Authority Website Link
Law	2019	Drug Administration Law	http://www.gov.cn/xinwen/2019-08/26/content_5424780.htm
Law	2019	Ordinances on Management of Human Genetic Resources	http://www.gov.cn/zhengce/2020-12/27/content_5574163.htm
Regulation	2015	Stem Cell Clinical Research Management Provisions (Trial)	https://www.nmpa.gov.cn/yaopin/ypfgwj/ypfgbmgzh/20150720120001607.html
Regulation	2020	Drug Registration Regulation	https://www.nmpa.gov.cn/directory/web/nmpa/zhuanti/ypzhcglbf/ypzhcglbfzhcwj/20200330180501220.html
Regulation	2020	Drug Manufacture Supervision Regulation	https://www.nmpa.gov.cn/yaopin/ypfgwj/ypfgbmgzh/20200330182901110.html
Regulation	2020	Registration Classification and Application Document Requirements of Biological Products	https://www.nmpa.gov.cn/yaopin/ypggtg/ypqtgg/20200630175301552.html
Regulation	2021	Provisions for Post-approval Changes of Medicinal Product (Trial)	https://www.nmpa.gov.cn/xxgk/ggtg/qtggtg/20210113142301136.html
Regulation	2021	Drug Development and Technical Review Communication Regulation	https://www.cde.org.cn/main/news/viewInfoCommon/b823ed10d547b1427a6906c6739fdf89
Regulation	2022	Drug Supply Supervision Regulation	https://gkml.samr.gov.cn/nsjg/fgs/202203/t20220322_340682.html
Regulation	2022	Provisions for the Management of Annual Report for Drug	https://www.nmpa.gov.cn/xxgk/fgwj/gzwj/gzwjyp/20220412172455115.html
Guideline	2003	Technical Guideline for Human Cell Therapy Research and Quality Control on Preparations	https://www.cde.org.cn/zdyz/domesticinfopage?zdyzIdCODE=62d75837caaf77d394058 90f6ade9204
Guideline	2003	Technical Guideline for Human Gene Therapy Research and Quality Control on Preparations	https://www.cde.org.cn/zdyz/domesticinfopage?zdyzIdCODE=65a2d7ba914ad1e0ccd72e6a952b0dc6
Guideline	2017	Technical Guideline for Research and Evaluation of Cell therapy Products (Trial)	https://www.cde.org.cn/zdyz/domesticinfopage?zdyzIdCODE=452c529b29963 82 97210fe4a1294eb31

(Continued)

TABLE 13E.1 (CONTINUED)
Relevant Key Regulation in China

Type	Year	Title of Regulation	Authority Website Link
Guideline	2020	Technical Guideline for Pharmaceutical Research and Evaluation of Gene Transduction and Modification System (Draft for Comments)	https://www.cde.org.cn/main/news/viewInfoCommon/fb6eec0d50516fb00f2c0657c2e23a59
Guideline	2020	Technical Guideline for Clinical Trials of Human Stem Cells and Their Derived Cell Therapy Products (Draft for Comments)	https://www.cde.org.cn/main/news/viewInfoCommon/1be11947797accf6c77ec01f32cc79509
Guideline	2021	Technical Guideline for Clinical Trials of Immunocell Therapy Products (Trial)	https://www.cde.org.cn/zdyz/domesticinfopage?zdyzIdCODE=cd15d9b4d5305683f507d15029e36895
Guideline	2021	Technical Guideline for Non-clinical Research of Gene-Modified Cell Therapy Products (Trial)	https://www.cde.org.cn/zdyz/domesticinfopage?zdyzIdCODE=b7dfbba537d5ecc30659d715f5045acb
Guideline	2021	Technical Guideline for Non-clinical Research of Gene Therapy Products (Trial)	https://www.cde.org.cn/zdyz/domesticinfopage?zdyzIdCODE=3c3eef7964f7950ca9a18b9ce095088c
Guideline	2021	Technical Guideline for Long-term Follow-Up Clinical Studies of Gene Therapy Products (Trial)	https://www.cde.org.cn/zdyz/domesticinfopage?zdyzIdCODE=948d4385437338fd7fd4919fd75f5f1a
Guideline	2022	Technical Guideline for the Clinical Risk Management Plan for Market Authorization Application of Chimeric Antigen Receptor T Cell (CAR-T) Therapeutic Products	https://www.cde.org.cn/zdyz/domesticinfopage?zdyzIdCODE=9c18eb2d5f9bb96423052d104e80665a
Guideline	2022	Technical Guideline for Pharmaceutical Research and Evaluation of Immunocell Therapy Products (Trial)	https://www.cde.org.cn/zdyz/domesticinfopage?zdyzIdCODE=adac7701a40bf0ad8691c409324a004c
Guideline	2022	Technical Guideline for Pharmaceutical Research and Evaluation of In Vivo Gene Therapy Products (Trial)	https://www.cde.org.cn/zdyz/domesticinfopage?zdyzIdCODE=c78a179af7edac00e6d8893dcdfef01f
Guideline	2022	Technical Guideline for Pharmaceutical Research and Evaluation of In Vitro Gene Therapy Products (Trial)	https://www.cde.org.cn/zdyz/domesticinfopage?zdyzIdCODE=c8d231a6cc6aee6eae2e49071696dc5

Regulatory Landscape in China

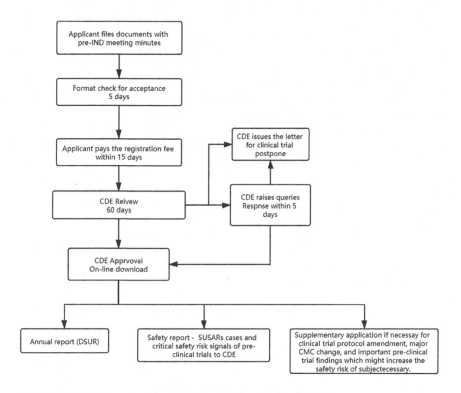

FIGURE 13E.2 Flowchart for the clinical trial application in China.

After the first clinical trial of a new drug is approved in China, the Development Safety Update Report (DSUR) shall be filed every year, and there are also requirements on safety reporting for the serious adverse events that happened in the drug clinical trial. For a new indication of the drug, another clinical trial application shall be filed to get approval again.

Market Authorization Application

In China, a drug market authorization application, also called a new drug application, shall be filed with NMPA to get the license of the drug before marketing any drug in China.

Figure 13e.3 contains the flowchart for the market authorization application in China.

In parallel with the CDE review, usually, a QC test on three batches of drug products is required, the test can be initiated either before or after the market authorization application is filed; based on the risk of the market authorization application, registration inspection on the research facility, the preclinical trial institutes, clinical trial institutes and even manufacture site may be conducted by CFDI.

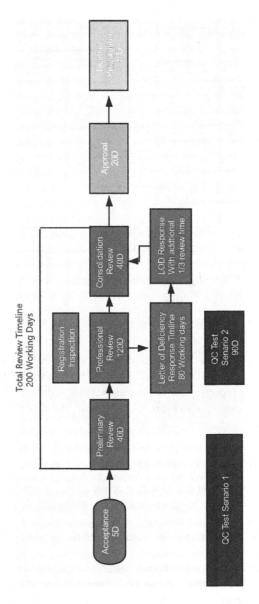

FIGURE 13E.3 Flowchart for the market authorization application in China.

APPLICATION DOCUMENT REQUIREMENTS

For the clinical trial application and marketing authorization application of therapeutic biological products, the applicant shall prepare the registration documents in accordance with ICH M4: Common Technical Documents (CTD) for the Registration Application of Human Drugs (hereinafter referred to as CTD). And there are specific requirements for regional information 3.2.R as follows.

3.2.R Regional information
3.2.R.1 Process verification

The process verification scheme and report shall be provided.

3.2.R.2 Batch record

For the clinical trial application, batch production and inspection records representing the process of the clinical trial samples shall be provided.

For the marketing authorization application, batch production and inspection records of the key clinical batches and at least three consecutive marketing scale validation batches shall be provided.

The testing reports of these batches shall be provided.

3.2.R.3 Analysis method validation report

The analysis method validation report shall be provided, including the typical chromatogram.

3.2.R.4 Stability chromatogram

The typical chromatogram of stability study shall be provided.

3.2.R.5 Comparability protocol (If applicable)
3.2.R.6 Others

So far, eCTD is still not mandatory in China, for both drug clinical trial application and marketing authorization application, eCTD submission is in trial period only.

ACCELERATED PROCEDURES

There are several procedures in China to accelerate drug development in China, including the breakthrough therapy procedure, conditional approval procedure, priority review and approval procedure and special review and approval procedure.

Table 13e.2 provides a summary of different accelerated procedures:

REGISTRATION FEES

For drug registration in China, a registration fee will be charged by NMPA for each submission according to the following drug registration pricing standards (Table 13e.3). Meanwhile, no fees will be charged for the communication meeting with CDE.

TABLE 13E.2
Summary on Different Accelerated Procedures to Accelerate Drug Development

Accelerated Procedures	Scope	When to Apply	Advantages
Breakthrough therapy procedure	Innovative new drug – for life-threatening disease or disease seriously affecting quality of life without effective treatment, or have significant advantages compared to current treatment methods with sufficient evidences	Before initiation of phase III clinical trial	1. Dedicated review team in China to guide the product development 2. Accelerated communication application procedure as type 1 3. Priority review and approval for market authorization application
Conditional approval procedure	1. Drugs indicated for serious life-threatening disease without effective treatment, or drugs with urgent needs in public health and the available clinical data can prove the efficacy and predict the clinical value 2. Vaccines with urgent needs to deal with emergent public health event or designed by health department, after evaluation it is shown␣the benefit is higher than risk	Before market authorization application – communication meeting with CDE for alignment first	1. Early approval – with conditions (post-market research requirements and timelines; risk management measures) 2. Priority review and approval for market authorization application
Priority review and approval procedure	1. Drugs with urgent medical needs in short supply, and new drugs and modified new drugs for treatment of serious infectious disease or drugs for rare disease 2. Pediatric drug (new product, dosage form, and strengths) with significant clinical value and meeting the physical characteristics of children 3. Vaccines with urgent needs for disease prevention or control 4. Drugs under breakthrough therapy procedure 5. Drugs under conditional approval procedure 6. Other situations demonstrated by NMPA	Along with market authorization application	1. Accelerated Review timeline for market authorization application: 130 working days 2. Registration inspection, QC test, as well as the Chinese generic name verification task will be arranged with priorities 3. Communication application for document supplementary raised by applicant – if agreed by CDE, documents can␣be filed without affecting review timeline 4. Accelerated NMPA administration approval – ten working days
Special review and approval procedure	Drugs needed for prevention of public health emergencies	At any time	1. Specific guidance team setting up by NMPA to guide the development and submission 2. The acceleration of all the procedures from acceptance to approval, such as acceptance within 24 hours, review within 15 working days

TABLE 13E.3
Example of Fees Charged by NMPA for Submissions

Drug Registration Pricing Standards			
			Unit: RMB 10,000 yuan
Project Category		Domestic Made	Import
New drug registration fee	Clinical trial	19.20	37.60
	Manufacturing/marketing	43.20	59.39
Generic drug registration fee	Manufacturing/marketing requiring no clinical trial	18.36	36.76
	Manufacturing/marketing requiring clinical trial	31.80	50.20
Supplementary application registration fee	Regular item	0.96	0.96
	Requiring technical review	9.96	28.36
Drug re-registration fee (once every five years)		Instituted by provincial price and fiscal department	22.72

POST-MARKET REQUIREMENTS

CONDITIONS SHOWN IN THE LICENSE

After the product approval, the market authorization holder (MAH) shall proactively carry out post-marketing studies to further verify the safety, efficacy and quality controllability of drugs and enhance ongoing management of marketed drugs.

Where the drug license and its attachments require the MAH to carry out related post-marketing studies, the MAH shall complete the studies within the prescribed timeline and submit a supplementary application, notification or report as required. The renewal submission will be affected if the MAH cannot meet the requirements in the drug license and its attachment.

ADVERSE EVENTS REPORTING

For domestic new drugs and imported drugs, within the first five years after the drug license is granted, all adverse reactions shall be reported; after five years, new and serious adverse reactions shall be reported. For domestic generic drugs, new and serious adverse reactions shall be reported.

The drug MAHs that discover or learn of a new or serious adverse drug reaction shall report it within 15 days and shall report any deaths immediately; other adverse drug reactions shall be reported within 30 days. Any follow-up information shall be reported promptly.

The drug MAH shall investigate any fatalities concerning which they receive information and shall endeavor to uncover the underlying facts, drug usage circumstances and occurrence and treatment of any adverse reactions, and shall complete and submit their investigative report within 15 days.

For serious adverse drug reactions attributed to imported drugs and domestic drugs used overseas (including those acquired through the self-reporting system, those discovered through post-marketing clinical studies and those reported in the literature), the drug MAH shall report within 30 days of receiving the notification.

PSUR

For domestic new drugs and imported drugs, a periodic safety update report (PSUR) shall be submitted every full year following the date on which the drug license was issued, up until the first re-registration; thereafter, PSUR shall be submitted once every five years. For domestic generic drugs, PSUR shall be submitted once every five years.

The aggregation time frame for a PSUR takes the date on which the drug license was issued as its starting point, and the submission date shall fall within 60 days of the data aggregation deadline.

The MAH shall also submit a pharmacovigilance annual report every year, before March 31 of each year; the report shall include the operation status of the drug adverse reaction monitoring system, the adverse reaction report status, the risk identification and control status and the post-market research status of the previous year.

Post-market Changes

For post-marketing changes to drugs, classified management shall be practiced depending on their risks to and the extent of their influence on the safety, efficacy and quality controllability of the drugs. Post-marketing changes are classified into changes subject to approval, notification and reporting.

MAHs shall fully assess and validate the impact of changes on the safety, efficacy and quality controllability of drugs and carry out related study work according to relevant regulations by reference to the relevant technical guidelines.

Annual Report

The MAH shall collect and summarize the information of the annual report for the drug of the previous natural year and report through the system of the annual report for the drug before April 30 of each year. For drugs approved for marketing in the current year, the MAH can combine the information in the annual report into the next year's report.

The contents of the annual report are divided into the public part and the product part.

(i) The public part, including the information of the holder, the overall situation of the products held, the overview of quality management, the construction and operation of the pharmacovigilance system, the acceptance of overseas entrusted processing and acceptance of the inspection by overseas regulatory authorities, etc.
(ii) The product part, including basic product information, production and sales, post-marketing research, change management and risk management, etc.

DRUG RECALL

The drug MAH shall analyze all data relevant to medicine safety, conduct investigations into possible safety hazards and immediately order recalls of all drugs with potential safety risks. In the case of recalls conducted from overseas by imported drug MAH, the MAH shall send a notice of termination of sale and usage for the intended drug to NMPA.

If the drug MAH decides to recall a medicine, it shall complete formulation and pre-implementation of the recall plan within 24 hours for Class I recall, within 48 hours for Class II recall and within 72 hours for Class III recall. The drug MAH shall notify the relevant drug management enterprises and organizations using the drug of the termination of sales and usage and report to the provincial or national drug administration authorities.

When the drug MAH initiates a recall, it shall submit a report of its investigation and recall plan to provincial or national drug administration authorities within one day for Class I recall, within three days for Class II recall and within seven days for Class III recall.

OTHERS

ADVERTISEMENT AND SALES PROMOTION

Drug MAHs, drug manufacturing enterprises, drug distribution enterprises and medical institutions shall not give or receive rebates or other improper gains in the procurement and sale of drugs. Drug MAHs, drug manufacturing enterprises, drug distribution enterprises or their agents shall not give monies or other improper gains under any pretext to the person(s)-in-charge, drug procurement staff, physicians, pharmacists, etc., of medical institutions using their drugs. The person(s)-in-charge, drug procurement staff, physicians, pharmacists, etc., of medical institutions, shall not receive monies or other improper gains under any pretext from drug MAHs, drug manufacturing enterprises, drug distribution enterprises or their agents.

Drug advertisements shall be subject to approval by the provincial drug administration authority. Drug advertisements shall not be published without approval.

The contents of drug advertisements shall be true and legitimate, shall be based on the package insert of the drug approved by NMPA and shall not contain false information. Drug advertisements shall not assert or guarantee efficacy or safety of the advertised drug; shall not make use of the name or image of state agencies, research institutes, academic institutions, industry associations or experts, academicians, physicians, pharmacists, patients, etc., for recommendation or endorsement. Non-drug advertisement shall not involve advertising of drugs.

Special Considerations for Overseeing CGT Products in China

Based on the special administrative measures for foreign investment access (negative list) (2021 edition) issued by the National Development and Reform Commission (NDRC) and Ministry of Commerce, it is prohibited to invest in the development and use of human stem cell, gene diagnosis and therapy for foreign companies in China.

The collection, preservation, utilization or exportation of China's human genetic resources are subjected to special administration by the Chinese government; human cells, tissues and gene information are considered human genetic resources. If companies would like to export human cells and tissues outside China, Human Genetic Resource Administration of China (HGRAC) approval shall be required. It may take several months to get the approval of HGRAC even when the companies can provide sufficient documents and rationales. Meanwhile, the importation of human cells and tissues into China is also controlled strictly due to epidemic prevention reasons.

Taking consideration of all of the aforementioned factors, it is necessary to build the production line in China for overseeing the CGT products, and it is also required to find a local partner for product development and marketing in China, either via patent licensing or via joint adventure.

Clinical Trial Management for Stem Cell Products

Before any clinical trial of stem cell products, the clinical trial institutes shall file record submission documents to the provincial health commission; after the review by both the provincial health commission and drug administration authority, NHC and NMPA will be notified to complete the record.

IITs for stem cell products needn't be approved by NMPA. Instead, the trial project shall pass the review by an institutional academic committee and ethics committee for approval, then the trial project documents shall be filed with the provincial health commission for the record; after the review by both the provincial health commission and drug administration authority, NHC and NMPA will be notified to complete the record.

REFERENCES

Drug Administration Law. (2019). http://www.gov.cn/xinwen/2019-08/26/content_5424780.htm

Drug Development and Technical Review Communication Regulation. (2021). https://www.cde.org.cn/main/news/viewInfoCommon/b823ed10d547b1427a6906c6739fdf89

Drug Manufacture Supervision Regulation. (2020). https://www.nmpa.gov.cn/yaopin/ypfgwj/ypfgbmgzh/20200330182901110.html

Drug Registration Regulation. (2020). https://www.nmpa.gov.cn/directory/web/nmpa/zhuanti/ypzhcglbf/ypzhcglbfzhcwj/20200330180501220.html

Drug Supply Supervision Regulation. (2022). https://gkml.samr.gov.cn/nsjg/fgs/202203/t20220322_340682.html

Ordinances on Management of Human Genetic Resources. (2019). http://www.gov.cn/zhengce/2020-12/27/content_5574163.htm

Provisions for Post-approval Changes of Medicinal Product (Trial). (2021). https://www.nmpa.gov.cn/xxgk/ggtg/qtggtg/20210113142301136.html

Provisions for the Management of Annual Report for Drug. (2022). https://www.nmpa.gov.cn/xxgk/fgwj/gzwj/gzwjyp/20220412172455115.html

Registration Classification and Application Document Requirements of Biological Products. (2020). https://www.nmpa.gov.cn/yaopin/ypggtg/ypqtgg/20200630175301552.html

Stem Cell Clinical Research Management Provisions (Trial). (2015). https://www.nmpa.gov.cn/yaopin/ypfgwj/ypfgbmgzh/20150720120001607.html

Technical Guideline for the Clinical Risk Management Plan for Market Authorization Application of Chimeric Antigen Receptor T Cell (CAR-T) Therapeutic Products. (2022). https://www.cde.org.cn/zdyz/domesticinfopage?zdyzIdCODE=9c18eb2d5f9bb96423052d104e80665a

Technical Guideline for Clinical Trials of Human Stem Cells and Their Derived Cell Therapy Products(Draft for Comments). (2020). https://www.cde.org.cn/main/news/viewInfoCommon/1be1194797accf6c77ec01f32cc79509

Technical Guideline for Clinical Trials of Immunocell Therapy Products (Trial). (2021). https://www.cde.org.cn/zdyz/domesticinfopage?zdyzIdCODE=cd15d9b4d5305683f507d15029e36895

Technical Guideline for Human Cell Therapy Research and Quality Control on Preparations. (2003). https://www.cde.org.cn/zdyz/domesticinfopage?zdyzIdCODE=62d75837caaf77d39405890f6ade9204

Technical Guideline for Human Gene Therapy Research and Quality Control on Preparations. (2003). https://www.cde.org.cn/zdyz/domesticinfopage?zdyzIdCODE=65a2d7ba914ad1e0ccd72e6a952b0dc6

Technical Guideline for Long-term Follow-up clinical Studies of Gene therapy Products (Trial). (2021). https://www.cde.org.cn/zdyz/domesticinfopage?zdyzIdCODE=948d4385437338fd7fd4919fd75f5f1a

Technical Guideline for Non-clinical Research of Gene-modified Cell Therapy Products (Trial). (2021). https://www.cde.org.cn/zdyz/domesticinfopage?zdyzIdCODE=b7dfbba537d5ecc30659d715f5045acb

Technical Guideline for Non-clinical Research of Gene Therapy Products (Trial). (2021). https://www.cde.org.cn/zdyz/domesticinfopage?zdyzIdCODE=3c3eef7964f7950ca9a18b9ce095088c

Technical Guideline for Pharmaceutical Research and Evaluation of Gene Transduction and Modification System (Draft for Comments). (2020). https://www.cde.org.cn/main/news/viewInfoCommon/fb6eec0d50516fb00f2c0657c2e23a59

Technical Guideline for Pharmaceutical Research and Evaluation of Immunocell Therapy Products (Trial). (2022). https://www.cde.org.cn/zdyz/domesticinfopage?zdyzIdCODE=adac7701a40bf0ad8691c409324a004c

Technical Guideline for Pharmaceutical Research and Evaluation of in vivo Gene therapy Products (Trial). (2022). https://www.cde.org.cn/zdyz/domesticinfopage?zdyzIdCODE=c78a179af7edac00e6d8893dcdfef01f

Technical Guideline for Pharmaceutical Research and Evaluation of in vitro Gene Therapy Products (Trial). (2022). https://www.cde.org.cn/zdyz/domesticinfopage?zdyzIdCODE=c8d231a6cc6aee6e6ae2e49071696dc5

Technical Guideline for Research and Evaluation of Cell Therapy Products (Trial). (2017). https://www.cde.org.cn/zdyz/domesticinfopage?zdyzIdCODE=452c529b299638297210fe4a1294eb31

13f Regulatory Landscape in Japan

Hazel Aranha
GAEA Resources Inc., Northport, USA

CONTENTS

Background and History 285
Regulatory Body in Japan 286
ASRM 288
    New Directions under ASRM 289
PMDA: Roles and Responsibilities 290
Standard and Priority Review 290
Accelerated or Conditional Approval 291
Breakthrough Therapy Designation 291
Cartagena Act for Gene Therapies 292
Conclusions 293
Websites 293
References 293

BACKGROUND AND HISTORY

In September 2014, the Japanese legislature enacted a suite of laws and amendments specifically designed to rationalize and streamline the regulation of regenerative medicine. These reforms were inspired in part by the 2012 'Yokohama Declaration' (Yokohama, 2014) issued by the Japanese Society for Regenerative Medicine (JSRM), which expressed concerns that the existing drug regulatory system was poorly suited to the timely and efficient premarket review and approval of cell-based therapeutics. The JSRM also noted that the lack of effective practice standards governing the use of cell-based interventions by private clinics had turned Japan into a 'therapeutic haven' for businesses to market unproven interventions to patients. The 2014 regulatory package included reforms designed to promote safety while expediting the regulatory review process and included the establishment of a new category of medical product 'regenerative medicine products'; the previous Pharmaceutical Affairs Law (PAL) became the Pharmaceuticals, Medical Devices, and Other Therapeutic Products Act (PMD Act). Around the same time, the government passed the Act on the Safety of Regenerative Medicine (ASRM), which established a review system for the use of regenerative medicine in private medical practices operating outside the national health insurance system to increase oversight and safety.

DOI: 10.1201/9781003285069-19

While some early efforts to enforce ASRM were made to maintain the safety of regenerative medicine, the governance framework for private practices offering regenerative procedures was less than optimal (Cyranoski, 2010). This highly permissive oversight system for cell-based procedures through the private system under ASRM represented a challenge: while entities seeking to gain market authorization for regenerative medicine products under PMD Act were required to conform to more stringent regulations, private practices were subject to weaker oversight of their promotional claims and pricing under ASRM.

Rogue actors, for profit, marketed therapies to those seeking respite for their conditions; many of the recipients of the therapies were from outside Japan (Charo, 2016). In August 2017, six individuals were arrested under suspicion of selling unproven stem cell interventions outside the scope of the ASRM. These arrests marked the first law enforcement action under the ASRM. This enforcement of the ASRM was particularly important given the ambiguous governmental regulation of the private sector (Sipp, 2015). The suspects reportedly injected patients with derivatives of cord blood or other perinatal tissue obtained from a private cord blood bank that had gone bankrupt in 2009. This incident highlighted the need for an appropriate balance between encouraging innovation and reining in exploitative business practices.

REGULATORY BODY IN JAPAN

Regulatory policies are developed and administered by the Ministry of Health, Labour and Welfare (MHLW); the Pharmaceuticals and Medical Devices Agency (PMDA) executes the policies, and certain authorities are delegated to local governments. This chapter discusses the current regulatory system and the delegation of responsibilities.

In Japan, it is necessary to comply with the regulations of the PMD Act to manufacture, import and sell drugs and medical devices. The MHLW is one of the central government ministries and has a central role in promulgating and administering regulations under the PMD Act. MHLW conducts operations and services related to pharmaceutical affairs administration while working with the PMDA and prefectural governments. Within MHLW, pharmaceutical affairs administration is managed primarily by the Health Policy Bureau (HPB), the Pharmaceutical and Food Safety Bureau (PFSB) and the Pharmaceutical Affairs and Food Sanitation Council (PFSC). Of these bureaus, PFSB grants licenses and enforces regulations for clinical trials, evaluations and post-market safety measures. The majority of these tasks are delegated to the PMDA, but MHLW is the ultimate decision-maker (Agency).

To summarize: the roles of each of these organizations are as follows:

i. MHLW develops and implements regulatory policies and punitive actions for violations, such as approval and administrative orders related to health-care products;
ii. PMDA conducts operations such as examinations, inspections, and data analysis before an administrative measure is implemented; and

iii. Prefectural governments have the authority to grant licenses as established under the PMD Act and related laws and to undertake pharmaceutical affairs administration in their jurisdictional prefectures.

PMDA's reviews and related services consist of various activities, such as "consultations" providing advice in relation to regulatory submission, Good Laboratory Practices (GLP)/Good Clinical Practices (GCP)/ Good Post-marketing Study Practice (GPSP) inspections to ensure the submitted data are in compliance with the ethical and scientific standards and GMP/QMS/GCTP inspections to ensure quality management of the manufacturing facility for the product submitted for approval.

The agency's major areas of focus can be categorized broadly into three parts: product review affairs, safety policy affairs and health damage relief affairs.

- As part of product review, their duties include premarket submission review, GMP/GMS/GCTP (good gene, cellular, and tissue-based products manufacturing practice) inspections for manufacturing plants and standard development of the Japanese Pharmacopoeia.
- Safety functions include collecting and organizing information on side effects from manufacturers or medical institutions; research and analysis work that contributes to safety measures through scientific analysis; consultation services regarding safety measures from manufacturers; responding to consultations from the general public; and information service related to the safety of pharmaceuticals, medical devices, regenerative medicine products and similar responsibilities.
- Health Damage Relief System for sufferers from adverse drug reactions; for sufferers from infections via biological products; consignment and loan affairs; commissioned benefits affairs; payment of benefits based on the law on special measures. These relief systems are funded by annual contributions from industry.

The aforementioned system has developed and evolved over the years (Konomi et al., 2015; Okada et al., 2015). The PMD Act of 2014 replaced the PAL of 1960, amended in 2006, within which gene therapy is defined as a form of regenerative medicine. The regulatory reform in 2019 added elements to the regulatory system to take advantage of an increasingly digitized society and digital technology. Digitization of information includes precautions and posting on the PMDA homepage (enforced in September 2021), establishment of a legal compliance system for marketing authorization holders and manufacturers (enforced in September 2021) and compliance with unique device identifier requirements (will be enforced in December 2022).

The MHLW revised the PAL and implemented the Law on Securing Quality, Efficacy and Safety of Products, including the Pharmaceuticals and Medical Devices Act in 2014 (Nagai, 2019; Nagai & Ozawa, 2016, 2017). "Regenerative medical products" were defined in the Act as follows: (i) products used by introducing them into human cells for gene therapy and (ii) processed cells used for the reconstruction/repair/formulation of human body structure/function or treatment/ prevention of a disease. Regenerative medical products include in vivo and ex vivo gene medical products and cellular medical products. Organ transplantation, hematopoietic stem cell transplantation and blood transfusion are not classified as regenerative medical products. Regenerative medical products are recognized as nonhomogeneous and a system to address at an early stage with conditions and time limits by estimating efficacy and confirming safety was introduced. Based on these contents, the name of the PAL was changed to "Law Concerning Quality, Effectiveness and Safety of Pharmaceuticals, Medical Devices, etc."

All new drug/regenerative medical product applications are submitted to the PMDA; this body reviews applications for drugs, medical devices and regenerative medicines, and prepares review reports in Japan. The MHLW grants marketing authorization.

ASRM

Under the ASRM (Takashima et al., 2021), regenerative medicine procedures are defined (as noted in the Ministry of Health, Labour and Welfare's Act on the Safety of Regenerative Medicine) as those involving human cells processed for use in (i) the reconstruction, formation, or repair of human bodily structures or functions or (ii) the treatment or prevention of human diseases. The second of these clauses suggests that nearly any medical use of human cells is encompassed under the ASRM, and the scope of the law is not limited to stem/progenitor cells but appears to extend to human cellular materials in general, including uses not ordinarily considered "regenerative" in nature (e.g., cytotoxic T cells used against tumor cells). This has significant implications for Japan, where non-regenerative cell-based interventions, such as so-called cellular immunotherapies were offered by private clinics since the late 1990s (Sipp, 2015). Takashima et al. provide an interesting commentary regarding the rationale for the choice of the term regenerative therapy; they discuss the highlights and differences between the reports published in 2010 and 2013 by the Governmental Committee for Institutional Framework of Regenerative Medicine (Takashima et al., 2021).

The ASRM applies to all clinical and research institutions using regenerative medicine and using cell culture and processing. Under ASRM, regenerative medicine is classified based on cell types included and degree of manipulation:

(i) Class I: High Risk – pluripotent/genetically modified/xenogeneic cells that are used in therapy, e.g., transplantation of retinal pigment epithelium sheets derived from autologous iPS cells in patients with age-related macular degeneration;
(ii) Class II: Medium Risk – autologous stem cells, that may be used in therapy and research, e.g., autologous bone marrow cell infusion therapy for liver cirrhosis; and
(iii) Class III: Low risk – autologous differentiated cells that may be intended for research, total therapy, activated lymphocyte therapy for malignancies.

Under ASRM, any medical institution that plans to offer regenerative medicine must undergo review by a Certified Committee for Regenerative Medicine (CCRM) or Certified Special Committee for Regenerative Medicine (CSCRM) for class III and class I/II regenerative medicine techniques, respectively. For class I (high-risk) procedures, the institution next submits its protocol to the MHLW to seek its approval. However, for class II interventions, prior authorization from the ministry is not required. The ministry maintains a record of CCRMs and CSCRMs. Of the CCRMs registered with MHLW, an equal number (~ 46%) are affiliated with either medical or dental schools, publicly funded research centers, professional societies or local government agencies, or they are directly affiliated with private clinics specializing in areas of medicine, such as cosmetic surgery, dentistry, or neurology, respectively. An additional five are independent bodies specializing in the review of regenerative medicine protocols under ASRM.

There is concern that CSCRMs that are empowered to review class I/II regenerative medicine techniques (high to medium risk) may not always provide the required oversight. Only one-third of the CSCRMs are registered at academic (. ac.jp) or government (. go.jp) domains, the rest list a private email (gmail.com, yahoo.co.jp addresses) or no contact email at all. The empowering of private CSCRMs to review plans for potentially risky procedures involving pluripotent or genetically modified cells (which goes against the International Society for Stem Cell Research's Guidelines for Stem Cell Research and Clinical Translation), may facilitate unethical interventions and unfortunately may have failed to silence the "therapeutic haven" alarm raised by JSRM in 2011. These deficiencies are being addressed by the MHLW.

Under the ASRM, a manufacturing license from the MHLW is required for specified cell-processing operations. The procedures vary depending on where the manufacturing/facilities are located (Japan or foreign countries, in a medical institute or other location). Applications fall into one of the following categories: (i) Accreditation is required in the case of manufacturers and facilities located in foreign countries. The application for accreditation and inspection, and site information (e.g., the blueprint and equipment for processing specific cell products), are submitted to the MHLW for review. In some cases, the PMDA will investigate the facility. (ii) Notification is required in the case of manufacturing at a medical institute located in Japan. The application for notification is submitted to the regional Bureau of Health and Welfare. An inspection is not required. (iii) Permission is required in the case of manufacturing at a nonmedical institution in Japan. The application for permission and inspection is submitted to the regional Bureau of Health and Welfare. In some cases, the PMDA will inspect the facility.

When medical institutions outsource manufacturing of cell-processing products, they have to select licensed manufacturers.

NEW DIRECTIONS UNDER ASRM

The current focus of the ASRM revisions is on the regulatory framework for in vivo gene therapy and risk classification and the scope of the exemptions within the ASRM. One of the latest partial revisions of the Act, in June 2020, is related to genome-editing technology, where gene-edited cells, in addition to gene-transferred cells, are incorporated into class I of the risk categorization.

As the field of regenerative medicines continues to progress, we will be faced with additional challenges that will require a better demarcation of what constitutes a medical or research treatment; the latter usually requires greater oversight and review. Open issues include standardization and authorization for cell processing and risk classification and oversight over research and unproven therapies. It will be necessary for us to reconsider the safety concepts and revisit the range of risk factors used to determine review classifications and clarify the path for unproven therapies to transition to evidence-based interventions.

PMDA: ROLES AND RESPONSIBILITIES

The PMDA reviews applications for drugs, medical devices and regenerative medicines and prepares review reports in Japan. Consistency in quality and a favorable benefit-risk balance must be established before any medicinal product can receive marketing authorization. Until about a decade ago, this focus on quality, safety and efficacy driven by evidence-based science, knowledge, and technology led to longer development timelines and therefore a delay in getting medicinal products approved and available for use in patients. Expedited regulatory review projects for innovative drugs and regenerative medical products were developed in the US, the EU and Japan. Each regulatory agency elaborated an original regulatory framework and adopted regulatory initiatives similar to other regulatory agencies. For example, the Food and Drug Administration (FDA) developed the breakthrough therapy designation, European Medicines Agency (EMA) introduced the priority medicines (PRIME) designation and the PMDA introduced the Sakigake designation. Note that, following the UK's withdrawal from the EU, the UK implemented its own scheme for expedited access, the innovative licensing and access pathway, or ILAP. Comparisons and differences between the expedited review processes are discussed elsewhere (Jokura et al., 2018; Nagai, 2019).

STANDARD AND PRIORITY REVIEW

In Japan, both standard review and priority review are available. In addition to orphan drugs, the MHLW designates medical products as priority review products based on the following requirements: (i) no standard therapy exists or there is superior clinical usefulness as compared with the existing products in terms of quality of life of patients, efficacy or safety, and (ii) it applies to serious diseases or orphan drug designation. Products granted conditional approval automatically enjoy priority review designation.

ACCELERATED OR CONDITIONAL APPROVAL

In Japan, the MHLW implemented the Law on Securing Quality, Efficacy and Safety of Products including the PMDA as the revision of the PAL in 2014 (Nagai, 2019; Nagai & Ozawa, 2016, 2017). The conditional and term-limited approval system for regenerative medical products was instituted in the Act. The conditional and term-limited approval is generally granted based on promising results of exploratory phase I/ II trials in terms of efficacy and safety. Sponsors must conduct post-marketing clinical studies to confirm the efficacy and safety, and resubmit applications for regular approval within a predetermined period (not more than seven years).

The conditional approval system for drugs is applicable and may be granted if all of the following requirements are met: (i) no standard therapy exists or superior clinical usefulness can be demonstrated as compared with the existing products in terms of quality of life of the patients, efficacy, or safety; (ii) it is applicable to serious diseases; (iii) it is difficult or it would take too long to conduct a confirmatory study; (iv) exploratory clinical studies can show the efficacy and safety of the drug; and (v) surveillance or clinical studies must be conducted as a post-marketing requirement.

BREAKTHROUGH THERAPY DESIGNATION

A unique feature of the Sakigake (meaning pioneer or forerunner in Japanese) designation is the necessity of the product being first developed in Japan, thus actively supporting the development of advanced-therapy medicinal products (ATMPs) by academia or research institutions.

The Strategy of Sakigake was announced as a pilot scheme in April 2015 and consisted of the Sakigake Designation System and the Scheme for Rapid Authorization of Unapproved Drugs. Following amendment of the PMD Act in 2019, Sakigake became permanently embedded in Japanese law as of 1 September 2020. As of December 2021, 25 drugs received Sakigake designation. Table 13f.1 summarizes the features and benefits for granting conditional approval under the Sakigake designation. This conditional approval requires that there are existing exploratory clinical studies that show efficacy and safety, and recognizes that an extended amount of time would be required to conduct confirmatory studies. As a post-marketing requirement, clinical studies were required; however, recent examples indicate that comparative studies may not be necessary and post-marketing surveillance is acceptable (Nagai, 2019). Table 13f.2 summarizes the significant advantages of the Sakigake designation in terms of conducting priority consultations, reduced waiting time, early assessment, and expedited reviews. An extended reexamination period may be available after consultation.

TABLE 13F.1
Breakthrough Therapy Designation in Japan (Sakigake designation): Features and Requirements

Features	Requirements
Prioritized consultation (reduced waiting time) Substantial preapplication consultation Expedited review (a target total review time of six months for drugs, devices and In-Vitro Diagnostics (IVDs), and a designated priority review) (total review time for Sakigake-designated regenerative medical products is not established) Assignment of a PMDA concierge An extended reexamination period	• No standard therapy exists or superior clinical usefulness is demonstrated as compared with existing products in terms of quality of life, efficacy or safety • Applicable to serious diseases, diseases with unmet medical needs • The medical product has been first developed in Japan, and a sponsor is planning to submit a marketing authorization application • Data from early phase clinical studies, nonclinical studies, and prominent effectiveness must be demonstrated

TABLE 13F.2
Conditional and Term-limited Approval (Sakigake designation): Benefits

Advantages of Application Using Sakigake Designation[1]

Prioritized consultation	Waiting time: 2 months →1 month
Substantialized preapplication consultation	De facto review before application
Prioritized review	Waiting time: 12 months →6 months* (*for new drug or new medical device)
Review partner	PMDA manager as a concierge
Substantial post-marketing safety measures	Extension of reexamination period

[1] Caveat: Necessity of product being first development in Japan is the original feature of the Sakigake designation

CARTAGENA ACT FOR GENE THERAPIES

Gene therapy products (in vivo and ex vivo) in Japan are regulated by the Cartagena Act on the Conservation and Sustainable Use of Biological Diversity Through Regulations on the Use of Living Modified Organisms (LMOs).

The Cartegena Act is the law that regulates the use of LMOs and provides rules to assess the effect of LMOs on biological diversity in advance and also the way to appropriately use LMOs. The Cartagena Protocol was adopted by the international community in January 2000 and entered into force in September 2003. In February 2004, the Cartagena Act became effective and the Cartagena Protocol entered into force in Japan. The Cartagena Act separates uses into two types and uses different approaches for evaluating each type: (i) Type 1 (Use under the open system): Uses for conveyance and cultivation for food, feed, etc., approved only when the LMOs are judged not to cause adverse effects on biological diversity; (ii) Type 2 (Use under the closed system): Uses in laboratories, factories, etc. Prior approval by the MHLW and minister of the environment is required when attempting Type 1 use to ensure the LMOs do not have a negative effect on the local environment. It is necessary for the sponsor to apply and obtain the approval of the MHLW regarding the Type 1 use at the medical institution before starting the clinical trial.[38]

Cells collected from a patient, genetically modified and returned back to the patient would fall under Type 2; however, when the cells, without modification, are returned to humans in clinical trials or clinical research, it is classified as a Type 1 use. In development of gene therapies, Type 2 use falls under the PMD Act for nonclinical research, and Type 1 use is applied to them before clinical use.

CONCLUSIONS

The existing regulations are evidence that the Japanese regulatory system is in line with international regulations that address regenerative medicine therapeutics. The government of Japan approved the second phase of its five-year health and medical strategy, which was implemented on 1 April 2020. The new strategy established six priority areas for further research and development, including pharmaceuticals; regenerative, cell and gene therapies; genomes and data platforms; medical equipment; and basic disease research. Japan is poised to grow as an investment and approval destination for multinationals developing regenerative treatments.

WEBSITES

MHLW PMD Act. Available online: https://www.mhlw.gp.jp/english/policy/health-medical/pharmaceuticals/dl/150407-01.pdf (accessed: 04 July 2022)

MHLW Strategy of Sakigake. Available online: http://www.mhlw.go.jp/english/policy/health-medical/pharmaceuticals/140729-01.html (accessed: 04 July 2022)

PMDA Organization. Available online: https://www.pmda.go.jp/english/about-pmda/outline/0003.html (accessed: 04 July 2022)

PMDA Reviews. Available online: https://www.pmda.go.jp/english/review-services/reviews/0001.html (accessed: 04 July 2022)

PMDA Regulatory Science Center Available online: http://www.pmda.go.jp/english/rs-sb-std/re/0003.html (accessed: 04 July 2022)

REFERENCES

Charo, R. A. (2016). On the road (to a cure?) – Stem-cell tourism and lessons for gene editing. *N Engl J Med*, *374*(10), 901–903. doi:10.1056/NEJMp1600891

Cyranoski, D. (2010). Korean deaths spark inquiry. *Nature*, *468*(7323), 485. doi:10.1038/468485a

Jokura, Y., Yano, K., & Yamato, M. (2018). Comparison of the new Japanese legislation for expedited approval of regenerative medicine products with the existing systems in the USA and European Union. *J Tissue Eng Regen Med*, *12*(2), e1056–e1062. doi:10.1002/term.2428

Konomi, K., Tobita, M., Kimura, K., & Sato, D. (2015). New Japanese initiatives on stem cell therapies. *Cell Stem Cell*, *16*(4), 350–352. doi:10.1016/j.stem.2015.03.012

Nagai, S. (2019). Flexible and expedited regulatory review processes for innovative medicines and regenerative medical products in the US, the EU, and Japan. *Int J Mol Sci*, *20*(15). doi:10.3390/ijms20153801

Nagai, S., & Ozawa, K. (2016). Regulatory approval pathways for anticancer drugs in Japan, the EU and the US. *Int J Hematol*, *104*(1), 73–84. doi:10.1007/s12185-016-2001-7

Nagai, S., & Ozawa, K. (2017). New Japanese regulatory frameworks for clinical research and marketing authorization of gene therapy and cellular therapy products. *Curr Gene Ther*, *17*(1), 17–28. doi:10.2174/1566523217666170406123231

Okada, K., Koike, K., & Sawa, Y. (2015). Consideration of and expectations for the pharmaceuticals, medical devices and other therapeutic products act in Japan. *Regen Ther*, *1*, 80–83. doi:10.1016/j.reth.2015.04.001

Sipp, D. (2015). Conditional approval: Japan lowers the bar for regenerative medicine products. *Cell Stem Cell*, *16*(4), 353–356. doi:10.1016/j.stem.2015.03.013

Takashima, K., Morrison, M., & Minari, J. (2021). Reflection on the enactment and impact of safety laws for regenerative medicine in Japan. *Stem Cell Reports*, *16*(6), 1425–1434. doi:10.1016/j.stemcr.2021.04.017

Yokohama, D. (2014). https://www.jesc.or.jp/Portals/0/center/activity/kokusai/H25_yokohama_en.pdf Accessed 04, july 2022.

13g Regulatory Landscape in India

Arun Bhatt
Clinical Research & Drug Development –
Consultant, Mumbai, India

Rajesh Jain
PharmaLex India Pvt. Ltd, Mumbai, India

CONTENTS

Introduction 295
Stem Cell Therapy Guidelines 296
    Classification of Stem Cells 296
    Review and Oversight Committees 297
    Ethical Considerations 297
    Scientific Considerations 298
Gene Therapy Guidelines 301
    GTP 301
    Classification of Gene Therapy 301
    Review and Oversight Committees 302
    Ethical Considerations 302
    Responsibility of the Stakeholders 303
    Scientific Considerations 304
    Regulatory Process for Clinical Trial Approval and New Drug
    Approval 307
Conclusions 309
References 309

INTRODUCTION

India is one of the emerging hot spots for the development of regenerative medicine applications such as stem cell-based therapies. A search of the Clinical Trial Registry of India showed 55 clinical trials registered till June 2022. Despite rapid progress in stem cell research, Stempeucel® – a human bone marrow–derived mesenchymal stem cells–based therapy – is the only product currently approved in India. Academic researchers, concerned about unregulated growth of unproven stem cell therapies, have felt the need to create a robust regulatory environment to promote scientific and ethical clinical trials and availability of effective innovative

DOI: 10.1201/9781003285069-20

stem cell–based treatments. In India, cellular therapies and human cells and tissues used for transplantation have been included in regulations as biological products. To facilitate the development of safe and effective stem cell treatments, the Indian Council of Medical Research (ICMR) and the Department of Biotechnology (DBT) published National Guidelines for Stem Cell Research (NGSCR) in 2007, which were revised in 2017.

Gene therapy is still in its infancy in India. In 2019, ICMR and DBT have developed National Guidelines for Gene Therapy Product Development and Clinical Trials (NGGTPDCT).

Although Indian guidelines refer to US Food and Drug Administration and European guidance documents, the regulatory documentation requirements and review processes are different. Details of Indian regulations are described in this chapter.

STEM CELL THERAPY GUIDELINES

NGSCR defines stem cells as undifferentiated cells with a capacity for self-renewal, proliferation, and differentiation into diverse types of functional cells. The guidelines provide a framework for monitoring mechanisms and regulatory pathways for basic, clinical research and product development. They also discuss procurement of gametes, embryos and somatic cells for derivation and propagation of any stem cell lines; their banking and distribution; international collaboration; exchange of cell/lines; and education for stakeholders and advertisement. These guidelines apply to all stakeholders who may be associated with both basic and clinical research involving any kind of human stem cells and their derivatives. The guidelines do not apply to the use of hematopoietic stem cells. The guidelines categorize research into permissible areas, restrictive areas and prohibited areas.

Cell or stem cell–derived products are new drugs as per New Drugs and Clinical Trials Rules 2019 (NDCTR). Regulatory requirements and NGSCR for the clinical development of stem cells are discussed below.

CLASSIFICATION OF STEM CELLS

Stem cells are classified as 'Somatic Stem Cells' (SSCs), and 'Embryonic Stem Cells' (ESCs).

Different levels of stem cell manipulation include:

- **Minimal Manipulation** refers to processing that does not alter the number or the biological characteristics and function of the cells or tissue relating to their utility for reconstruction, repair or replacement. Processing includes simple isolation/separation, washing, centrifugation and suspension in culture medium/reagents, cutting, grinding, shaping, overnight culturing without biological and chemical treatment, decellularization and cryopreservation for a period not exceeding 72 hours.

Regulatory Landscape in India

- **Substantial or More than Minimal Manipulation** is ex vivo alteration in the cell population (enhancement or depletion of specific subsets), expansion, cryopreservation, or cytokine-based activation, but one not expected to result in alteration of cell characteristics and function.
- **Major Manipulation** refers to the genetic and epigenetic modification of stem cells, transient or permanent or of cells propagated in culture leading to alteration not only in their numbers but also in biological characteristics and function.

Review and Oversight Committees

NAC-SCRT and IC-SCR, constituted as per Annexure I, should ensure that the review, approval, and monitoring processes of all research projects in the field of stem cell research are conducted in compliance with the national guidelines. Table 13g.1 summarizes the responsibilities per committee when reviewing, approving and monitoring research projects.

Ethical Considerations

The investigator/institution, the sponsors and the IEC should follow the ethical principles of the National Ethical Guidelines for Biomedical and Health Research Involving Human Participants, 2017.

TABLE 13G.1
Review Committees for Stem Cell Research

Committee	Responsibility
National Apex Committee for Stem Cell Research and Therapy (NAC-SCRT)	• Constituted by Department of Health Research (DHR), Govt. of India • Serve as an advisory body to promote and facilitate stem cell research in India • Perform a comprehensive review of the therapeutic use of stem cells and formulate policies to curb unethical practices • Review specific controversial or ethically sensitive issues referred to the committee • Review annual reports of IC-SCR to ensure compliance with national guidelines and ethical practices
Institutional Committee for Stem Cell Research (IC-SCR)	• Independent functioning body at the institutional level that registered with NAC-SCRT • Approval for undertaking any institutional stem cell research and clinical trials
Institutional Ethics Committee (IEC)	• Review and approval of clinical trial protocol, informed consent and other documents

Scientific Considerations

Scientific aspects of stem cells as therapeutic options include:

- **Prohibited fields for clinical research/trials**
  - Human germ-line gene therapy and reproductive cloning
  - In vitro culture of intact human embryos
  - Xenogeneic cells
  - Xenogeneic-human hybrids
  - Genome-modified human embryos, germ-line stem cells or gametes for developmental propagation
  - Implantation of human embryos after in vitro manipulation into uterus in humans
- **Chemistry, Manufacturing and Control (CMC) requirements**: The guidelines highlight the need for (1) detailed description of degree of manipulation(s) required for cell processing, (2) thorough documentation of all processes (3) requirements for aseptic production in facility. CMC should include procedures for:
  - Cell Collection/Processing/Culture Conditions
  - Cell Culture
  - Final Cell Harvest
  - Process Timing and Intermediate Storage
  - Final Formulation

Processing stem cells or their products/derivatives should be done in a Central Drugs Standard Control Organization (CDSCO)-certified and licensed facility complying with Good Manufacturing Practice (GMP) requirements.

- **Quality control and quality assurance (QC/QA)**

For all clinical trials, rigorous QC/QA for product development is critical and should be compliant with requirements as per Schedule M of the Drugs and Cosmetics Act, 1940. Institutions or companies involved in clinical trials using stem cells and/or their derivatives should prepare detailed Standard Operating Procedures (SOPs) on the development and manufacturing processes, to provide reproducibility of well-characterized, clinical-grade cells that meet the desired standards of quality.

QA/QC release criteria for stem cell–derived products as per NGSCR Annexure VI include:

1) Product Identity
2) Cellular Component
3) Non-cellular Components or the Active Substance
4) Product Purity
5) Impurities
6) Viability

7) Potency
8) Tumorigenicity

Release criteria specifications for the final product should be provided in the format prescribed in the NGSCR.

- **Labeling and packaging**
  - For an investigational product for clinical trial, the label should be: "Caution: New Drug – Only for Investigational Use."
  - To reduce the chance of potential mix-ups, the label should contain (1) the date of manufacture, (2) storage conditions, (3) expiry date and time, (4) product name and (5) two non-personal patient identifiers.
  - For autologous cells intended for autologous use the label should be: "FOR AUTOLOGOUS USE ONLY" and "NOT EVALUATED FOR INFECTIOUS SUBSTANCES" if donor testing and screening are not performed.
- **Shipping and transport**
  - The stability studies should demonstrate that integrity, sterility and potency of the product are maintained under the proposed shipping conditions.
  - If the final product is delivered in a frozen state to the clinical trial site, the product description should include the process of shipping and data to show that the product can be thawed with consistent results.
- **Preclinical studies**
  - All laboratory procedures, e.g., bio-distribution, safety and toxicity studies should be conducted under aseptic conditions in a CDSCO-certified GMP and Good Laboratory Practice facility (GLP) for human applications.
  - For preclinical studies on animals, the laboratory should have GLP certification from the Department of Science and Technology (DST).
  - Preclinical studies can be allowed only after approval from IC-SCR, the Institutional Animal Ethics Committee (IAEC), and the Committee for the Purpose of Control and Supervision of Experiments on Animals (CPCSEA).
  - For preclinical studies involving human tissue, approval from IEC is necessary.
  - Duration of safety study might be longer as compared to standard single-dose studies for chemical entities since the infused cells/biological entities may induce long-term effects.
  - Risks for tumorigenicity should be thoroughly assessed before initiation of the clinical trial.
- **Clinical trial conduct**
  - Stem cell administration to humans outside the purview of clinical trials is not allowed. Commercial therapeutic use of stem cells is prohibited.

- An institution or laboratory developing or processing stem cells for human use should have National Accreditation Board for Testing and Calibration (NABL) accreditation for all laboratory procedures required for product development.
- CDSCO approval is mandatory even if the products are not intended for market authorization.
- Import of any type of stem cells and/or their products/derivatives for clinical trials requires a license from CDSCO and approvals from the regulatory authority of the country of origin.
- For international collaborative projects, prior approval of the Health Ministry Screening Committee (HMSC) is necessary.
- Clinical trials are permitted only in institutions/hospitals having IC-SCR registered with NAC-SCRT and IEC registered with CDSCO.
- The investigator or sponsor should obtain approval of CDSCO for clinical trials after obtaining clearance from both IC-SCR and IEC of the trial site.
- All investigators participating in a clinical trial should obtain approvals from their own IC-SCR and IEC.
- Clinical trials should be conducted using only clinical-grade cells.
- Each institution should have a list of investigators conducting stem cell research and ensure that NGSCR and NDCTR are followed.
- The protocol should be designed carefully as per NGSCR Annexure II and NDCTR.
- Prior to procurement of biological material for isolation of stem cells, voluntary informed consent from the donor should be video recorded and documented.
- The patient information sheet (PIS) and the informed consent form (ICF), prepared as per NDCTR, should have prior approval of IC-SCR and IEC.
- The clinical trial participant should be informed about specific aspects of stem cell research, such as:
  - Status on the application of stem cells in the disease
  - Experimental nature of the proposed clinical trial
  - Possible short- and long-term risks and benefits
  - Irreversibility of the intervention
  - Source and characteristics of stem cells and degree of ex vivo manipulation, if any
  - IC-SCR and IEC approvals
- The follow-up period should be two years or longer depending on the type and source of cells used, the intended clinical application and the age and gender of the recipient.
- Data Safety Monitoring Board (DSMB) should be set up prior to conducting the clinical trial.
- The institution/investigator should maintain medical and study records of clinical trial participants for at least 15 years.

GENE THERAPY GUIDELINES

NGGTPDCT provides guidance for research, development and clinical trials of Gene Therapy Products (GTPs) in India. All GTPs being developed for human applications must adhere to these guidelines and should be used only under the purview of well-defined and approved clinical trials.

Gene therapy is defined as the process of introduction, removal or change in content of an individual's genetic material with the goal of treating the disease and a possibility of achieving a long-term cure. Gene therapy encompasses all processes wherein a nucleotide sequence with or without its regulatory elements required for correction of a deleterious or defective genotype or phenotype is introduced.

GTP

This is a biologic that could introduce alterations in the genome, including ex vivo gene-modified/edited cells/tissues/organs, to achieve a therapeutic outcome.

GTPs include:

- In vivo and ex vivo molecular therapeutics such as:
    - Recombinant viral vectors
    - Non-viral vectors
    - Microbial/bacterial vectors
    - Oncolytic viruses
    - Modifications resulting from the use of CRISPR and other similar technologies
    - ShRNA and siRNA
    - Ex vivo genetically modified cells
    - Soluble/particulate/emulsion/nano-based interventions containing any form of genetic material/nucleic acid
    - Combination of any of the above with other cellular/non-cellular drugs and/or devices for the purpose of treatment, augmentation, or vaccination
    - DNA vaccines where the final product is a nucleic acid

As GTPs are new drugs, academic trials of GTPs cannot be conducted. Regulatory requirements and guidelines relevant to approval of clinical trials of GTP are discussed in the following section.

CLASSIFICATION OF GENE THERAPY

- Germ-line gene therapy, applied to germ line or gametes, which can be transmitted vertically across generations, is prohibited in India due to ethical and social considerations.

- Somatic cell gene therapy affects the targeted cells/tissue/organs in the patient and is not passed on to subsequent generations. This includes:
  - ex vivo approach: Cells from an individual are genetically modified/corrected outside the body followed by transplantation into the same or a different individual. In clinical trials involving ex vivo gene therapy, if the cells from the human participant e.g., stem cells or bone marrow cells, are harvested and genetically modified with or without expansion in the laboratory, the final modified cell is considered GTP. In ex vivo approach, modified cells are conferred new properties via the introduction of genetic material or modification of existing genes, e.g., expression of novel chimeric antigen receptors on immune cells to target cancers. Stem cells expressing wild-type copies of mutated genes or corrected for mutations using gene editing are considered GTPs.
  - ex vivo approach: Gene-modified cells are transplanted with or without expansion in vitro. Cells/stem cells that are genetically modified through any technique involving the introduction of nucleic acid sequences to differentiate into organ-specific cell types to regenerate various organs are considered GTPs. Such genetically manipulated stem cells would require additional approvals as per NGSCR.
  - in vivo approach: Gene is delivered directly to target cells/tissues/organs in the patients. Gene augmentation, replacement of a mutated copy of the gene by a healthy copy, correction of mutation by gene editing methods, silencing of a dominant mutation, altering the expression of genes by affecting transcription (transcription factors, epigenetic modulators) or splicing (exon skipping) are all considered GTP.

REVIEW AND OVERSIGHT COMMITTEES

The scientific and ethical concerns for GTP clinical trials require a thorough evaluation of the profound effect that genes can exert on living cells by conferring novel properties and functions, and can cause serious harms such as teratogenicity, excessive immune activation, introduction of unwanted mutations or unwanted host immune response. These need a robust review and oversight from Gene Therapy Advisory and Evaluation Committee (GTAEC), Review Committee on Genetic Manipulation (RCGM), Institutional Biosafety Committee (IBSC) and IEC. Table 13g.2 summarizes the responsibilities for each committee involved in research and clinical trials.

ETHICAL CONSIDERATIONS

All stakeholders involved in the development of GTPs should ensure the protection of rights, safety, dignity and fundamental freedom of the participant. This includes all processes of obtaining human tissues/cells for research, diagnosis and planning and conduct of clinical trials. They should follow the ethical principles

TABLE 13G.2
Review Committees for GTP Research and Clinical Trials

Committee	Responsibility
GTAEC	• Review all GTP clinical trial applications and provide recommendations to CDSCO and provide inputs to Cell Biology–Based Therapeutic Drug Evaluation Committee (CBBTDEC). • Set up standards in coordination with the CDSCO, for safety, efficacy, quality control, procedures for GTP and its licensing/approval. • Review and approval of scientific, clinical and ethical content of clinical trials. • Collect all the data related to GTP clinical trials that would help in improving the quality and conduct of a clinical trial and their comparison and harmonization with international data. • Formulate policies to inculcate scientific and ethical practices among stakeholders.
RCGM	• Approval of GTP and its components to be used and procedures • Approval of import/export and exchange of all recombinant DNA products.
IBSC	• Oversee and establish institutional procedures planned for GTP development, production and preclinical testing for the safety of the GTP.
IEC	• Review and approval of the clinical trial protocol, informed consent and other documents.

of the National Ethical Guidelines for Biomedical and Health Research Involving Human Participants, 2017.

RESPONSIBILITY OF THE STAKEHOLDERS

- **Investigators/Institution**
  - Each institution should have a list of investigators conducting gene therapy trials and should ensure that national guidelines, regulations and best practices are followed.
  - The GTPs should be processed in an approved GLP and GMP facility as per the Drugs and Cosmetic Act, 1940.
  - The institute should have SOPs for development, production, storage and disposal of the GTPs or its components as per the Regulations and Guidelines on Biosafety of Recombinant DNA Research and Biocontainment 2017.
  - The institute should perform proper risk assessment to ensure containment measures are in place prior to safe operation to ensure product/personnel protection.
  - Exchange of indigenously developed GTPs or their components with investigators/institutions within the country or outside the country requires a Memorandum of Understanding (MoU) or Material Transfer Agreement (MTA) with the other investigator/institute and prior approval of Ethics Committee (EC) and IBSC.

- For international collaborative projects/global clinical trials, prior approval of HMSC is necessary.
- Institutions conducting development of GTPs or clinical trials of GTPs involving stem cells should follow NGSCR.
- **EC**
  - The EC must ensure that the SOPs for research and clinical trials for GTP comply with the national guidelines.
  - Biological material from humans can be obtained only from clinics/hospitals that have an EC.
- **Sponsors**
  - Government agencies/other sponsors facilitating gene therapy trials must ensure that the project submitted for financial support is approved by GTAEC and CDSCO, and preclinical animal studies have been approved by IBSC and RCGM.
  - The clinical trial project team should include a GTP expert or collaborate with the GTP producer who can provide guidance about the GTP characteristics relevant to clinical management and patient safety.
  - The sponsor should organize specialist training and ensure the availability of certain drug classes, e.g., IL6 receptor antagonists, to treat common and severe immediate post-treatment immune reactions.
  - If the route of administration is novel or very invasive, the clinical team should have the capability to perform the same and minimize risk to subjects as well as manage unforeseen complications.
  - The clinical team also must have procedures and mechanisms in place to provide long-term care to patients in a GTP trial.

SCIENTIFIC CONSIDERATIONS

These include:

- selection of appropriate gene delivery vector/modality for the disease/tissue target,
- design of the expression cassette to ensure clinically relevant expression levels,
- specificity of gene expression to prevent unwanted side effects or off-target effects and
- minimizing immune reactions of the host.
- **CMC**

All details of the manufacture, storage and transportation of each individual component of the final GTP should be described in an approved Product Specification File and should adhere to GMP requirements.

- CMC information for GTP should include all the details for:
  1) Vector Components: Vector Sequence and Maps, Vector Sequence Analysis

Regulatory Landscape in India

  2) Cellular Components and Reagents for GTP Production: Autologous and/or Allogeneic Cells
  3) Banking of GTPs
  4) Maintenance of GTP Stock
  5) Working Cell/GTP Bank
  6) Reagents
  7) Excipients
- Product manufacturing (GMP-grade) should describe all critical procedures followed for:
  1) Vector Production/Purification
  2) Preparation of Ex Vivo Gene-Modified Autologous or Allogeneic Cells
     a. Method of Cell Collection/Processing/Culture Conditions
     b. Ex Vivo Gene Modification
     c. Irradiation of Ex Vivo Gene-Modified Cells
     d. Final Harvest
  3) Process Timing and Intermediate Storage
  4) Final Formulation
  5) Product Testing
  6) Microbiological Testing
     a. Sterility Testing (Bacterial and Fungal Testing)
     b. Mycoplasma Testing
     c. In Vitro Viral Testing
     d. In Vivo Viral Testing
     e. Species-Specific Virus Testing
     f. Retrovirus Testing
     g. Adenovirus Testing
     h. Adeno-associated Virus Testing
  7) Final GTP
     a. Identity
     b. Purity – Leftover Contaminants, Pyrogenicity/Endotoxin
     c. Potency
- **GMP Guidelines**
  1) Organization and Personnel: All personnel should have relevant scientific qualifications and should be trained in aseptic procedures, testing for contaminants and GTP production processes. Access in GTP production should be limited to authorized personnel. They should wear protective apparel covering the whole body to protect the product from contamination.
  2) Infrastructure: Design and construction of the facility should follow the guideline for (a) areas for manufacturing production and storage of GTP, (b) instrumentation, (c) tissue and cell culture, (d) waste disposal, (e) process data, (f) facility validation and (g) sanitation.
  3) Production and Process Controls: Written protocols for production and process control designed to assure quality and purity should be available.

4) QC Unit: It should approve each batch of GTP produced to ensure that (a) sterility and GMP are maintained for all critical processes and (b) all materials used are as per the standards. SOPs should be established to ensure the quality, purity, potency and identity of the final GTP stocks.

- **Preclinical Studies**
  - If the GTP has been previously authorized for clinical use elsewhere for the same indication, preclinical evaluation may be expedited.
  - Expedited preclinical evaluation is not applicable to GTPs (a) that are developed for first clinical use, (b) that are a modification of previously used GTP, (c) that require preclinical testing as a part of lot-release criteria of clinical-grade GTP or (d) for which outcome data is only available from a "similar" GTP but not the same GTP.
  - Imported GTPs or their modified variants must undergo preclinical animal model studies with due approval of RCGM.
  - Preclinical studies should be conducted in suitable non-human vivo and in vitro models.
  - Efficacy testing should focus on (a) vector and the nucleic acid cargo, (b) route of administration, (c) pharmacokinetics and (d) dose-response.
  - Multiple routes of administration must be studied in disease-specific preclinical models to determine their relative efficacies.
  - Safety testing should include toxicity studies, monitoring off-target effects, bio-distribution and environmental risk assessment studies.
- **Clinical Trials Planning and Conduct**
  - A first-in-human clinical trial of a GTP of foreign origin or its modified variants is not allowed in India.
  - Clinical trials of GTP are mostly single-arm trials. Hence, historical data and natural history of the disease must be obtained prior to patients being considered for enrollment.
  - The sponsors/investigators must provide information from genetic tests for all clinical trial participants to establish correct diagnosis and patient selection.
  - Trial sponsors must provide sequencing data from manipulated cells and broad gene expression or protein profiling to address concerns about off-target effects during the trial phases.
  - As therapeutic potential of the GTP may be limited by adverse immune reactions, all potential trial participants should be tested for pre-existing neutralizing antibodies to the vectors used in the GTP.
  - The protocol should be designed as per Appendix II of guidelines and NDCTR, considering all unique and critical aspects such as selection criteria, disease severity/staging, co-morbidities, prior immune or recombinant protein therapy, dose and route of administration, unique study designs tailored to small population samples, Quality of Life and risk evaluation and mitigation procedures.

Regulatory Landscape in India

- The trial design should include detailed clinical phenotyping and companion diagnostic and monitoring tests throughout the trial to enable collection of extensive data on efficacy, immune reactions, adverse events, biomarkers, phenotypic data. These will be essential for subsequent studies, approval of market authorizations and development of related GTP.
- Since the risks are typically high, healthy volunteers/placebo groups should not be included in trials.
- Limited patient pool of inherited, familial, rare diseases may not permit large-scale clinical trials.
- Each patient should be requested to complete disease-related questionnaires, read the PIS and sign ICF.
- Patient and caregiver/relative should be counseled and provided with a detailed description of risk and benefit parameters.
- Patient can withdraw from the trial at any time before the administration of GTP. However, once GTP is administered, s/he should remain part of the clinical trial.
- The investigators should exercise extreme caution while increasing the number of enrolled patients, as the effects of GTP could be serious and far-reaching.
- Enrollment of vulnerable populations should follow relevant regulations and guidelines.
- Each trial can use only one GTP vector type for delivery.
- Immune suppression is a critical component to be considered in clinical trial design.
- Efficacy endpoints should take into consideration, GTP expression, bioactivity of the GTP, biochemical, enzymatic, sequencing-based end points for gene transfer and clinical improvement.
- Clinical development plans should include a detailed monitoring plan to protect the trial participants against any unforeseen adverse events and serious adverse events.
- The trial sponsors should develop monitoring tests for potential innate and adaptive immune responses likely to occur in the administered patients.
- DSMB should monitor the compliance during the trial.
- As GTP introduces new genetic elements or alterations. Long-term follow-up of five to ten years is strongly recommended.
- All records about GTPs and clinical trials must be maintained for at least 15 years.

REGULATORY PROCESS FOR CLINICAL TRIAL APPROVAL AND NEW DRUG APPROVAL

The clinical trial approval or new drug approval process for stem cell-based therapy or GTP is as prescribed in NDCTR 2019.

For clinical trial approval, the following documents/information are required:

- Chemical and pharmaceutical information (Second Schedule Table 1)
- Protocol (Third Schedule Table 2)
- Case Record Form
- Investigator's Brochure (Third Schedule Table 7)
- Undertaking by the Investigator (Third Schedule Table 4)
- Informed Consent Document and ICF (Third Schedule Table 3)
- EC approval from committee with CDSCO (Third Schedule Table 1)
- Sponsor's undertaking to provide compensation in the case of clinical trial related injury or death
- For new chemical entity and global clinical trial:
  a. Assessment of risk versus benefit to the patients
  b. Innovation vis-à-vis existing therapeutic option
  c. Unmet medical need in the country.
- For Global clinical trial:
  a. Justification for undertaking clinical trials in India
  b. Number of participants from India

For marketing authorization, the following documents/information as per Second Schedule Table 13g.1 are needed:

1. Introduction: A brief description of the drug and the therapeutic class to which it belongs
2. Chemical and pharmaceutical information
3. Animal pharmacology
4. Animal toxicology
5. Phase I Human or Clinical pharmacology (Report format Third Schedule Table 6)
6. Phase II Therapeutic exploratory trials (Report format Third Schedule Table 6)
7. Phase III Therapeutic confirmatory trials (Report format Third Schedule Table 6)
8. Special studies, if relevant
9. Regulatory status in other countries
10. Prescribing Information (Third Schedule Table 8)
    1. Proposed full prescribing information
    2. Drafts of labels and cartons
11. Samples and testing protocol/s
12. New chemical entity and global clinical trial:
    1. Assessment of risk versus benefit to the patients
    2. Innovation vis-à-vis existing therapeutic option
    3. Unmet medical need in the country
13. Copy of license to manufacture any drug for sale granted by state licensing authority

Permission of CDSCO is needed before the manufacture or import of an investigational new drug for clinical trial.

The application for clinical trial approval to CDSCO can be done only after all other approvals from different committees are in place. CBBTDEC reviews applications for clinical trials and marketing authorization for stem cells and recommendations to CDSCO to approve or reject an application.

CONCLUSIONS

Research and development of stem cells and GTPs is still evolving in India. Overall regulatory approach and requirements for data and documentation follow guidelines from developed economics. However, the need for reviews and oversight by institutional and government committees seems to be a mechanism to ensure that the highest standards of science and ethics are maintained during research and development of high-risk innovative therapies. This complex regulatory milieu will become less challenging for the researchers and sponsors as the country gains more experience in understanding and monitoring the risks and benefits of advanced therapies.

REFERENCES

Central Drugs Standard Control Organization Drugs and Cosmetics Act 1940. Available from: https://cdsco.gov.in/opencms/export/sites/CDSCO_WEB/Pdf-documents/acts_rules/2016DrugsandCosmeticsAct1940Rules1945.pdf Last accessed on 14 Jun 22

Central Drugs Standard Control Organization. New Drugs and Clinical Trials Rules 2019. Available from: http://www.cdsco.gov.in/opencms/opencms/Pdf-documents/NewDrugs_CTRules_2019.pdf. Last accessed on 26 Mar 2019

Indian Council of Medical Research Medical Research and Department of Biotechnology. National Guidelines for Gene Therapy Product Development and Clinical Trials. Indian Council of Medical Research, New Delhi, November 2019. Available from: https://main.icmr.nic.in/sites/default/files/guidelines/guidelines_GTP.pdf. Last accessed on 14 Jun 22

Indian Council of Medical Research Medical Research and Department of Biotechnology. National Guidelines for Stem Cell Research. Director General Indian Council of Medical Research New Delhi, October 2017. Available from: https://main.icmr.nic.in/sites/default/files/guidelines/Guidelines_for_stem_cell_research_2017.pdf. Last accessed on 14 Jun 22

Indian Council of Medical Research National Ethical Guidelines for Biomedical and Health Research Involving Human Participants. Available from: http://www.icmr.nic.in/sites/default/files/guidelines/ICMR_Ethical_Guidelines_2017.pdf. Last accessed on 16 Oct 2017

Viswanathan S, Rao M, Keating A, Srivastava A. (2013, August). Overcoming challenges to initiating cell therapy clinical trials in rapidly developing countries: India as a model. *Stem Cells Transl Med*, 2(8), 607–13. doi: 10.5966/sctm.2013-0019. Epub 2013 Jul 8. PMID: 23836804; PMCID: PMC3726140

14 Avoiding Pitfalls during Advanced Therapy Development

Kirsten Messmer
Agency IQ, Garner, USA

CONTENTS

Introduction 311
    Chemistry, Manufacturing and Control 312
    Nonclinical Studies 314
    Clinical Studies 315
    Approval and Beyond 318
Conclusion 318
References 319

INTRODUCTION

Advanced therapy development has seen rapid growth. The number of products under development is growing almost exponentially. Many global regulators have issued guidance for cell and gene therapies in general, but also addressing specific technologies such as gene editing and Chimeric Antigen Receptor (CAR)-T cell therapy development. However, advanced therapy product development poses specific challenges that can occur at any stage of development and may involve various functions (e.g., regulatory, clinical trial team, commercial). It is pertinent to strategically assess any potential issues before they develop and find appropriate solutions. Of course, not all problems can be foreseen.

Nonetheless, developers of advanced therapies should consider the entire process of product development, marketing authorization, pricing/reimbursement and providing the product to patients at the start of the program. This forethought will facilitate resolving future issues efficiently to avoid program delays such as clinical holds or extensive questions (or even rejection) during the marketing authorization application assessment. It is important to involve a cross-functional team in developing a regulatory strategy for product development. It is equally important to ensure the strategy is frequently updated as product development progresses, new knowledge is gained and additional guidelines from regulatory authorities and/or ethics committees may become available.

This chapter provides an overview of some potential issues throughout the various development stages but is not all-inclusive.

CHEMISTRY, MANUFACTURING AND CONTROL

Chemistry, manufacturing and controls are one of the most important aspects of advanced therapy development. Since the products are inherently more complex, their characterization, manufacturing process, attributes to control and specifications to apply should be determined early and appropriately controlled. Ensuring that the advanced therapy is consistently of good quality, safe and effective is paramount. It is not wise to consider 'fixing' any gaps later, as these can swiftly turn into costly delays through clinical holds or issues during the marketing authorization application. Requests for additional studies by regulatory authorities to ensure product quality should be strictly avoided.

The differentiation between active/drug substance and drug product can be somewhat complicated for cell and gene therapies. Questions such as is the vector, the cells or a combination the active substance must be determined early as they will play a key role during development of the quality control strategy. If in doubt, it would be prudent to consult with the regulatory agencies involved to ensure alignment.

Products that use cells from human donors such as cell therapies or CAR-T cell gene therapies and others require donor eligibility determination. The initial donor testing, eligibility determination and procurement of the source material must be mapped out early. Donor testing and eligibility determination in large part prevent the transfer of adventitious agents and transmissible diseases. However, the prevalence of disease will vary by region and with that the testing panel required by the regulatory authority. Even within the EU, the test requirements may vary by individual Member States. The testing strategy needs to be determined early and should potentially be aligned with the regulatory authorities involved. This buy-in on the testing panel is particularly important if minimization of the required agents to test for is desired. Additionally, facilities that procure somatic cells and/or tissue-based products will require a tissue establishment license and/or register that must be in place once procurement starts. These logistics and needed certifications should be laid out early to avoid delays.

Scalability of the manufacturing process is another consideration that should be addressed early. Developers should ask, how large does each batch need to be and how early could I potentially reach this batch size? Vector-based gene therapies and allogeneic cell therapies lend themselves to larger batch sizes. It is important to consider the scale-up early. Generally, regulatory authorities require that clinical data needs to be obtained with the intended commercial product. Quality and nonclinical investigations may be acceptable even if they use a smaller-scale manufacturing process. However, clinical studies need to be conducted with drug products obtained through the planned commercial process. The sooner the final commercial manufacturing process can be established, the better. Generally, comparability studies would be needed if there are differences in the manufacturing process for the product used in Phase I to the product used in Phase III and intended for commercialization after authorization. This creates the risk of additional regulatory agency questions.

Autologous cell therapies and cell-based gene therapies potentially can only be manufactured in small batches and may need to consider some advanced manufacturing technologies. Point-of-care manufacturing (i.e., the product is manufactured in close proximity to the recipient) may be an option for autologous cell therapies. However, reproducibility of the manufacturing process becomes a concern to address if the advanced therapy is manufactured at the treatment site outside of a dedicated manufacturing facility. The manufacturing process developed during product development must consistently produce a safe and effective product that meets the applicable quality standards.

Aseptic manufacturing controls and viral safety assessment need to be considered since most advanced therapies will not be able to undergo terminal sterilization. A sterility-release testing strategy should be implemented early. The European Commission recently released the updated Good Manufacturing Practice guideline Annex I (EU, 2020) on sterile product manufacturing, which was subsequently adopted by the Pharmaceutical Inspection Co-operation Scheme PIC/S scheme (PIC/S, 2022) making it widely applicable although not legally binding since it is a guideline.

Gene therapy developers using viral vectors such as the adeno-associated virus should keep an eye out for potential additional guidance requiring further product testing and characterization. Although this viral vector was one of the earliest used for gene therapy development, the US Food and Drug Administration (US FDA) held a two-day advisory committee meeting (FDA, 2021) last year to address various potential safety issues. A large part of the debate centered around the presence, effect and/or function of empty capsids (i.e., capsids that do not contain the target vector). Since not much is known about their potential impact on safety, there may be additional requirements for characterization of the gene therapy product with respect to full and empty capsids, as well as those that only contain a fragment of the intended vector.

Cell and gene therapy products must be appropriately characterized to establish identity, correctness of any gene vectors involved, purity and any impurities present. The manufacturing process should involve steps to adequately clear any impurities from the product, but the remaining impurities must be determined. Product potency is often an additional factor that might be difficult to demonstrate. However, regulatory authorities generally are stringent on potency assessments. The development and use of an acceptable potency assay should be implemented early in the development program.

Setting specifications and ranges for controls can be particularly difficult for cell therapies and cell-based gene therapies. This type of therapy uses human source materials which naturally have a greater variability. Therefore, the ranges are wider, but they must still be appropriate to ensure a safe product of good quality is consistently manufactured. Developers should plan early to generate large enough batches and sufficient information per batch to support product quality information. Regulatory authorities may accept wider ranges and specifications, but they must be justified and supported by development data.

Chemistry, manufacturing and control deficiencies are often the reason for red flags and additional questions during the marketing authorization application assessment. The deficiencies may be resolved if additional data is available from the developer. However, it is more likely that data is missing, requiring additional studies to be conducted, which of course costs extra time. The need for additional studies to obtain missing data would likely lead to a complete response letter from the US FDA or a negative opinion from the European Medicine Agency's (EMA) Committee for Medicinal Products for Human (CHMP). Although many of these considerations seem obvious, they should be evaluated very early during development and built into the development plan and strategy to ensure adequate data supporting the consistent manufacture of a quality product with a commercial process that is safe and effective.

Cell and gene therapy developers should also keep up to date with ongoing workshops and advisory committee meetings since they may give rise to additional testing requirements in the future. Being aware of these potential requirements may provide an advantage through early implementation of additional testing. The US FDA two-day advisory committee meeting (FDA, 2021) to discuss safety issues related to gene therapies using adeno-associated viruses serves as one critical example. These meetings can provide a good gauge of the thinking of regulators and industry on specific issues that may be implemented in future additional or updated guidance.

Nonclinical Studies

Nonclinical studies are a key part of new drug development and are necessary before human studies are performed. However, standard nonclinical studies may not be possible due the specific nature of advanced therapies. The standard pharmacokinetic/pharmacodynamic and various toxicology studies may not provide information that would allow prediction of the response in humans. Considerations for nonclinical studies depend on the characteristics of the new therapeutic product and its primary mode of action and therefore can vary widely. The following are a few examples of challenges that can be encountered. However, the list is not exhaustive due to the varied nature of advanced therapies and the rapid emergence of new technologies.

Available animal models for nonclinical proof-of-concept studies should be carefully evaluated and their suitability justified. Generally, an animal model should as closely as possible mimic the human disease and disease progression. Similarly, the animal model should show a similar response to any therapeutic intervention. However, particularly for gene therapies, there may not be an adequate animal model. The genetic defect leading to disease in humans may not have the same effect in animals, or perhaps the gene may not even be present in a similar form. *Ex vivo* data and/or *in vitro* data may need to be generated to obtain the necessary nonclinical data. However, this potential hurdle should be considered very early in development to ensure a suitable model, whether it be *in vitro*, *ex vivo* or an animal model, is available to conduct nonclinical studies.

The distribution of cell and gene therapy throughout various tissues may give rise to additional safety and toxicity concerns. Cell and gene therapies generally aim to target specific tissues in which they exert the desired effect. However, these therapeutics may also distribute to nonspecific tissues where they could lead to immune, inflammatory or other adverse responses. As an example, gene editing-based products may lead to off-target effects (i.e., unintended effects) that need to be investigated. Developers should consider the potential effect on safety and toxicity of off-target alterations early in the development program.

Nonclinical studies will also be a pertinent tool in evaluating if the scale-up strategy was appropriately designed. Although the amounts of the therapeutic product may not be significantly more than that used for initial proof-of-concept studies, the product used in nonclinical studies should be representative of that used in clinical studies. Therefore, the ability to scale up the manufacturing process of nonclinical study material needs to be carefully considered. Comparability studies or a repeat of the nonclinical studies may be needed if a change in the manufacturing process is needed to generate enough clinical therapeutic product later in the clinical development program. Of course, either of these options would add time and cost to the development program.

Developers should consider discussing the available and chosen model(s) as well as nonclinical study design with regulatory authorities to ensure suitability before starting studies. Regulatory authorities offer opportunities to discuss innovative products at a very early developmental state (e.g., INTERACT meetings in the US FDA, 2020; Innovation Task Force, EMA, 2020; meetings in the EU). These meetings can provide an early indication of regulatory expectations for developers. Additionally, regulatory agencies gain insights into highly innovative products early on, which supports understanding during further regulatory activities in the development program.

CLINICAL STUDIES

There are numerous guidance documents addressing clinical studies and clinical study considerations for gene and cell therapies. In general, cell and gene therapy products are tested through the standard Phase I to Phase III program like traditional treatments before approval can be considered. However, there are additional factors and/or requirements implemented by national legislation, which may not necessarily be covered by guidance documents.

Clinical studies are generally regulated at the national level and each country may have different requirements. It is important to keep in mind, that although cell and gene therapy-based products must be approved through the European Commission, clinical trials with these products in the EU are still regulated at a national level and require approval by each Member State's regulatory agencies as well as ethics committees where the trial is to be conducted.

National regulators may have very strict requirements on participant selection and eligibility criteria. It will be critical to ensure all national guidelines and requirements are met to ensure approval of the clinical trial. Additionally, clinical trial participant selection for gene therapies may require genetic testing to ensure the potential participant has the appropriate mutation, deletion or addition for the proposed gene therapy to be effective. National requirements vary widely regarding the need for counseling to support genetic testing (e.g., absolute requirement or recommended). This is particularly important if the genetic testing could reveal additional genetic diseases that were previously not considered. It will be important to determine the need, content and timing for genetic counseling to ensure appropriate measures are put in place to ensure trial approval and an efficient start of participant enrollment.

Several countries (e.g., EU Member States, Australia) require additional approval by statutory bodies for genetically modified organisms, or GMOs. A GMO (EU, 2001) is "an organism, with the exception of human beings, in which the genetic material has been altered in a way that does not occur naturally by mating and/or natural recombination." The approval of clinical trials using GMOs may further be complicated if the country distinguishes between GMOs for "contained use" and those for "deliberate release." In the EU, according to the applicable directives, deliberate release (EU, 2001) means "any intentional introduction into the environment of a GMO or a combination of GMOs for which no specific containment measures are used to limit their contact with and to provide a high level of safety for the general public and the environment," while contained use (EU, 2009) is "any operation in which micro-organisms are genetically modified" or manipulated "for which physical barriers, or a combination of physical barriers together with chemical and/or biological barriers, are used to limit their contact with the general population and the environment." Some countries may consider all gene therapies as "deliberate release" or make the distinction between the two mechanisms. The terminology for these two conditions may also vary globally. For example, Australia makes the same distinction, but the terminology is changed to dealings involving or not involving deliberate release. Based on considerations of whether the GMO is contained or not, approvals from additional national authorities may be required as appropriate. This could add additional time to the compilation of the application package, approval time lines and overall burden of preparing the clinical trial application. However, particularly taking the impact on approval time lines into account will be critical to ensure clinical trial initiation in all planned countries according to the development plan.

Another question to ask during the selection of trial sites: Is there adequately trained personnel available and does the infrastructure allow handling and administration of cell and gene therapies? Cell and gene therapies are not ordinary biologics and require specific conditions to be viable therapeutic products. This is particularly important for cell therapy-based products, which likely require ensuring certain time lines and temperature requirements are met to ascertain sufficient viable cells in the manufactured batch and can be administered to the patient. Consider also that perhaps specific administration requirements may need to be met (e.g., intrathecal injection), which may require specific training of the personnel conducting the administration and adequate facilities.

Traditionally, clinical trial sites were selected by investigator experience and patient availability. However, this paradigm may not work for cell and gene therapies. There is a shift toward accounting for geographic distances and logistics between the experimental product manufacturing site to the trial site, availability of appropriate equipment at the trial site and adequately trained staff to perform the procedure. Time lines and trial feasibility will be largely influenced by the specific advanced therapy investigated, the material involved, and the skills required. Take for example the shelf life of a particular cell therapy: The final product must have sufficient numbers of viable cells in order to be effective and/or safe. The maximum time for isolation, manufacture into the desired product and administration will depend on the specific cell therapy product. A shorter shelf life will require manufacturing closer to the clinical trial site and may even involve manufacturing at the point of care. Staff may need to use specialized equipment such as laminar flow hoods and special centrifuges to handle starting material and/or final product. The advanced therapy may need to be stored under cryogenic conditions requiring that staff on the research team is appropriately trained in handling that equipment. The interrelationship of manufacture and administration sites, facility and personnel training requirements is an extremely important consideration that should be addressed early in the development program to ensure cross-functional alignment, the quality of the product and safety of the trial participant.

Many advanced therapies are initially developed in the academic research environment and may have been under development for years before progressing to clinical development and approval considerations. Records may only exist in paper format for some of these products. However, regulatory agencies are moving toward mandatory electronic submission which may necessitate the conversion of these existing records. There are two factors to consider: (a) time needed to convert all records to electronic files and potentially 'catch up' on submission and (b) completeness of information in the files. Consider an investigational program that spans one or two decades. During that time, investigators and staff may have changed, which may also have influenced the detail of information provided. It is important to assess the completeness of records and develop a 'conversion plan' for products that may have initial records that are paper based.

Long-term follow-up also deserves a special consideration as the required time frame is generally much longer than for traditional chemical or biological products. During that time patients may move, including to different countries. Therefore, it will be important to plan necessary follow-up visits and ensure that the most convenient options are available for trial participants. Budgetary planning may be required considering potential cross-border moves to ensure all participants can be followed up as required by the trial protocol. Additionally, emerging information may require a different follow-up schedule. Although planning for such events is almost impossible, general consideration should be given to such a situation. Follow-up procedures and contingencies should be developed as early as possible.

Approval and Beyond

The pricing and reimbursement strategy must be developed much earlier for advanced therapies than for other types of medicines. A couple of years ago, I asked two regulatory affairs professionals during a dinner discussion what they thought was most important to consider early during an advanced therapy development program. They both indicated that surprisingly, pricing and reimbursement stand out, as they need to be considered at the latest during Phase II, according to their judgment. Typically, most product candidates fail during clinical development due to a lack of safety and/or efficacy. Therefore, it would appear counterintuitive to consider the 'post-marketing strategy' at an early development stage. However, advanced therapies are extremely expensive products, and authorities making decisions on reimbursement of these products are increasingly looking at alternative pricing models. The regulatory authorities and decision-makers recognize the value of cell and gene therapies as they can often be curative. However, the cost may be the last stumbling stone to commercial success due to the high cost.

Developers of advanced therapies should explore options to consult with pricing/reimbursement authorities to ensure all evidence of effectiveness is gathered during the development program. There are opportunities to hold joint advice meetings with the EMA and health technology assessment bodies in Europe. Such meetings may provide a valuable resource to ensure a development program that addresses regulatory requirements, as well as those needed for market entry.

CONCLUSION

Development programs for advanced therapies need to evaluate a wider range of factors when designing the development plan and strategy due to their highly innovative nature. The development plan and strategy should also be updated frequently as new information becomes available. The risk of any issues or challenges occurring during the development program should be assessed early and mitigation options determined. However, these will likely change throughout the development program. A good rule of thumb for advanced therapies seems to be "more is better" – more controls, more considerations addressed, more information available, etc. There are a significant number of information requests during advanced therapy assessment from the US FDA. It would be prudent for developers to try and pre-empt these requests as well as any development challenges through due diligence and a well-planned development program.

Some factors to think about include:

- Early agency interactions should be considered to ensure a development program efficiently fulfills regulatory requirements.

Avoiding Pitfalls during Advanced Therapy Development

- Chemistry, control and manufacturing are an absolute must to get right and provide sufficient information. Most issues at the time of product approval arise in this area.
- Scale-up and/or scale-out should be considered very early since the intended commercial product must be comparable to the product used during product development.
- Discuss clinical models with regulators to ensure acceptability for regulatory decision-making.
- Account for extra work to develop the clinical strategy taking into account additional approval requirements, particularly for GMOs, clinical trial site experience regarding personnel and available equipment as well as logistics of any movements between sites required for the advanced therapy product.
- Ensure a commercial strategy is implemented early, as the final cost of advanced therapies is high and may require innovative reimbursement strategies.
- The development strategy requires a cross-functional team to address all aspects to ensure a time- and cost-efficient development program.
- Consider reviewing approval documents for approved advanced therapies to gain an understanding of questions asked by regulatory agencies during the approval process.

REFERENCES

EMA. (2020). Innovation in medicines | European Medicines Agency (europa.eu)

EU. (2001). Directive 2001/18/EC of the European Parliament and of the council of 12 March 2001 on the deliberate release into the environment of genetically modified organisms and repealing council directive 90/220/EEC. CL2001L0018EN0030010.0001.3bi_cp 1..1 (europa.eu)

EU. (2009). Directive 2009/41/EC of the European Parliament and of the council of 6 May 2009 on the contained use of genetically modified micro-organism. *Journal of the European Union*. Directive 2009/41/EC of the European Parliament and of the Council of 6 May 2009 on the contained use of genetically modified micro-organisms (Recast)Text with EEA relevance (europa.eu)

EU. (2020). Annex 1. Manufacture of sterile products. European Union. (europa.eu)

FDA. (2020). Initial targeted engagement for regulatory advice on CBER products (INTERACT). INTERACT Meetings | FDA.

FDA. (2021). Cellular, Tissue, and Gene Therapies Advisory Committee September 2–3, 2021 meeting announcement. Cellular, Tissue, and Gene Therapies Advisory Committee September 2–3, 2021 Meeting Announcement - 09/02/2021 - 09/03/2021 | FDA.

Pharmaceutical Inspection Convention Pharmaceutical Inspection Co-Operation Scheme (PIC/S). (2022). Revised ANNEX 1 (Manufacture of sterile medicinal products) to guide to good manufacturing practices for medicinal products. PS/INF 26/2022 (Rev. 1). PIC/S Secretariat, 14 rue du Roveray, CH-1207 Geneva. 4737 (picscheme.org).

15 Going Forward – Existing and Evolving Technologies (CRISPR, mRNA, siRNA)

Humberto Vega-Mercado
Bristol Myers Squibb, Cell Therapy Operations, Global Manufacturing Sciences and Technology, Warren, USA

Hazel Aranha
GAEA Resources Inc., Northport, USA

CONTENTS

Introduction 321
Clustered Regularly Interspaced Short Palindromic Repeats (CRISPR) –
Historical Review and Future 321
mRNA – Historical Review and Future 322
siRNA – Historical Review and Future 325
Final Remarks 328
References 330

INTRODUCTION

The advances in cell and gene therapies in recent years are in part the result of increased research involving cell biology, antiviral immunity, DNA and RNA functionalities, and the immune system responses across multiple research disciplines. In this chapter, we will focus on existing and evolving technologies that will continue to change the development of therapies (e.g., CRISPR, mRNA, siRNA).

CLUSTERED REGULARLY INTERSPACED SHORT PALINDROMIC REPEATS (CRISPR) – HISTORICAL REVIEW AND FUTURE

CRISPRs are repeating DNA sequences in the genomes of prokaryotes (e.g., bacteria) initially identified in E. coli in 1987 when a series of repeated sequences interspersed with spacer sequences were cloned while analyzing a gene responsible for the conversion of alkaline phosphatase (Ishino et al., 1987). Similar patterns were reported in other bacteria and halophilic archaea (Mojica et al., 1993, 1995).

Mojica et al. (2005) and Pourcel et al. (2005) reported that the spacer regions between the repeat sequences are homologous to sequences of bacteriophages, prophages, and plasmids, and they pointed out that the phages and plasmids do not infect host strains harboring the homologous spacer sequences in the CRISPR. It was suggested that the CRISPRs triggered the capture of foreign invading DNA, unique antiviral system, to create a memory of past viral infections from the same virus (Mojica et al., 2005 and Pourcel et al., 2005; Han and She, 2017). In summary, when a virus infects a cell, the spacer sequences in CRISPR arrays are transcribed to generate short CRISPR RNA (crRNA), which guides the CRISPR-associated sequence (Cas) protein to cleave complementary DNA or RNA viral sequences.

These discoveries culminated in the development of genome editing tools based on the CRISPR-Cas9 and other class II CRISPR-Cas systems as described by Travis (2015), Han and She (2017) and Ishino et al. (2018).

More recently, Carroll (2021) defines genome editing as the use of programmable DNA modifying agents that can be engineered to precisely attack genomic sites chosen at will. There are three principal platforms in this arena: zinc-finger nucleases (ZFNs), transcription activator-like effector nucleases (TALENs) and CRISPRa. CRISPR has dominated as a genome editing tool due to its simplicity since a single, constant protein is needed for all applications by designing one new guide RNA using Watson-Crick base pairing rules. CRISPR has revolutionized genetics research by making it possible to generate intentional, targeted genomic changes in essentially any sequence in any organism (Carroll 2021). Table 15.1 provides the reader with additional references on CRISPR, ZFN and TALEN (e.g., initial work and experiments to applications).

The use of CRISPR has transcended the academic world to become part of the industry. The reader can find in the market a variety of ready-to-use kits and solutions (e.g., software, reagents) to design, build, deliver, screen and validate CRISPR applications (Thermo-Fisher Sci. Inc., 2022; Bio-Rad, 2022; TriLink Biotechnologies, 2022). A typical microbiology laboratory setting is commonly used to handle and manipulate the reagents during gene editing activities.

mRNA – HISTORICAL REVIEW AND FUTURE

The current knowledge and use of messenger RNA (mRNA) and lyposomes are based on work dating back to the 1960s as summarized by Dolgin (2021):

a. mRNA Work
  i. Discovery of the mRNA in 1960
  ii. mRNA synthesized in lab in 1984
  iii. mRNA used as treatment in rats in 1992

TABLE 15.1
List of References on CRISPR, ZFN and TALEN

Year	Articles
1989	Rudin, N, Sugarman, E., & Haber, J. E. (1989). Genetic and physical analysis of double-strand break repair and recombination in *Saccharomyces cerevisiae*. *Genetics*, 122, 519–534.
1994	Rouet, P., Smih, F., & Jasin, M. (1994). Introduction of double-strand breaks into the genome of mouse cells by expression of a rare-cutting endonuclease. *Mol Cell Biol*, 14, 8096–8106.
1995	Choulika, A., Perrin, A., Dujon, B., & Nicolas, J-F. (1995). Induction of homologous recombination in mammalian chromosomes by using the I-*Sce*I system of *Saccharomyces cerevisiae*. *Mol Cell Biol*, 15, 1968–1973. Segal, D. J., & Carroll, D. (1995). Endonuclease-induced, targeted homologous extrachromosomal recombination in *Xenopus* oocytes. *Proc Natl Acad Sci USA*, 92, 806–810.
1996	Kim, Y-G., Cha, J., & Chandrasegaran, S. (1996). Hybrid restriction enzymes: Zinc-finger fusions to *Fok*I cleavage domain. *Proc Natl Acad Sci USA*, 93, 1156–1160.
2000	Smith, J., Bibikova, M., Whitby, F. B., Reddy, A. R., Chandrasegaran, S., & Carroll, D. (2000). Requirements for double-strand cleavage by chimeric restriction enzymes with zinc finger DNA-recognition domains. *Nucleic Acids Res*, 28, 3361–3369. Rong, Y. S., & Golic, K. G. (2000). Gene targeting by homologous recombination in *Drosophila*. *Science*, 288, 2013–2018.
2001	Bibikova, M., Carroll, D., Segal, D. J., Trautman, J. K., Smith, J. Kim,...Y-G. (2001). Stimulation of homologous recombination through targeted cleavage by chimeric nucleases. *Mol Cell Biol*, 21, 289–297.
2002	Makarova, K. S. et al. (2002). A DNA repair system specific for thermophilic Archaea and bacteria predicted by genomic context analysis. *Nucleic Acids Research*, 30(2), 482–496. Bibikova, M., Golic, M., Golic, K. G., & Carroll, D. (2002). Targeted chromosomal cleavage and mutagenesis in *Drosophila* using zinc-finger nucleases. *Genetics*, 161, 1169–1175.
2003	Bibikova, M., Beumer, K., Trautman, J. K., & Carroll, D. (2003). Enhancing gene targeting with designed zinc finger nucleases. *Science*, 300, 764. Porteus, M. H., & Baltimore, D. (2003). Chimeric nucleases stimulate gene targeting in human cells. *Science*, 300, 763.
2005	Urnov, F. D., Miller, J. C., Lee, Y-L., Beausejour, C. M., Rock, J. M., ... Augustus, S. (2005). Highly efficient endogenous gene correction using designed zinc-finger nucleases. *Nature*, 435, 646–651.
2006	Makarova, K. S. et al. (2006). A putative RNA-interference-based immune system in prokaryotes: Computational analysis of the predicted enzymatic machinery, functional analogies with eukaryotic RNAi, and hypothetical mechanisms of action. *Biology Direct*, 1, 7. Beumer, K., Bhattacharyya, G., Bibikova, M., Trautman, J. K., & Carroll, D. (2006). Efficient gene targeting in Drosophila with zinc finger nucleases. *Genetics*, 172, 2391–2403.
2008	Beumer, K. J., Trautman, J. K., Bozas, A., Liu, J-L., Rutter, J., & Gall, J. G. (2008). Efficient gene targeting in Drosophila by direct embryo injection with zinc-finger nucleases. *Proc Natl Acad Sci USA*, 105, 19821–19826.

(Continued)

TABLE 15.1 (CONTINUED)
List of References on CRISPR, ZFN and TALEN

Year	Articles
2013	Shah, S. A. et al. (2013). Protospacer recognition motifs: mixed identities and functional diversity. *RNA Biology*, 10(5), 891–899. Beumer, K. J., Trautman, J. K., Mukherjee, K., & Carroll, D. (2013). Donor DNA utilization during gene targeting with zinc-finger nucleases. *G3 (Bethesda)*, 3, 657–664. Cho, S. W., Kim, S., Kim, J. M., Kim, & J. S. (2013). Targeted genome engineering in human cells with the Cas9 RNA-guided endonuclease. *Nat Biotechnol*, 31, 230–232. Cong, L., Ran, F. A., Cox, D., Lin, S., Barretto, R., ... Habib, N. (2013) Multiplex genome engineering using CRISPR/Cas systems. *Science*, 339, 819–823.
2014	Doudna, J. A., & Charpentier, E. (2014). Genome editing. The new frontier of genome engineering with CRISPR-Cas9. *Science*, 346, Article 1258096.
2015	Morange, M. et al. (2015). What history tells us XXXVII. CRISPR-Cas: The discovery of an immune system in prokaryotes. *Journal of Biosciences*, 40, 221–223.
2017	Barrangou, R., & Horvath, P. (2017). A decade of discovery: CRISPR functions and applications. *Nat Microbiol*, 2, 17092. Shmakov, S., Smargon, A., Scott, D., Cox, D., Pyzocha, N., ... Yan, W. (2017). Diversity and evolution of class 2 CRISPR-Cas systems. *Nat Rev Microbiol*, 15, 169–182.
2018	Sharma, P. K. et al. (2018). Tailoring microalgae for efficient biofuel production. *Frontiers in Marine Science*, 5, 382. Urnov, F. D. (2018). Genome editing B.C. (before CRISPR): Lasting lessons from the "old testament." *Crispr J*, 1, 34–46.
2019	Kotagama, O. W. et al. (2019). Era of genomic medicine: A narrative review on CRISPR technology as a potential therapeutic tool for human diseases. *Biomed Research International*, 2019. Simon, A. J., Ellington, A. D., & Finkelstein, I. J. (2019). Retrons and their applications in genome engineering. *Nucleic Acids Res*, 47, 11007–11019.
2020	Anzalone, A. V., Koblan, L. W., & Liu, D. R. (2020). Genome editing with CRISPR-Cas nucleases, base editors, transposases and prime editors. *Nat Biotechnol*, 38, 824–844.

    iv. mRNA tested as cancer vaccine in 1995
    v. First clinical trial of mRNA vaccine for babies in 2013
  b. Lipid-based delivery systems
    i. Fist liposomes produced in 1965
    ii. Liposomes used for drug delivery in 1971
    iii. Liposome used for vaccine delivery in 1974
    iv. Lipid nanoparticles for delivery of DNA in 2001
    v. Scalable method for manufacturing lipid nanoparticles in 2005
    vi. Approval of first drug with lipid nanoparticles in 2018
  c. mRNA and Lipids
    i. Fist liposome-wrapped mRNA in 1978
    ii. Delivery of synthetic mRNA in liposome to human cells in 1989
    iii. Liposome-wrapped mRNA delivered to mice in 1990
    iv. First mRNA vaccine tested for influenza (mice) in 1993

Going Forward – Existing and Evolving Technologies

      v. First mRNA vaccine in lipid nanoparticles (mice) in 2012
      vi. First clinical trial of mRNA vaccine (influenza) in lipid nanoparticles in 2015
      vii. COVID-19 mRNA-based vaccine authorized (emergency) in 2020
    d. Commercial Research & Development (R&D) using mRNA
      i. Merix Biosciences founded in 1997
      ii. CureVac founded in 2000
      iii. BioNTech founded in 2008
      iv. Novartis and Shire create mRNA divisions in 2008
      v. US Defense Adv. Research Project Agency – funding mRNA vaccine research in 2012

Focusing on mRNA and liposomes, let's start with the work by Dimitriadis (1978a, 1978b, 1978c) who reported the entrapment of ribonucleic acids in large unilamellar liposomes. Those macromolecules were protected from ribonucleases and can be isolated both intact and biologically active (Dimitriadis, 1978c). Fusion of these liposomes yields RNA-filled cells. Dimitriadis (1978a) presented evidence of introduced rabbit reticulocyte 9S mRNA into mouse spleen lymphocytes and directed the synthesis of globin.

Then in 1984, Krieg and Melton (Krieg & Melton, 1984; Melton et al., 1984) described the synthesis of eucaryotic messenger RNAs. Injection into the cytoplasm of frog oocytes and addition to wheat germ extracts show that these synthetic RNAs did function efficiently as messenger RNAs. They confirmed that mRNA is essential for translation in injected oocytes and showed that most of the flanking region, including the poly A tail, can be deleted without the abolition of protein synthesis. Although they saw synthetic mRNA mainly as a research tool for studying gene function and activity, the method remains currently in use while enabling the production of large amounts of mRNA and consequently protein from any cDNA clone. Table 15.2 provides the readers with additional scientific literature on the subject.

The work on mRNA has resulted, as described earlier and predicted by Geall et al. (2012), in the rapid development and successful manufacture of new commercial drug products, more recently the COVID-19 vaccine (Kremsner et al., 2021). This technology is expected to continue expanding in use not just in vaccines but also in other drug products and therapies.

siRNA – HISTORICAL REVIEW AND FUTURE

Fire et al. (1998) discovered the RNAi mechanism while working on the gene expression in the nematode *Caenorhabditis elegans*. They investigated the requirements for structure and delivery of the interfering RNA and found that double-stranded RNA was substantially more effective at producing interference than either strand individually. After injection into adult animals, purified single strands had at most a modest effect, whereas double-stranded mixtures caused potent and specific interference. Also, they noticed the need for few molecules

TABLE 15.2
List of References on mRNA

Year	Articles
1970	Hill, M., & Huppert, J. (1970). Fate of exogenous mouse DNA in chicken fibroblasts *in vitro*: Non-conservative preservation. *Biochim. biophys. Acta*, 312, 26–35.
	Graessmann, A. (1970). Mikrochirurgische zellkerntransplantation bei säugetierzellen. *Expl Cell Res.*, 60, 373–382.
1974	Furusawa, M., Nishimura, T., Yamaizumi, M. & Okada, Y. (1974). Injection of foreign substances into single cells by cell fusion. *Nature* 249, 449–450.
1975	Loyter, A., Zakai, N., & Kulka, R. G. (1975). "Ultramicroinjection" of macromolecules or small particles into animal cells. A new technique based on virus-induced cell fusion. *J. Cell Biol.*, 66, 292–304.
1976	Stacey, D. W., & Allfrey, V. G. (1976). Microinjection studies of duck globin messenger RNA translation in human and avian cells. *Cell* 9, 725–732.
1978	Ostro, M. J., Giacomoni, D., Lavelle, D., Paxton, W., & Dray, S. (1978). Evidence for translation of rabbit globin mRNA after liposome-mediated insertion into a human cell line. *Nature* 274, 921–923.
1989	Malone, R. W., Felgner, P. L. & Verma, I. M. (1989). Cationic liposome-mediated RNA transfection. *Proc. Natl Acad. Sci. USA*, 86, 6077–6081.
	Malone, R. W. (1989). mRNA Transfection of cultured eukaryotic cells and embryos using cationic liposomes. Focus, 11, 61–66.
1990	Wolff, J. A., Malone, R. W., Williams, P., Chong, W., Acsadi, G., Jani, A., & Felgner, P. L. (1990). Direct gene transfer into mouse muscle in vivo. *Science*, 247, 1465–1468.
1993	Jirikowski, G. F., Sanna, P. P., Maciejewski-Lenoir, D., & Bloom, F. E. (1993). Reversal of diabetes insipidus in Brattleboro rats: intrahypothalamic injection of vasopressin mRNA. *Science*, 255, 996–998.
1993	Martinon, F., Krishnan, S., Lenzen, G., Magne, R., Gomard, E., Guillet, J. G., Levy, J. P., & Meulien, P. (1993). Induction of virus-specific cytotoxic T lymphocytes in vivo by liposome-entrapped mRNA. *Eur. J. Immunol.*, 23, 1719–1722.
1995	Conry, R. M., LoBuglio, A. F., Wright, M., Sumerel, L., Pike, M. J., Johanning, F., Benjamin, R., Du, D., & Curiel, D. T. (1995). Characterization of a messenger RNA polynucleotide vaccine vector. *Cancer Res.*, 55, 1397–1400.
1996	Boczkowski, D., Nair, S. K., Snyder, D., & Gilboa, E. J. (1996). Dendritic cells pulsed with RNA are potent antigen-presenting cells in vitro and in vivo. *Exp. Med.*, 184, 465–472.
1998	Karikó, K., Kuo, A., Barnathan, E. S., & Langer, D. J. (1998). Phosphate-enhanced transfection of cationic lipid-complexed mRNA and plasmid DNA. *Biochim. Biophys. Acta*, 1369, 320–334.
1999	Karikó, K., Kuo, A., & Barnathan, E. (1999). Overexpression of urokinase receptor in mammalian cells following administration of the in vitro transcribed encoding mRNA. *Gene Ther.*, 6, 1092–1100.
	Smerdou, C., & Liljeström, P. (1999). Non-viral amplification systems for gene transfer: Vectors based on alphaviruses. *Curr Opin Mol Ther*, 1, 244–251.
2000	Hoerr, I., Obst, R., Rammensee, H. G., & Jung, G. (2000). In vivo application of RNA leads to induction of specific cytotoxic T lymphocytes and antibodies. *Eur. J. Immunol.*, 30, 1–7.
2004	Karikó, K., Ni, H., Capodici, J., Lamphier, M., & Weissman, D. J. (2004). mRNA is an endogenous ligand for Toll-like receptor 3. *Biol. Chem.*, 279, 12542–12550.
2005	Karikó, K., Buckstein, M., Ni, H., & Weissman, D. (2005). Suppression of RNA recognition by Toll-like receptors: the impact of nucleoside modification and the evolutionary origin of RNA. *Immunity*, 23, 165–175.
	Jeffs, L. B., Palmer, L. R., Ambegia, E. G., Giesbrecht, C., Ewanick, S., & MacLachlan, I. (2005). A scalable, extrusion-free method for efficient liposomal encapsulation of plasmid DNA. *Pharm. Res.*, 22, 362–372.

(Continued)

TABLE 15.2 (CONTINUED)
List of References on mRNA

Year	Articles
2007	Probst, J., Weide, B., Scheel, B. Pichler, B. J., Hoerr, I., Rammensee, H. G., & Pascolo, S. (2007). Spontaneous cellular uptake of exogenous messenger RNA in vivo is nucleic acid-specific, saturable and ion dependent. *Gene Ther.*, 14, 1175–1180.
2010	Warren, L., Manos, P. D., Ahfeldt, T., Loh, Y. H., Li, H., Lau, F., Ebina, W., Mandal, P. J., Smith, Z. D., Meissner, A., Daley, G. Q., Brack, A. S., Collins, J. J., Cowan, C., Sclaeger, T. M., & Rossi, D. J. (2010). Highly efficient reprogramming to pluripotency and directed differentiation of human cells with synthetic modified mRNA. *Cell Stem Cell*, 7, 618–630.
2011	Graham, B. S. (2011). Biological challenges and technological opportunities for respiratory syncytial virus vaccine development. *Immunol Rev*, 239, 149–166.
	Rappuoli, R., Mandl, C. W., Black, S., & De Gregorio, E. (2011). Vaccines for the twenty-first century society. *Nat Rev Immunol*, 11, 865–872.
2012	Weiss, R., Scheiblhofer, S., Roesler, E., Weinberger, E., & Thalhamer, J., (2012). mRNA vaccination as a safe approach for specific protection from type I allergy. *Expert Rev Vaccines*, 11, 55–67.
	Weiss, R., Scheiblhofer, S., Roesler, E., Weinberger, E., & Thalhamer, J. (2012). mRNA vaccination as a safe approach for specific protection from type I allergy. *Expert Rev Vaccines*, 11, 55–67.
2021	Aldrich, C., Leroux-Roles, I., Huang, K. B., Bica, M. A., Loeliger, E., Schoenborn-Kellenberg, O., Walz, L., Leroux-Roels, G. von Sonnenburg, F., & Oostvogels, L. (2021). Proof-of-concept of a low-dose unmodified mRNA-based rabies vaccine formulated with lipid nanoparticles in human volunteers: A phase 1 trial. *Vaccine*, 39, 1310–1318.

of injected double-stranded RNA required per affected cell, suggesting that there could be a catalytic or amplification component in the interference process.

Elbashir et al. (2001) showed that 21-nucleotide siRNA duplexes specifically suppress expression of endogenous and heterologous genes in different cell lines. They used siRNA to specifically silence the expression of different genes: siRNA against reporter genes coding for sea pansy (*Renilla reniformis*) and two sequence variants of firefly (*Photinus pyralis*, GL2 and GL3) luciferases. Also, they reported results in work with mammalian cells (e.g., NIH/3T3, monkey COS-7 and Hela S3); the reporter genes were more strongly expressed but the specific suppression was less complete. In summary, after transfection with the siRNAs, there was a significant reduction in the gene expression level of all the silenced genes, and the silencing effects were directly proportional to the level of their endogenous gene expression.

The production of naturally occurring siRNAs is catalyzed by the Dicer enzyme from long double-stranded RNA (dsRNA) and small hairpin RNAs (Bernstein et al., 2001). It was suggested that 21-nucleotide siRNA duplexes provide a new tool for studying gene function in mammalian cells and may eventually be used as gene-specific therapeutics (Elbashir et al., 2001).

Bruno (2012) pointed out that an important property of siRNA to be controlled during the drug discovery process is its potential off-target effect, which limits its specificity. Specifically, proper selection of excipients for the formulation of siRNA is key since the molecule is relatively big, heavily charged and susceptible to degradation in the body fluids; therefore, the delivery vehicle has to provide protection as well as enable cell penetration and release, as discussed by Bruno (2011).

Saw and Song (2020) highlighted the breakthroughs in siRNA therapeutics with the first us Food and Drug Administration (US FDA)–approved RNAi therapeutics Onpattro (Patisiran, a treatment of Hereditary Transthyretin Amyloidosis (hATTR) or transthyretin familial amyloid polyneuropathy (TTR-FAP) or familial amyloid cardiomyopathy (TTR-FAC), a rare, progressive and fatal disease) and effective siRNA delivery system focusing on siRNA nanocarrier in clinical trials. The latter is critical, as one hurdle in achieving potent siRNA therapeutics did not lie on the design of potent siRNA, but rather on their administration at their target as biologically active siRNAs.

Delivery systems are divided into viral or nonviral techniques.

e. Viruslike nanoparticles:
   These are proved to be highly efficient in delivering siRNA but safety concerns around long-term infusion of viral vectors make them less attractive versus nonviral vectors (Saw & Song, 2020)
f. Nonviral vectors:
   These are namely lipid-based nanoparticles (Saw et al., 2017; Zatsepin et al., 2016) and polymeric nanoparticles (Xu et al., 2016, 2017). Delivery systems are reported elsewhere (see Table 15.3 for a summary of references): encapsulation, adsorbed onto nanoparticle surface, complexed with condensation agent; conjugated with other biomacromolecules, or being modified to increase siRNA stability.

The work on siRNA has resulted, as described earlier, in the development and a handful of potential commercial drug products as RNAi became a go-to technique (Saw & Song, 2020). It is expected that there will be a continued increase in successful translation rates from laboratories to clinics in the upcoming years as technology continues to improve. This technology is expected to continue expanding in use of not just vaccines but also in other drug products and therapies.

FINAL REMARKS

In this handbook, we have focused on cell and gene therapies developed in recent years that have reached or are near commercial manufacturing stage. Starting with a review of the basic research performed early in the 20th century that laid the foundations for the development of highly effective novel treatments for diseases with unmet medical need, we discussed the challenges and safety concerns that were seen in earlier trials of gene therapy and went on to scientific progress and efforts to improve and ensure the safety of gene therapy products. A brief look at the controversial subject of genome editing technologies highlighted the need for

TABLE 15.3
List of References on siRNA (Stability and Delivery Systems)

Year	Articles
2006	Choung, S., Kim, Y. J., Kim, S., Park, H. O., & Choi, Y. C. (2006). Chemical modification of siRNAs to improve serum stability without loss of efficacy. *Biochem Biophys Res Commun*, 342, 919–927.
2009	Elbakry, A., Zaky, A., Liebl, R., Rachel, R., Goepferich, A., & Breunig, M. (2009). Layer-by-layer assembled gold nanoparticles for siRNA delivery. *Nano Lett*, 9, 2059–2064.
	Jeong, J. H., Mok, H., Oh, Y. K., & Park, T.G. (2009). siRNA conjugate delivery systems. *Bioconj Chem*, 20, 5–14.
	Terrazas, M., & Kool, E. T. (2009). RNA major groove modifications improve siRNA stability and biological activity. *Nucl Acids Res*, 37, 346–353.
2012	Kenski, D. M., Butora, G., Willingham, A. T., Cooper, A. J., Fu, W., Qi, N., Soriano, F., Davies, I. W., & Flanagan, W. M. (2012). siRNA-optimized modifications for enhanced *in vivo* activity. *Mol Ther – Nucleic Acids*, 1, e5.
2014	Steinbacher, J. L., & Landry, C. C. (2014). Adsorption and release of siRNA from porous silica. *Langmuir*, 30, 4396–4405.
	Tagalakis, A. D., Lee, D.H.D., Bienemann, A. S., Zhou, H., Munye, M. M., Saraiva, L., McCarthy, D., Du, Z., Vink, C. A., Maeshima, R. (2014). Multifunctional, self-assembling anionic peptide-lipid nanocomplexes for targeted siRNA delivery. *Biomaterials*, 35, 8406–8415.
2015	Beloor, J., Zeller, S., Choi, C.S., Lee, S.K., & Kumar, P. (2015). Cationic cell-penetrating peptides as vehicles for siRNA delivery. *Therap Deliv*, 6, 491–507.
2017	Dana, H., Chalbatani, G. M., Mahmoodzadeh, H., Karimloo, R., Rezaiean, O., Moradzadeh, A., Mehmandoost, N., Moazzen, F., Mazraeh, A., ... Marmari, V. (2017). Molecular mechanisms and biological functions of siRNA. *Inter J Biomed Sci*, 13, 48–57.
	Dang, C.V., Reddy, E. P., Shokat, K.M., & Soucek, L. (2017). Drugging the "undruggable" cancer targets. *Nat Rev Cancer*, 17, 502–508.
	Gertz, M. A. (2017). Hereditary ATTR amyloidosis: burden of illness and diagnostic challenges. *Am J Managed Care*, 23, S107–S112.
	Schluep, T., Lickliter, J., Hamilton, J., Lewis, D. L., Lai, C. L., Lau, J. Y., Locarnini, S. A., Gish, R. G., & Given, B. D. (2017). Safety, tolerability, and pharmacokinetics of ARC-520 injection, an RNA interference based therapeutic for the treatment of chronic hepatitis B virus infection, in healthy volunteers. ClinI Pharm Drug Dev, 6, 350–362.
	Nair, J.K., Attarwala, H., Sehgal, A., Wang, Q., Aluri, K., Zhang, X., Gao, M., Liu, J., Indrakanti, R., ... Schofield, S. (2017). Impact of enhanced metabolic stability on pharmacokinetics and pharmacodynamics of GalNAc-siRNA conjugates. *Nucl Acids Res*, 45, 10969–10977.
2018	Springer, A. D., & Dowdy, S. F. (2018). GalNAc-siRNA conjugates: leading the way for delivery of RNAi therapeutics. *Nucl Acid Therap*, 28, 109–118.
	Kosmas, C. E., Muñoz Estrella, A., Sourlas, A., Silverio, D., Hilario, E., Montan, P. D., & Guzman, E. (2018). Inclisiran: A new promising agent in the management of hypercholesterolemia. *Diseases*, 6, 63.
2019	Eisenstein, M. (2019, October 16). Pharma's roller-coaster relationship with RNA therapies. *Nature*, 574 (7778), S4–S6.

temperance of scientific innovations with considerations related to ethical, legal and social issues. In the handbook, we discuss the significant benefits and risks related to gene editing and summarize bioethical highlights and current thinking related to some key questions. Chapters 3 to 11 focus on overviews of manufacturing processes, facilities, equipment and basic requirements for successful development, design and implementation of cell and gene therapies. The discussion includes a risk-based approach for advanced therapy medicinal products as described by the European Medicines Agency; overview of manufacturing processes and analytical methods for cell and gene therapies, design and construction of facilities and equipment; validation, verification, qualification, overview of contamination control; training requirements for the workforce supporting manufacturing and testing activities; and basic elements for storage and distribution of these therapies. Finally, we provide an overview of the regulatory landscape across major markets (e.g., USA, Canada, Europe, India, China, Japan, Australia, New Zealand, Malaysia, Singapore and South America (Brazil and Argentina)). We close this edition of the handbook with this chapter where we provide an overview of key tools and technologies that are providing additional options to the field of cell and gene therapies: CRISPR, mRNA and siRNA.

Our search for better ways to treat patients will continue. Current available technologies have made possible genome modification and editing at the somatic and germline levels. Our study of epigenomics, facilitated by these advances will serve as a conduit to personalized 'regulomes'. Individual -omes as well as the integrated profiles of multiple –omic information will continue to be explored and are expected to be invaluable for preventive measures and precision medicine, for transforming medical care from traditional symptom-oriented diagnosis and treatment of diseases toward disease prevention and early diagnostics.

REFERENCES

Bio-Rad Labs., Inc (2022). CRISPR gene editing kits. CRISPR Gene Editing Kits I Bio-Rad

Bernstein, E., Caudy, A.A., Hammond, S.M., & Hannon, G.J. (2001, January). Role for a bidentate ribonuclease in the initiation step of RNA interference. *Nature*, 409(6818), 363–412. 6. Bibcode:2001Natur.409..363B.

Bruno, K. (2011). Using drug-excipient interactions for siRNA delivery. *Advanced Drug Delivery Reviews*, 63, 1210–1226.

Bruno, K. (2012). Ten years of siRNA – a clinical overview. Eur. Pharm. Rev., 17(3).

Carroll, D. (2014). Genome engineering with targetable nucleases. *Annual Review of Biochemistry*, 83(2014), 409–439.

Carroll, D. (2021). A short, idiosyncratic history of genome editing – ScienceDirect. *Gene & Genome Editing* (Vol. 1). Elsevier. https://doi.org/10.1016/j.ggedit.2021.100002

Dimitriadis, G.J. (1978a). Translation of rabbit globin mRNA introduced by liposomes into mouse lymphocytes. *Nature*, 274, 923–924.

Dimitriadis, G.J. (1978b). Entrapment of ribonucleic acids in liposomes. *FEBS Letters*, 86, 289–293.

Dimitriadis, G.J. (1978c). Introduction of ribonucleic acids into cells by means of liposomes. *Nucleic Acids Research*, 5, 1381–1386.

Dolgin, E. (2021). The tangled history of mRNA vaccines. *Nature*, 597, 318–324.

Elbashir, S.M., Harborth, J., Lendeckel, W., Yalcin, A., Weber, K., & Tuschl, T. (2001). Duplexes of 21-nucleotide RNAs mediate RNA interference in cultured mammalian cells. *Nature*, 411, 494–498.

Fire, A., Xu, S.Q., Montgomery, M.K., Kostas, S.A., Driver, S.E., and Mello, C.C. (1998). Potent and specific genetic interference by doublestranded RNA in Caenorhabditis elegans. *Nature*, 391, 806–811.

Geall, A.J., Verma, A., Otten, G.R., Shae, C.A., Hekele, A., Banerjee, K., Cu, Y., Beard, C.W., Brito, L.A., Krucker, T., O'Hagan, D.T., Singh, M., Mason, P.W., Valiante, N.M., Dormitzer, P.R., Barnett, S.W., Rappuoli, R., Ulmer, J.B., & Mandl, C.W. (2012). Nonviral delivery of self-amplifying RNA vaccines. *Proceedings of the National Academy of Sciences of the USA*, 109, 14604–14609.

Han, W. & She, Q. (2017). Chapter one – CRISPR history: Discovery, characterization, and prosperity. *Progress in Molecular Biology and Translational Science*, 152, 1–21.

Ishino, Y., Shinagawa, H., Makino, K., Amemura, M., & Nakata, A. 1987. Nucleotide sequence of the *iap* gene, responsible for alkaline phosphatase isozyme conversion in *Escherichia coli*, and identification of the gene product. *Journal of Bacteriology*, 169, 5429–5433.

Ishino, Y., Krupovic, M., & Forterre, P. (2018). History of CRISPR-Cas from encounter with a mysterious repeated sequence to genome editing technology. *Journal of Bacteriology*, 200, 7.

Kremsner, P.G., Mann, P., Kroidl, A. Leroux-Roels, I., Schindler, C., Gabor, J.J., Schunk, M., Leroux-Roels, G. Bosch, J.J. Fendel, R. Kreidenweiss, A. Velavan, T.P., Fotin-Mleczek, M., Mueller, S.O. Quintini, G., Schonborn-Kellenberger, O., Vahrenhorst, D., Verstraeten, T., Alves de Mesquite, M. Walz, L., Wolz, O.O., Oostvogels, L., & CV-NCOV-001 Study Group. (2021). Safety and immunogenicity of an mRNA-lipid nanoparticle vaccine candidate against SARS-CoV-2. Wien Klin Wochenschr, 133(17–18), 931–941.

Krieg, P.A., & Melton, D.A. (1984). Functional messenger RNAs are produced by SP6 in vitro transcription of cloned cDNAs. *Nucleic Acids Research*, 12, 7057–7070.

Melton, D.A., Krieg, P.A., Rebagliati, M.R., Maniatis, T., Zinn, K., & Green, M.R. (1984). Efficient in vitro synthesis of biologically active RNA and RNA hybridization probes from plasmids containing a bacteriophage SP6 promoter. *Nucleic Acids Research*, 12, 7035–7056.

Mojica, M.J., Juez, G., & Rodríguez-Valera, F. (1993). Transcription at different salinities of *Haloferax mediterranei* sequences adjacent to partially modified PstI sites. *Molecular Microbiology*, 9, 613–621.

Mojica, F.J., Ferrer, C., Juez, G., & Rodríguez-Valera, F. (1995). Long stretches of short tandem repeats are present in the largest replicons of the archaea *Haloferax mediterranei* and *Haloferax volcanii* and could be involved in replicon partitioning. *Molecular Microbiology*, 17, 85–93.

Mojica, F.J.M., Díez-Villaseñor, C., García-Martínez, J., & Soria, E. (2005). Intervening sequences of regularly spaced prokaryotic repeats derive from foreign genetic elements. *Journal of Molecular Evolution*, 60, 174–182.

Pourcel, C., Salvignol, G., and Vergnaud, G. (2005). CRISPR elements in *Yersinia pestis* acquire new repeats by preferential uptake of bacteriophage DNA, and provide additional tools for evolutionary studies. *Microbiology*, 151, 653–663.

Saw, P.E., Park, J., Jon, S., & Farokhzad, O.C. (2017). A drug-delivery strategy for overcoming drug resistance in breast cancer through targeting of oncofetal fibronectin. *Nanomedicine: Nanotechnology, Biology and Medicine*, 13, 713–722.

Saw, P.E. & Song, E.W. (2020). siRNA Therapeutics: A clinical reality. *Science China Life Sciences*, 63(4), 485–500.

Thermo-Fisher Sci. Inc. (2022). Invitrogen: Complete cell engineering solutions: CRISPR genome editing tools. Complete cell engineering solutions (thermofisher.com). Thermo Fisher. Sci. Inc.

Travis, J. (2015). Science's 2015 breakthrough of the year: CRSPR genome-editing technology shows its power. Amer. Assoc. Advanc. Sci. Science, 350 (6267), 1456–1457.

Xu, X., Saw, P.E., Tao, W., Li, Y., Ji, X., Yu, M., Mahmoudi, M., Rasmussen, J., Ayyash, D., ... Zhou, Y., (2017). Tumor microenvironment-responsive multistaged nanoplatform for systemic RNAi and cancer therapy. *Nano Letters*, 17, 4427–4435.

Xu, X., Wu, J., Liu, Y., Yu, M., Zhao, L., Zhu, X., Bhasin, S., Li, Q., Ha, E., ... Shi, J. (2016). Ultra-pH-responsive and tumor-penetrating nanoplatform for targeted siRNA delivery with robust anti-cancer efficacy. *Angew Chem Int Ed Engl.*, 55(25), 7091–7094.

Zatsepin, T.S., Kotelevtsev, Y.V., & Koteliansky, V. (2016). Lipid nanoparticles for targeted siRNA delivery-going from bench to bedside. *IJN*, 11, 3077–3086.

Glossary

Acceptance criteria: Condition, numerical limits, ranges or measures for acceptable results.

Action limit: A limit that is established for a performance attribute. If exceeded, it indicates the analytical procedure is trending outside of expected performance. The response to exceeding an action limit should be pre-defined and, at a minimum, trigger an investigation.

Active Pharmaceutical Ingredient: Chemical or biological compound or entity used in the formulation of drug products with known therapeutic effect.

Alert limit: An established limit for a performance attribute that is more stringent than an action limit, and, therefore, it provides an early warning of a potential trend away from normal operating conditions when exceeded. The response to exceeding an alert limit should be pre-defined but is usually less comprehensive to exceeding an action limit.

Analytical Method Transfer: Planned activities used to communicate, provide and confirm the proper implementation of new analytical methods and required equipment at a location.

API: Any substance or mixture of substances used in manufacturing that becomes an active ingredient in the drug product. APIs furnish pharmacological action or other direct effects in the diagnosis, cure, mitigation, treatment, or prevention of disease, or affect the structure and function of the body.

Appearance: Visual inspection that evaluates the color and clarity of the product.

Biannual cGMP Inspection: Periodic regulatory inspection of a facility to assess the cGMPs.

Chimeric Antigen Receptor (CAR): Fabricated receptor that is delivered to target cells (T Cells) via a viral vector. The CAR will allow the T Cell expressing it to identify and attack specific cells expressing the antigen of interest.

Chain of Identity: Mechanism to monitor and maintain the identity of each and every material associated with a patient 100% of the time during production and testing.

Chemistry, Manufacturing and Controls (CMC): Term used to describe the section of a drug application (e.g., FDA New Drug Application) which details the pharmaceutical development, stability, manufacturing process and controls used in the production of a drug.

Clean room: Environmentally controlled production area where airborne contamination is carefully monitored and controlled.

Commissioning: The steps to verify that a new or modified asset can meet its design intent.

Critical Process Parameters (CPP): A process parameter whose variability has an impact at least on one critical quality attribute. The CPP should be monitored and controlled to ensure the process produces the desired quality.

Critical Quality Attribute (CQA): A physical, chemical, biological or microbiological property that should be within a pre-defined range or limit to ensure the desired product quality.

CRISPR: Repeating DNA sequences in the genomes of prokaryotes (e.g., bacteria) initially identified in E. coli in 1987 when a series of repeated sequences interspersed with spacer sequences.

Design Qualification: Documented verification that the proposed design of the facilities, equipment, or systems are suitable for the intended purpose.

Drug Master File (DMF): A DMF is a submission to the FDA that the holder may use to provide confidential detailed information about facilities, processes, or articles used in the manufacturing, processing, packaging, and storing of one or more human drugs. The submission of a DMF is not required by law or FDA regulation. A DMF is submitted solely at the discretion of the holder. The information contained in the DMF can also support other applications (such as an IND, an NDA, or an ANDA), or another DMF.

Drug Products: Finished products containing an active pharmaceutical ingredient at a pre-determined concentration or dose for the prevention or treatment of a medical condition.

Drug Substance: Active ingredient required to formulate the drug product.

Environmental Monitoring: Program to periodically monitor the environmental conditions of the facility.

Failure Mode and Effect Analysis: Method used for risk assessment considering the mode of failures and the impact of those failures on the quality of the products and the safety of patients.

Good Laboratory Practice: GLP regulations were put into effect in 1981 to ensure adequate quality control for nonclinical (animal) studies and to provide an adequate degree of consumer protection. The regulations specify minimum standards for the proper protocol and conduct of safety testing and contain sections on facilities, personnel, equipment, standard operating procedures, test and control articles, quality assurance, records and reports, and laboratory disqualification

Good Manufacturing Practice or Current Good Manufacturing Practice: cGMP for finished pharmaceuticals (21 CFR 312.211) ensures that drugs meet the Food, Drug, and Cosmetic (FD&C) Act requirements for safety, and meet the identity, strength, quality and purity characteristics that they purport or are represented to possess

Good Tissue Practice or Current Good Tissue Practice: cGTP requirements govern the methods used in, and the facilities and controls used for, the manufacture of human cells, tissues, or cellular or tissue-based

Glossary

products (HCT/Ps) in a way that prevents the introduction, transmission, or spread of communicable diseases by HCT/Ps. Communicable diseases include, but are not limited to, those transmitted by viruses, bacteria, fungi, parasites, and transmissible spongiform encephalopathy agents.

Identity: Confirmation that the intended product has been manufactured.

Installation Qualification: Documented evidence of the satisfactory installation of a piece of equipment satisfying the user requirement specification (URS).

Mechanism of Action: The specific interaction of a drug that produces its pharmacological effect.

Mode of Action: The functional and/or anatomical changes that result from the administration of a drug. Represents an intermediate level of complexity between the mechanism of action and physiological outcome.

Operation Qualification: Documented evidence of the satisfactory operation of a piece of equipment satisfying the functional requirement specifications (FRS).

Performance Qualifications: Documented evidence of the satisfactory operation of a piece of equipment satisfying the process-specific conditions.

Potency: It is a quantitative measure that is indicative of the intended biological activity of the active ingredient(s) (21 CFR 600.3(s) and 21 CFR 610.10).

Pre-licensing Inspection: Regulatory inspection of a new facility with no previous regulatory activities.

Pre-approval Inspection: Regulatory inspection of a facility where a new product or process is introduced.

Process Performance Qualifications: Documented evidence that the process performs as described in the procedures and batch record and the product satisfies the quality attributes as defined in the acceptance criteria and release parameters.

Process validation: Documented evidence that the process, operated within established parameters, can perform effectively and reproducibly to make an ingredient or product meeting its predetermined specifications and quality attributes.

Purity: Assessments of product-related purity (e.g., intended active ingredient(s) and unintended cellular impurities that are either modified or unmodified) and process-related impurities (e.g., residual activation reagent, residual growth factors), whether or not impurities impact the safety and efficacy of the product (21 CFR 600.3(r)).

Quality assurance: The planned and systematic activities that ensure a food, drug, or device will be processed and produced with consistency, meeting all analytical and performance specifications within and between batches. QA primarily involves (1) review and approval of all procedures related to production and maintenance, (2) review of associated records, and (3) auditing, performing, and evaluating trend analyses.

Quality control: The steps taken during the generation of a product or service to ensure it meets requirements and the product or service is reproducible. QC usually involves (1) assessing the suitability of incoming components, containers, closures, labeling, in-process materials, and the finished products; (2) evaluating the performance of the manufacturing process to ensure adherence to proper specifications and limits; and (3) determining the acceptability of each batch for release.

Quality system: Formalized business practices that define management responsibilities for organizational structure, processes, procedures, and resources needed to fulfill product and service requirements, customer satisfaction, and continual improvement. Under a quality system, it's normally expected that the product and process development units, the manufacturing units, and the QU will remain independent.

Quality unit: A group organized within an organization to promote quality in general practice by ensuring the various operations associated with all systems are appropriately planned, approved, conducted, and monitored. The QU has the authority to create, monitor, and implement a quality system

Raw material: "Raw material" means any ingredient intended for use in the manufacture of a drug substance or drug product, including those that may not appear in that product. A RM can be either an active or inactive ingredient.

Safety: Testing for the level of transduction, sterility, mycoplasma, pyrogens, such as endotoxin, and adventitious agents, such as replication-competent lentivirus.

Strength: It is the concentration of the active ingredient(s) in the product, which is used for dosing.

Technology Transfer: Planned activities used to communicate, provide and confirm the proper implementation of new processes and equipment at a location.

Validation Master Plan (VMP): VMP is the foundation of a company's Good Manufacturing Practice validation program. It's a comprehensive plan that should include process validation, facility and utility qualification and validation, equipment qualification, cleaning, and computer validation.

Index

Pages in *italics* refer figures and pages in **bold** refer tables.

A

Abecma®, **8**, 10, **51**, 137
accelerated approval, 61, 203, 222–223; *see also* conditional approval
acceptance criteria, xix, xxi, 333, 335
 container, 128–129
 endotoxin, 126
 identity, 108
 performance, 90, 92–98, 113, 149, 171, 205
 sterility, 123
 test, 89, 119, 132, **141**, 142
ACCESS Consortium, 237
accreditation, xx, 53, 175–181
 of organization, 239, 248, 249, 289, 300
accuracy
 analytical procedures, 14, 82, 90–94, 103, 106, 111, 113, 115–117, 121, 126
acridine orange (AO), **114**
Act on the Safety of Regenerative Medicine (ASRM), 285–286
action limit, 100, **145**, 333
active pharmaceutical ingredient (API), xix, 136, 151, 173, 237, 333, 334
acute lymphoblastic leukemia (ALL), 3, **6**, 137
Adeno-associated virus (AAV)
 therapy, **185**, 194–195
 vector, xix, 2, 5, **6–7**, **31**, 36–37, 76, 105, 161, **164**, 167, **185**, 194–195
adenosine deaminase (ADA) deficiency, 1, 2, **6**
adenovirus, 28, 36, 37, 155, 166, 167, 305
Adli, M., 21, 24
Advanced Therapy Medicinal Products (ATMP)
 definition, 197, 209
 distribution, 186
 regulatory framework, 185, 195–196, 209, 225, 231
 risk-based-approach, 27–33, 330
 training, 186
adventitious agents, 83, 87–88, 95, 123, 153, 155–156, 160–161, 167–168, 171, 312, 336
adverse
 event, 2, 15, **19**, 28, 307
 event reporting, 235, 243, 251, 253–255, 260, 267, 269, 275, 279
 reaction, 44, 267, 279, 280
agarose gel electrophoresis, 103, 106
 EtBr stained, 44, 45

Agenda 2030 for sustainability, 23
AIDS, 15, **17**; *see also* HIV
air handling, xx, xxi, 67–69, 70, 72–73, 139, 144, **146**
airlocks, 69, 71, 73
Ajayi, O.O., 157, 172
alert limit, 100, 169, 333
Alexandra Beumer Sassi, 203, 205, 206, 207
Ali, S., 154, 173
Alipogene tiparvovec, *see* Glybera®
Allan Bream, 202, 207
Allogeneic cell therapies, 2, 49, 51–52, 83, 104, **158**, 161, 163–165, **169**, 183, 193
 distribution logistics, **184–185**, 186, 189
 regulatory framework, 217, **220**, 256, 305
Alnabhan, R., 57, 58, 60
Alofisel™, **51**
Amariglio, N., 28, 33
Amendola, M., 4, 11
American College of Medical Ethics and Genomics (ACMG), xix, 21
American Type Culture Collection (ATCC), 167
Amicus® Separator, 52
7-Aminoactinomycin (7-AAD), **114**
Amirache, F., 101, 131
Amoasii, L., 15, 16, 24
analytical method transfer, *see* technology transfer
Anderson, E.S., 45
Anderson, W.F., 1, 10, 11
anion Exchange Chromatography, *see* Chromatography
ANMAT (Administración Nacional de Medicamentos, Alimentos y Tecnología Médica), 226–227, 230
 Disposición 3602/2018227, 227, 230
annexin V, **114**
anthrax, **18**, 35
antibiotic resistance, 16, **17**, 35
ANVISA (Agência Nacional de Vigilância Sanitária)
 regulatory, 226, 228–231
apheresis, xix, *52*, 66, 70–71, 164, 165, 189
 collection systems, *52*
apoptosis, 58, 85
Aranha. H., xv, xvii, 11, 13, 14, 24, 153, 154, 155, 157, 172, 173, 285, 321

337

Argentina
 regulatory framework, 225–226
Argentinian Pharmaceutical Laboratories
  Industrial Chamber, 225–226
ASEAN, 246, 248, **249**, 259
aseptic
 operations/manipulations, 40, 49, 66, 69, 75, 138, 142–144, **146**, 147, 153, **154**, 164, 168–169, 199, 298, 299, 305, 313
 process simulation (APS), 146
assisted reproductive technology (ART), 13, 20
Association for the Advance of Blood & Biotherapies (AABB), xix, 53
Association of Clinical Research Professionals (ACRP), 180
ASTM E2935, 96, 131
Atidarsagene autotemcel, *see* Libmeldy™
Audit, 167, 170, 202, 252, 336
Ausbiotech, 175, 181, 234, 241
Australia New Zealand Clinical Trials Registry (ANZCTR), 234
Australian Public Assessment Report (AusPAR), 236, 238
Australian Register of Therapeutic Goods (ARTG), 235–238
Australian Regulatory Guidelines for Biologicals (ARGB), 237, 241
Australian Regulatory Guidelines for Prescription Medicines (ARGPM), 235–236, 241
Autologous cell therapies, **6–9**, 10, 28, 49, **50–51**, 52, 65, 70, 83, 115, 137, **158–159**, 161, **162**, 163–165, **169**, 313
 distribution logistics, 183, 184, **184**, **185**, 186–187
 regulatory framework, 217, 227, 256, 268, 299, 305
AutoMACS® Pro Separator, 53–*54*
Avery, O.T., 1, 10, 11
axicabtagene ciloleucel, **6**, 238; *see also* Yescarta

B

bacteriophages, 1, 322
barcodes, 193
Barone, P., 155, 157, 170, 172
barriers, 66, 144, 316
Baseman, H.S., 148, 151
Bashor, C.J., 49, **50–51**, 60
batch fermentation, *see* Fermentation
Baum, C., 4, 11, **39**
B-cell maturation antigen (BCMA), **8**, **9**, 137
Bennett, E.P., 105–106, 131
Betibeglogene autotemcel, *see* Zynteglo™

Bhardwaj, A.J., 104, 131
Bhatt A., xvii, 295
bioburden, 43, 169, **170**
biohazardous waste, 144
biological License Application (BLA), xix, 90, 97, 109, 157, *190*, 218, 222
biological safety cabinets (BSC), 40, 73, 144, 168
biomarkers, 16, 307
BioNTech, 325
Bio-Rad, 322, 330
Bioreactors, 40, 65, 73–74, 77, 156, 168
Biosafety, 37, 168, 194, 228–229, 303
Blaese, R.M., 2, 10, 11
Bohne, J., 36, **38–40**, 41, 46
Boldt, J., 18, 24
bone marrow
 cells, 1, 14, 164, 244, 296
 cell infusion, 1, 244, 288, 295
Borah, P., 102, 103, 131
bovine spongiform encephalopathy agent, 166
bovine viral diarrhea virus, 166
Brazil, regulatory framework, 203, 207, 225–230, 237, 330
Brenner, S., 35, 46
Brexucabtagene autoleucel, 238; *see also* Tecartus
Breyanzi®, **9**, 10, **51**, **118**, 137
Bridges, W., 180, 181
Brita Salzmann, 202, 207
Brokowski, C., 21, 24
Bruno, K., 327, 330
Bruton's tyrosine kinase (BTK) inhibitor, **8**
Bukrinsky, M.I., 37, 46
Bulaklak, K., 49, **50–51**, 60
Buning, H., 37, **39**, 46
Bushman F., 101, 131

C

Calcein AM, **114**
Cámara Industrial de Laboratorios Farmacéuticos Argentinos, (CILFA), 226
Cameron, P., 107, 131
Canada, regulatory framework, 237, 248
Canavan, J.B., 57, 58, 60
cancer, 10, 15, 36, 118, **214**, 215, 302
 treatment, 16, **17**, 219
 vaccines, 256, **324**
capsids, empty, 313
 shuffling, 37
Carcinoembryonic antigen (CEA), **51**
Carmen, J., 117, 131
Caroll, R.G., 36, **38–40**, 60, 104

Index

Caroll, S.M., 36, **38–40**, 46
Carroll, D., 21, 24, 131, 322, **323**, 324, 330
Carroll, P.J., 322, 331
CAR-T cell therapy
    logistics demands, 184, 189
    regulatory framework, 218–**220**, 234–235, 241, 270, **274**, 283–284, 312
    side effects, 10; *see also* chimeric antigen receptor (CAR)
Cartagena Protocol, 194, 196, 292
Carvykti™, **9**, 10, 137
Cas genes (Cas 9, Cas 12, Cas 13), 14, 15; *see also* CRISPR-*Cas*9
Castella, M., 167, 172
Catapult Program, 180, 181
CD4+ directed CAR T cells, 2, **9**, **17**, 53, 57
CD8+ directed CAR T cells, 2, **9**, 53, 57, 84
CD19 directed CAR T cells, **6–9**, 49–51, 57, 60, **118**
CD34+ directed CAR T cells, 3, **6–8**, **118**
cell banking system, *41*, **158**, 258
cell counters, 73, 76, 78
cell lines, general, 15, 40, 128, 161, **162**, **170**, 327
    Chinese hamster ovary (CHO), 155–157
    human, 155, 167, 170
    stem cells, 296
cell separation, 53, 56, 61, 73, 76, 290, 298
Center for Biologics Evaluation and Research (CBER), 207, 218, 221, **223**, 319
Central Drugs Standard Control Organization (CDSCO), 299, 300, **303**, 304, 308–309
centrifugation, 42, 53, 56, 296
Certified Committee for Regenerative Medicine (CCRM), 289
Certified Special Committee for Regenerative Medicine (CSCRM), 289
chain of custody, xix, 53, 183, 184, 186, 192, 195
chain of identity, xix, 53, 65, 66, 140, 333
Chakraverty, A., 175, 181
Challener C., 204, 207
Chandrasegaran, S., 46, **323**
Chang, H.Y., 15, 26
change control, 76
    impact assessment, 147
Charo, R.A., 286, 293
Charpentier, E., 14, 24, 25, 324
Chemistry, Manufacturing and Controls (CMC), xix, 333
    deficiencies, 248, 289, 314
    regulatory requirements, 86, 88, 94–95, 108, 123, 186, 196–197, 200, 201, 203–206, **220**, 246, 258–259, **261**, 262, 297–298, 304

Chiarle, R., 107, 131
Chile, regulatory framework, 225
Chimeric antigen receptor (CAR), xix, 10, 49, **51**, 61–62, 83, 102, 109, 131–132, 172, **184**, **220**, 234, 270, 302, 311, 333
China, xviii, 5, 13
    regulatory framework, 136, 203, 269–283, 330
Chinese Hamster Ovary (CHO), *see* cell lines
Chisholm, O., xvii, 175, 233
Chlorine, 69
Chromatography, 149, 157, 164, 174
    anion-exchange, *41*, 42
    hydrophobic interaction, *41*, 42
    size exclusion, *41*, 43
Ciltacabtagene autoleucel, **9**; *see also* Carvykti™
cirrhosis, 288
Clean room, 40, 66, 70, 71, 76, 333
Cleaning and Sanitization, 67–68, 139, 143
Clemson, M., 42, 46
clinical trials, 5, 16, **17**, 87, 90, 106, 108–109, 113, 127, 135, 167, **179**, 189, 193, 198, 219, 234–235, 243, 295
    notification (CTN), 234–235, 251–252
    Phase I, 86, 94–95, 101–102, 111, 114, 154
    Phase II/III, 95–96, 117
    regulatory oversight, 212, **214**, **220**, 221, 225–230, 234, 245–247, 251, 253, 264, 270, 272, **274**, 286, 293, 296, **297**, 298–309, 315–316, 328
CliniMACS™ Cell Separator, 53, *55*, 73
clustered regulatory interspaced short palindromic repeats (CRISPR), 13, 14, *see* CRISPR
COBE Spectra® Apheresis System, 52
Cobrinik, D., 37, 47
Code of Federal Regulations (CFR)
    21 CFR 210, 168
    21 CFR 211, 89, 109, 168
    21 CFR 312, 86, 95–96, **223**
    21 CFR 600, 86, 87, 113, 168, 173, 218, **223**
    21 CFR 1271.90, 164, **223**
Cohn, R.D., 15, 26
collection kit, 161
Colombia, 225
Colosimo, A., 46
Commissioning, 71, 76, 333
Committee for Advanced Therapies (CAT), 212–213
Committee for Medicinal Products for Human Use (CHMP), 212–213
Comparability studies, 87, 88, 106, 107, **200**, **214**, 219, 277, 312, 315
    analytical method, 96, 100, 112, 113, 117, 122, 136

Competency, **162**
  workforce, 175–176, 180
compressed gases, 65, 76, 140, 162
conditional approval, 205, 216, 277, **278**, 285, 290, 291, 294
conjugation, 1, 98, 110
Consortium on Adventitious Agent Contamination in Biomanufacturing (CAACB), 155, 157, 170
container
  closure, 75, 83, 129, 202, 206, 248, 258, 261
  content, 83, 128–130
  label, 205
  shipping, 193
  volume, 45
  waste, 144
contamination control, 67, 172, 330
  regulatory approach, 153–160, 168
  risk-based approach, 153–160, 161–168, 172
Continued process verification, 130, 150
contract manufacturing organizations (CMO), 65, 75–78
Control of Drugs and Cosmetics Regulations (CDCR), 256, 266, 267
Cooney, A.L., 2, 11
Cord blood, **50**, **51**, **118**, 164, 165, 244, 286
Cornelis, M.C., 16, 24
corrective actions and preventive actions (CAPAs), xix, 76, 96, 154, 156, 168, 267
cosmid, 36
Council for International Organisations of Medical Sciences (CIOMS), 253
courier service, 159, 183, 187
Couturier, M., 46
COVID-19, 14, 219, **220**, 325
Creative Biolabs, 43, 46
CRISPR/Cas9
  gene editing tool, 13–18, 83, 102, 104, 106, 132, 321, 322, 324
  bioethics of use, 20–24
Cristina Mota, 203, 207
Critical Process Parameters (CPP), **200**, **201**, 206, 334
Critical Quality Attribute (CQA), 82–88, 100, 105, 111, 113–114, 116, 128–130, 334
  process understanding, 200, **201**, 206
Crohn's disease, 57
Crosetto, N., 107, 131
Cryogenic storage and distribution, 183–192
cryopreservation, 58–59, 66, 71, 76, 123, **162**, 296–297
Cundell, T., 154, 172
CureVac, 325

Cuzin F., 35, 46
Cyranoski, D., 286, 293
cystic fibrosis, 15, 16, **17**, 25
Cytiva, 43, 46, 58, 59, 60
Cytokine Release Syndrome (CRS), xix, 10
Cytometry Methods, 82, 109, 111

D

Dabrowska M., 106, 131
dangerous goods, 185, 194, 195, 196
data collection, 43, 59, 69, 76, 97, 147, 150
Davis-Pickett, M., xvii, 81
de Lecuona I., 15, 21, 24
Declaration of Helsinki, 13
Dellaire G., 14, 25
DeMaster, L., 81
densitometry, 44
Department of Biotechnology (DBT), 122, 136, 296
Depil S., 51, 57, 58, 60
detoxification, **17**, **18**
diffuse large B-cell lymphoma (DLBCL), xix, **6**, **7**, **9**, 137
dimethyl sulfoxide (DMSO), 77, 186
Dimitriadis A.G., 325, 330
Diphtheria toxin, 16, **17**
Directive 2001/83/EC, 4, 29, 33, 196, 210, 211, **214**, 233
Directive 2001/83/EC, 5, 28, 210
Directive 2009/120/EC, 4, 11, 33, 196, 211, **223**
Directive 2009/120/EC, 4, 33
Division of Cellular and Gene Therapies (DCGT), 218
Dong B., 37, 46
donor
  eligibility, 164, 206, 219, 312
  testing, 218, 299, 312
dosage, 36, 186, 187, 193, 230, 261, **278**
  form, 43, 128, 247, **278**
  units, 83, 129
Doudna, J.A., 14, 24, 25, 324
droplet digital PCR (ddPCR), 103–106
Drug Control Authority (DCA), 256
drug master file (DMF), 334
drug product, xix, xx, 58, 59, 65, 66, 70, 71, 75, 85, 86, 90, 91, 94, 108, 137, 205, 312, 325, 328, 333, 334, 336
Drug Registration Guidance Document (DRGD), 257
drug substance, 86, 91, 108, 137, 200, 205, 312, 333, 334
Dubin, G., 157, 172
Duchenne muscular dystrophy, 15, 16, **17**, 24, 61
Dynabeads®, 53, 58, 61

Index

E

Egypt
    regulatory, 203
eLabels, 193
Elbashir, S.M., 327, 331
electronic reporting, 192, 193, 248, 249, 317
El-Karidy, A.E.H., 83, 131
Elsner, C., 36, **38–40**, 41, 46
Elverum, K., 43, 46, 52, 59, 61
Embryonic Stem Cells (ESC), **32**, 228, 296
Endotoxin, 42, 44, 83, 126, 336
environmental monitoring, xix, xx, 67, 69–70, 88, 135, 142–143, 146, 168, 334
Environmental Protection Agency (EPA), 240
epigenomics, 23, 330
episome
    definition, 35
    development, 37
Epstein-Barr virus (EBV), xix, 167
Equipment
    design considerations, 65, 73–75
Escherichia coli (E.coli), 41, **38–39**, 155, 321
Esvelt, K.M., 16, 24
European Group for Blood and Marrow Transplantation (EBMT), xix, xx, 53
European Medicines Agency (EMA), 5, 10, 27–30, 84, 127, 144, 194–195, 200, 211–223, 228–230, 248, 258, 260, 266, 290, 314–315, 318–319
European Public Assessment Report (EPAR), xx, **6–9**, 217
European Society for Blood and Marrow Transplantation (EBMT), 53
excipients, 43, 77, 86, 229, 305, 328
Exondys 51®, **50**
expedited review, 203, 290–**292**
Extractables, 75, 206

F

Facility and Equipment
    design, 65–78, 161, 227
    qualification, 135–160
Facility Fit Assessments, 136, 138–139
FACS WOLF-G2, *56*
Factor VIIII (FVIII), xx, **51**
factory acceptance tests, 78
Failure Mode and Effect (FMEA), 138, 147, 159, 160, 334
Falkow, S., 46
familial amyloid cardiomyopathy (TTR-FAC), 328
Fasshauers, S., 243, 255

FDA, xxi, 5, 10, 27, 49, 136, 139, 144, 148–149, 194, 318, 333, 334
    approvals, **50–51**, 328
    environmental requirement, **145**
    oversight, 28, 86, 97, 101, 117–**118**, 155, 313–315
    regulatory framework, 61, 63, 131, 186, 196, 199, 200, 205, 217–224, 230–237, 248, 258, 260, 266–267
feasibility
    hybrid vectors, 36
    logistics, 187, 195
    site selection, 187, 317
Federal Register (FR), xix, 5, 217, 235, 241
    58 FR 53248, 5
fermentation, *41*
    batch, 41, 42, 45, 270
Ferreira L.M.R., 49, 57, 58
fill-and-finish, *41*, 43, 45, 128–129, 163, 189
Firth, A., 15, 16, 25
Fischer, A., 11, 28, 33
Fisher, F., 81
flow cytometry, 54, 73, 82, 102, 109–111, 115–116, 124
fluorescence-activated cell sorting (FACS), xx, 53–*55*, *56*
Food and Drug Administration (US FDA), *see* FDA
Foundation for the Accreditation of Cellular Therapy (FACT), xx, 53
Fredcricq, P., 35, **38–40**, 46
freezers, 76, 78, 195
Freitas F.C.P., 102, 131
Fresenius Kabi, 52, *55*, 56
Friedmann, T., 10, 11
Fukasawa, T., 48
functional requirement specifications (FRS), 78, 335

G

gammaretroviruses, 3, 36
Gantz, V.M., 15, 20, 25
Garcia L.A.O., 228, 231
G-CON, 71, *72*
Geall, A.J., 325, 331
gel electrophoresis, 44, 103
    SYBR gold-stained, 45
Gendicine™, 5, 269
Gene Therapy Advisory and Evaluation Committee (GTAEC), 302, **303**
Gene Therapy Medicinal Product (GTMP), 4, 28, 194, 210, **214**

genetically modified organism (GMO)
  ecosystem impact, 20–21
  risk, 185, 194, 228–229, 239–241, 251, 316, 319
genetically modified organisms (GMO), 4, 5, 20, 194, 239, 261, 316, 319
genomic safe harbors (GSHs), 4, 11
Germline gene therapy, 15, 16, 20, 22, 23, 83, **214**, 301, 330
Gersbach, C.A., 49, **50–51**
Giannoukos, G., 108, 132
Gibson, M., 76, 136, 151
Gintuit™, **50**, **118**
Glybera®, 5, **6**, **9**, 10
GMP Inspection, 243, 252, 255, 266, 333
Golay, J., 154, 172, 333
Gomes, K.L.G., 229, 230, 231
Good Clinical Practices (GCP), 186, 196, 213, 234, 247, 259, 287
Good Distribution Practice (GDP), 199, 247, 267
Good Laboratory Practice (GLP), 213, 287, 299, 303, 334
Good Manufacturing Practice or Current Good Manufacturing Practice (GMP), 155, 160, 218–219, 227, 334
  ATMP-related, 186, 196, 197, 213, 252, 254, 259
  facilities, 74, 88, 139, 155
  regulatory review, 78–79, 137, 150–151, 173, 186, 213, 218–219, 227, 227, 298, 319
Good Post-marketing Study Practice (GPSP), 287
Good Tissue Practice or Current Good Tissue Practice (GTP), 259, 260, 266, 268, 335
Gorsky I., 148, 151
Gowning, 66–68, 70–73, 138–139, 143–144, **162**
Graft-versus-host disease (GvHD), **51**, 104
Greene, M., 16, 25
Griffith, F., 1, 11
Guedan S., 84, 85, 132
Guidance for Human Somatic Cell Therapy and Gene Therapy, 83, **220**
Guidelines for Advanced Therapy Development
  EMA, **214**
  FDA, **220**

H

Hacein-Bey-Abina, 3, 11, 28, 33
Haematopoietic progenitor cell (HPC), **51**
Haematopoietic stem cell (HSC), **51**

Haig, N.A., 81
Hammond, A., 20, 25
Han, W., 322, 331
Harada, K., 46
Harvesting, *41*, 42, 58–59, 71
Hastie, E.V., 37, 46
Hayes, W., 36, **38**, 46
Hazardous Substances and New Organisms Act, 240
He Jiankui, 13, 15
Health Canada, 237, 248
Health Ministry Screening Committee (HMSC), 300
Health Policy Bureau (HPB), 286
Health Products Act (HPA), 253
Health Science Authority (HAS), Singapore, 245–254
Heelan, B., xvii, 1, 27
Heidaran M., 201, 205
hematopoietic progenitor cells (HPC), 83
hematopoietic stem cell transplantation, 288
hematopoietic stem cells, **7**, 10, 37, 256, 296
heparin, 53, 166
Hepatitis B virus, xx, 16, **18**, 157, 165, 167
Hereditary Transthyretin Amyloidosis (hATTR), 328
High Throughput, Genome-Wide Translocation Sequencing (HTGTS), 107
HIV, xx, 13, 15, **17**, 36, 37, 100, 101, 167
Hollak, C.E., 154, 173
Holoclar™, **50**
Host cell proteins (HCPs), 44
Howe S.J., 3, 11
HPV, 167
HSNO Act, 241
Hu, J.F., xvii, 81
Huber, A., 40, 41, 42, 46
human cells, tissues, or cellular or tissue-based products (HCT/Ps), 335
human embryonic stem cells (hESC), 30, **32**, 228
Human leukocyte antigen (HLA), **51**
human research ethics committees (HREC), 234
HVAC, xx, 78
Hydrophobic interaction chromatography (HIC), *see* Chromatography

I

Ibrahim, M., xvii, 81
Idecabtagene vicleucel, **8**; *see also* Abecma
Ijantkar A., 81
IL2RG gene, 2, 3
Imlygic®, **6**, 10
India, regulatory framework, xvii, 295–309, 330

Index

Indian Council of Medical Research (ICMR), 296
Indonesia
  regulatory framework, 203
informed consent, 16, **18**, **297**, 300, **303**, 308
In-Process Testing, Analytical methods, 81
insertional mutagenesis, 2, 3, 4, 11, **214**
Installation Qualification (IQ), xix, xx, 78, 97, 115, 140, 335
Institutional Biosafety Committee (IBC), 239, 302, **303**
Institutional Committee for Stem Cell Research (IC-SCR), 297
Institutional Ethics Committee (IEC), 297, 302, **303**
Institutional Review Board (IRB), 219, 252
Instituto Nacional Central Único Coordinador de Ablación e Implante (INCUCAI), 226
INTERACT meetings, 219, **220**
interferon(IFN)-gamma, 57, 117
Interleukin (IL)
  IL-2, 3, 57, 108
  IL-4, 3
  IL-7, 3, 58
  IL-9, 3
  IL-15, 3, 58
  IL-21, 3
International Air Transport Association (IATA), 194, 196
International Clinical trials Registry Platform, 234
International Council for Harmonization (ICH)
  Q1A, 92
  Q2(R1), 91–94, 96–97, 100, 108–109, 132
  Q3B(R2), 120
  Q5A(R2), 156, 173
  Q5C, **200**, 258
  Q6B, 88
  Q7, 168, 173
  Q8(R2), 87–88, 136
  Q9(R1), 88, 136, 159, 173, 206
  Q10, 88, 136
  Q11, 88, 137, 159
  Q12, 151, 205
  Q14, 88–89, 91, 95–96
  S8, 120
  S12, **214**, **220**, 294
International Declaration on Human Genetic Data, 23
International Federation of Associations of Pharmaceutical Physicians (IFAPP), 180
International Society for Cell & Gene Therapy (ISCT), 53

International Society for Stem Cell Research, 289
International Society of Blood Transfusion (ISBT), xx, 53, 128
International Society of Cell and Gene Therapy (ISCT), xx, 53, 180, 181
International Society of Pharmaceutical Engineering (ISPE), xx, 65, 67, 79, 131, 136, 144, **146**, 151, 207
Investigational new drug (IND), xx, 86, 176, 186, 194, 205, **214**, 218–**220**, 228, 230, 247, 266, 299, 309, 317
iPS cells, 256, 288
irradiation, 75, **162**, 305
Ishikawa-Fishbone approach, 159–160
Ishino, Y., 322, 331
Israel, regulatory review, 203, 237
Ivins, B.E., 35, **38–40**, 45, 47

J

Jacob, F., 35, **38–40**, 41, 42, 46
Jain, R., xvii, 295
Japan, regulatory framework, 123, 126, 136, 229–231, 285–294, 330
Japanese Society for Regenerative Medicine (JSRM), 285
Jesse Gelsinger, 2, 12
Jimenez, L., 154, 173
Jinek, M., 14, 25
Johnston. C., 1, 11, 40
Joint Accreditation Committee of ISCT and EBMT (JACIE), xxi, 53
Jones, C.T., 180, 181
Jordan, B., 15, 25
June C.H., 57, 60

K

Kaeuferle T., 107, 132
Kandavelou, K., 46
Karanu, F., 117, **118**, 132
Kasia Averall, 199, 207
Katti A., 104, 132
Kelly, W.J., 42, 46
Kimberley Buytaert-Hoefen, 202, 206, 207
Koch, R., 35, 45, 47
Kong, S, 42, 47
Konomi, K., 287, 294
Koo, 144
Koo Lily, 199
Koo, T., 15, 16, 25
Kooijman, M., 28, 33
Kornberg, A., 47

Kotterman, M.A., 37, 47
Kralik. P., 103, 132
Kremsner, P.G., 325, 331
Krieg, P.A., 325, 331
Kumar, S.R., 4, 11
Kuriyan, A.E., 28, 33
Kyle Zingaro, 201, 207
Kymriah®, **6**, 10, **50**, **51**, **118**, 137, 238

L

L'Abbate, A., 36, **38–40**, 47
Labeling and packaging, 299
Latin America, 188, 225
Law on Securing Quality, Efficacy and Safety of Products, 290
Lazzarotto, C.R., 107, 132
Leachables, 75, 206
Lederberg, J., 1, 12, 35, **38–40**, 47
Lee. S., 37, 47
Lei, 226
Leukapheresis, 52, 59; *see also* apheresis
Lewis, P., 37, 40, 47
Libmeldy™, **8**, 10
life cycle management, 82–83, 96–97, 109, 151, 243, 249
Limberis, M.P., 37, 47
Linearity, 82, 93
lipoprotein lipase deficiency (LPLD), 5
liposomes, 324–**326**, 330
liquid nitrogen (LN2), 59, 188–189, *191*
Lisocabtagene maraleucel, **9**, 10; *see also* Breyanzi®
Little, S.F., 35, **38–40**, 45, 47
Liu W., 52, 60
Liu, J., **323**, **329**
living modified organisms (LMOs), 2–3, 194, 292, 293
LOVO® cell processing system, **56**, 73
Luxturna®, **7**, 10, 237, 240
lyophilization, 43
lyoprotectants, 43

M

MACI™ Ch5
Madsen, S.D., xvii, 81
Magnetic-activated cell sorting (MACS), 53
Mahmood, A., 154, 173
Mairhofer, J., 36, 47
Major histocompatibility complex (MHC), **51**
Malaria, 15, **18**
Malaysia, regulatory framework, 255
mantle cell lymphoma (MCL), 137

Marculescu, R., 81
Marketing Authorization Application (MAA) dossier
risk-based approach application (RBA), 27, 28–30, **31–32**, 216, 308
Marriott, N.G., 68, 69, 79
Massive parallel sequencing (MPS), 157
MasterControl™, 147
MasterControl™ document control system, 147
material used in, 26, 202–203
in village community, 62
Mayo Clinic, 176
McGreal R., 180, 181
McLaughlin, L.A., 68, 79
mechanism of action (MoA), 84
Mechanism of Action, 335
Medical tourism, **19**, 22
Medicine and Healthcare Product Regulatory Agency (MHRA), xx, 136, 237, 248
Melanoma, **6**, 10
Melton, D.A., 325, 332
Memi, F., 14, 25
Merix Biosciences, 325
Messenger RNA (mRNA), 104, 321–327
Messmer, K., xvii, 209, 311
metachromatic leukodystrophy (MLD), **8**, 10
Mexico, **188**, 225
Microfluidic Cell Sorter WOLF G2, 54
MilliporeSigma, 42, 47
Milone, M.C., 60, 61, 101, 132
Miltenyi Biotec, 53
minimal Manipulation, 211, 245, 296, 297
Ministry of Health, Labour and Welfare (MHLW), 286–291, 293
Mizrahy, H., xvii, 225
Miyoshi, H., 36, 37, 47
Mock U., 106, 132
Mode of Action, xx, 81, 82, 85, 87, 113, 116, 117, 210, 218, 314, 335
Modlich, U., 4, 11
Moffat, J., 37, 47
Mojica, F.J., 322, 331
Mojica, M.J., 322, 331
Molina, R., 15, 25
monoclonal antibodies, xiii, 28, 91, 118, 155–157, **158–159**
MONTRIUM™ document control system, 147
Muller, O., 18, 24
multiplex genome editing, 15
multi-product facilities, 199
Munis, A.M., 36, 47
murine leukemia virus, 36
Murine minute virus (MMV), 155
Murugan, K., 15, 25

Index

Mycoplasma, xx, 83, 87–88, 123–124, 155, 167, 171, 95, 105, 205, 305
    detection, 44, 124–125
Myelin basic protein (MBP), **51**
myelodysplasia, 3, 12
myocardial infarction, 16

N

Nagai. K., 287, 290, 294
Nain V., 104, 131
Nanoparticles, 53, 324
NASEM, 180
Naso, M.F., 37, **38–40**, 47
National Accreditation Board for Testing and Calibration (NABL), 300
National Apex Committee for Stem Cell Research and Therapy (NAC-SCRT), 297
National Ethics Commission, 228
National Guidelines for Gene Therapy Product Development and Clinical Trials (NGGTPDCT), 296
National Guidelines for Stem Cell Research (NGSCR), 296
National Health and Family Planning Commission (NHFPC), 271
National Health and Medical Research Council (NHMRC), 234
National Health Commission (NHC), 271
National Healthcare Security Administration (NHSA), 271
National Human Genome Research Institute (NHGRI), 36, 47
National Institute for Innovation on Manufacturing Biopharmaceuticals (NIIMBL), 180, 199
National Medical Products Administration (NMPA), 271
National Pharmaceutical Regulatory Agency (NPRA), 256
National Statement on Ethical Conduct in Human Research, 234
National Technical Commission on Biosafety – CTNBio, 228–229
Natkin, E., 45
New drug application (NDA), 132, 176, 246–247, *263*, 271, 275, 295, 299, 307, 334
New South Wales (NSW), 175, 176, 181, 233
next-generation sequencing (NGS), 106
Novartis, 137, 325
Novick, R.P., 47
NSW Health Commercialization Training Program, 176
NZ Medicines and Medical Devices Safety Authority, 240

O

O'Doherty, U., 101, 132
O'Sullivan, G.M., 233, 241
Ochratoxin, 16, **17–18**
Office of the Gene Technology Regulator (OGTR), 239
Office of Tissues and Advanced Therapies (OTAT), 218
Off-target effects; *see also* loss-of-function mutations, gain, 15
Oh Stevens, S., 199
Okada, K., 287, 294
Olcott, D., 180, 181
Oliveira, P.H., 36, **39**, **40**, 47
Onasemnogene abeparvovec, *see* Zolgensma
oncogenesis, 3, 15, **38**, **40**, 44, 101, 102
oncolytic viruses, **214**, **220**, 301
Onodera, M., 2, 11
on-target editing, 15
Open circular plasmids, 43
Operation Qualification (OQ), xxi, 78, 115, 136, 140, 335
ornithine transcarbamylase (OTC) deficiency, xxi, 2, 28, 33
orphan drug designation, 215, 219, **223**, 237, 290
Ota, M., 105, 132
Ozawa, K., 287, 290, 295

P

Papapetrou, E.P., 4, 11
Parenteral Drug Association (PDA), 79, 151, 168
Parker, T., 42, 47
Parkinson's disease, 16, **17**
parvovirus, 37
patents, **19**, 21, 25
    licensing, 270, 282
Patil, S., 5, 11
Pavani, G., 4, 11
PCR-restriction fragment length polymorphism (PCR-rFLP), 105
pegylated ADA (PEG-ADA), 1, 2
perforin, 84
periodic safety update report (PSUR), 280
peripheral blood stem cells (PBSCs), 164, 244
Perricaudet, M., 2, 12
Pharmaceutical Affairs and Food Sanitation Council (PFSC), 286
Pharmaceutical and Food Safety Bureau (PFSB), 286
Pharmaceutical Inspection Co-operation Scheme (PIC/S), 227, 238, 245
Pharmaceutical Regulatory Information System (PRISM), 252

Pharmaceuticals, Medical Devices, and Other Therapeutic Products Act (PMD Act), 230, 285–286
Pharmacovigilance Risk Assessment Committee (PRAC), 215
Pizzi, L.T., 180, 181
Plasmids, 5, 35, 36, 76
  colicinogenic (Col), 35
  manufacturing process, 37, 40–43
Plasticizers, 75
Plavsic, M., 154, 173
Point-of-care manufacturing, 199, 313
post-marketing requirement, 216, 223, 247, 279–281, 287, 291–292, 318
Pourcel, C., 322, 331
Powell, J.C., 81
Pre-licensing Inspection, 202, 335
Prescription Drug User Fee Act (PDUFA), 222
PRIME designation, 215, 290
Process Validation, 335
  Life Cycle, 142
Process, Method and Facility Changes, 146
Process-Related Impurities, 119–122
Product release, analytical methods, 81–130, 154, 163, 205
Professional Society for Health Economics and Outcomes Research, 180
Project Orbis, 237
prokaryotes, 14, 36, 321
Propidium Iodide (PI), **114**
prostate-specific antigen (PSA), 16, **17**
proteasome inhibitor, **8**, **9**, 10, 137
proto-oncogene, 3, **38**
Provenge, **118**
Public Health Service (PHS) Act, 5, 83, 218, 283; *see also* 42 USC §262
Putnam A.L., 57, 58, 60, 62
Pyrogens, 83, 336

Q

Qiu, Y., 154, 173
QR code, 193
Quaak, S.G.L., 43, 48
Quality agreements, 202
Quality by Design (QbD), xxi, 136, 207
Quality risk management (QRM), xxi, 136, 159, 200, 206

R

Ramanan, V., 16, 25
Ranucci, C.S., 157, 173
Raper, S.E., 28, 30, 33
Raposo, V.L., 15, 25

regenerative medicine, 176, 197, **220**–222, 233–234, 285–295
Regenerative Medicine Advanced Therapy (RMAT) designation, 221
Regulatory Affairs Professionals Society (RAPS), 180
Regulatory framework, India, 307–308
reimbursement, 211, 217, 233–234, 271, 311, 318–319
Replication Competent Retrovirus (RCR), 83, 95, 127–129, 336
Restrictive access barrier systems (RABS), xxi, 140, 168
retrovirus
  endogenous, 156, 157, 167, **220**, 305
  replication competent (RCR), xxi, 83, 95, 100, 127
  vector, 1–3, **6**–**8**, 36–37, 94, 100, 167, **220**
Review Committee on Genetic Manipulation (RCGM), 302, **303**
Ricchi, M., 103, 132
Ricciardi, A.S., 104, 132
Rietveld A., 199
Risk assessment and risk management plan (RARMP), 240
Risk Assessment, 14, 24, 240, 303, 334
  adventitious agents, 153, 156, 159–160
  environmental, **214**, 261, 306
  environmental monitoring (EM), 168
  environmental program, 69
  facility fit, 136
  impurity, 83, 89, 119–121
  regulatory, 215, 219, 240
  shipping and distribution, 185, 194
  tech transfer, 136
  tools, 147, 159–160
Robbins, J.J., 81
Robertson, J.A., 15, 25
Roblin, R., 10, 11
Robustness, 82, 91
Rodia, C.N., 81
Rogers, S., 1, 11
Roig, J., 52, 62
Rose, J.A., 37, 48
Rubino, M, 155
Russell, D.W., 36, **38**, 48

S

Saayman, S., 15, 25
Sakigaki designation, 291, **292**
Sale of Drugs Act, 256, 257, 267
Salsman, J., 14, 25
Sambrook, J., 36, 48
Samulski, R.J., 37, **39**, 46

Index

Sandoval-Villega, N., 104, 132
Sanger sequencing, 105–106
Saudi Arabia, regulatory framework, 203
Saw, P.E., 327, 328, 331
Schaffer, D.V., 37, 47
Schambach A., 4, 11, **39**
Schleef & Schmid, 43
Schleef, M., 43, 48
Schmidt, T., 43, 48
Schmitt, S., xviii, 136, 197
Schneider, C., 4, 12
Sealers, 40, 73, 76, 78
self-inactivating (SI) viral vectors, 4, **39**, 47, 133
SEPAX™, 73
Serviços e Informações do Brasil, 226
severe combined immunodeficiency (SCID), *see* X-linked severe combined immunodeficiency
She, Q., 322, 331
Sherkow, J.S., 21, 25
Shimizu, N., 36, **38–40**, 48
Shinwari, Z.K., 21, 25
Shipping and transport, 59, 70, 73, 76–77, 159, 161, 186, 202, 205–206, 299
Shire, 325
short hairpin RNA (shRNA), 37, 301
Shrivastava D., 81
Shukla, A., 154
sickle cell anemia, 15, 23
Silva, F., 42, 48
Silva, G., 104, 133
Sindusfarma, 226, 231
Singapore, regulatory framework, 203, 237, 243–254
Singh S., xviii, 81
single guide RNA (sgRNA), 104, 105
single use technologies, 59, 66, 71, **158**, **170**
Sipp D., 288, 294
small interfering RNA (siRNA), 301, 321, 325, 327–332
Sobecky, P.A., 48
Somatic cell gene therapy, 28, 201, 211, 301–302
Sonstein, S.A., 180, 181
South Korea, regulatory framework, 203
Spectra Optia® Apheresis System, 52
Spherox™, **50**
Spinraza®, **50**
Stanford Online, 180
State Administration for Market Regulation (SAMR), 271
steam sterilization, 69, 75
Stein, S., 3, 4, 12
stem cells, **6–8**, 28, 30, **32**, 27, **50–51**, 164, 228, 233, 286

centers/facilities, 191, 199
regulatory, **214**, 256, 269, **273–274**, 282, 288, 289, 295–300, 302–304, 307, 309
Stempeucel®, 295
StemSpan SFEM, 58
Stephens, P.J., 36, **38**, 48
Sterile filtration, 53
sterile welding and sealing, 40
Sterility, 83, 105, 205, 229, 289, 336
testing, 45, 70, 87, 95, 97, 123–124, 167–168, 172, 305–306, 313
Stolberg, S.G., 2, 12
Stonier, P.D., 180, 182
Strimvelis®, **6**, 10
summary of product characteristics (SmPCs), xxi, 10, 216
supplier qualification, 68, 202
surrogate endpoints, 203, 221–222
Swiss-Medic, 136, 237
Switzerland, regulatory framework, 237
Szybalska, E.H., 1, 12
Szybalska, W., 1, 12

T

Taghdisi, S.M., 16, 25
Taiwan, regulatory framework, 203
Takashima K., 288, 294
Talimogene laherparepvec, *see* Imlygic®
Tatum, E., 35, **38–40**, 47
Taylor, S.C., 103–104, 133
Tecartus®, **8**, 10, **51**, 137, 238
technology transfer, 135–137, 336
temperature loggers, 188
Therapeutic Goods Administration (TGA), 234, 235, 237
Thermo-Fisher, 58, 322
Tisagenlecleucel, **6**, 238; *see also* Kymriah®
tissue engineered product (TEP), 28, 210, 223
Townsend, B.A., 22, 23, 26
TRACKWISE®, 147
transcription activator-like effector nuclease (TALEN), xxi, 4, 14, 104, 131, 322, **323–324**
Transduction, 1, **7**, 10, 36, 36, 37, 45–47, 57, 59, 61, 336
control, 82, 87, 94, 100–102, 114, 128, **274**
transfection efficiency, 43–45, 327
transformation, 1, 2, 11, 12, 15, 22, 36, 40, 101
Translocation, 3, 15, 106–107
transthyretin familial amyloid polyneuropathy (TTR-FAP), 328
Travis, J., 322, 322
TriLink Biotechnologies, 322
Tsai, S.Q., 107, 133

Turchiano, G., 108, 133
Type 1 diabetes (T1D), **51**, **326**
Typical CAR T Cell Product, 81
    Design, 84–85

U

umbilical cord blood, 164, 244
Union Register, 216
Universal Declaration on Bioethics and Human Rights, 23
US Defense Adv. Research Project Agency, 325
US Pharmacopeia (USP)
    <1>, 45, 123, 124
    <63>, 124, 125
    <71>, 45, 123, 124
    <85>, 126
    <697>, 128
    <788>, 45, 130
    <790>, 130
    <791>, 130
    <797>, 163, 173
    <905>, 129
    <1010>, 96
    <1071>, 124
    <1151>, 128–129
    <1207>, 129
    <1223>, 124, 125
    <1226>, 95, 97
User Requirement Specifications (URS), xxi, 78, 335
UV spectrophotometry, 43–44

V

Vaccine, 4, 16, 18, 27, 167, 210, **220**, 328
    cancer, 256, 324
    COVID-19, 14, 325
    DNA, 301
    lipid nanoparticle, 324, 325, 330
    mRNA, **326**–327, 330
    recombinant, 244
    Rotavirus, 157
    urgent needs, **278**
    yellow fever, 241
Valeich J.E., 81
Validation, 4, 76, 82, 205
    master plan (VMP), xxi, 336
    method and systems, 45, 74, 88–91, 94–97, 99, 124–125, 135–139, 147–150, 154, 161, **204**, 229, 258, 277
    process, **19**, 23, **32**, 78–79, 147–151, 163, 200, 206, 266, 277
Validation, Verification and Qualification Considerations
Van Craenenbroeck, K., 36, 48

van der Loo, 167
Van Roy, N., 36, **38–40**, 48
Vassy, J.L., 18, 26
VEEVA™ document control system, 147
Vega-Mercado H., xv, xvi, xviii, 35, 49, 65, 79, 135, 148, 151, 321
Victoria, J.G., 157, 174
viral safety, 154, 156, **164**, 172
    assessment, 313
Viral vectors
    copy number (VCN), 101–103, 109
    design, 3, 37, 41
    dose control, 82, 101
    manufacturing process, 40–45
Vouillot, L., 106, 133
Vyondys, 51

W

Walther, W., 43, 48
Wang, J., xviii, 81
Wang, K.C., 15, 26
Wang, X., 26, **40**
Watanabe, T., 35, **38–40**, 48
Wee, S.L., 13, 26
Wei Zexi, 270
Weinberg, J.B., 37, 48
Welders, 40, 73, 76, 78
White, J.A., 180, 182
Whitman, M., 43, 46, 52, 29, 61
Wienert B., 107, 133
WOLF-G2, 54, *56*
Wollman, E.L., 35, 46
Wong, T.W.Y., 15, 26
Wright, G., **50**, 63
Wright, J.F., 167, 173
Wu X., 81
Wyles, S.P., 175, 176, 182

X

xenogeneic cells, 83, **214**, 217, 288, 298
X-linked chronic granulomatous disease (X-CGD), xix, 3, 12
X-linked severe combined immunodeficiency (SCID), 1, 2, 3, **6**, 10, 11, 28
Xu H., 57, 58, 63, 328
Xu, S.Q., 328, 331
XURI Cell Expansion Systems, 58

Y

Yan, W., 107, 133, **324**
Yang, B., 157, 174
Yang, W., 16, 26
Yang, Z., 106, 133

Yeh, P., 2, 12
Yescarta®, **7**, 10, **50**, **51**, 118, 238, 270
Yokohama Declaration, 285
Yuan, X., 16, 26

Z

Zatsepin, T.S., 328, 332
Zhang, K., xviii, 269

zinc-finger nuclease (ZFN), xxi, 3, 4, 14, 104, 131, 322, **323–324**
Zinder, N.D., 1, 12
Zisman A., 68, 79
Zobel, A., xviii, 183
Zolgensma®, **7**, 10
Zou, Q., 20, 26
Zufferey, R., 101, 133
Zynteglo™, **7**, 10

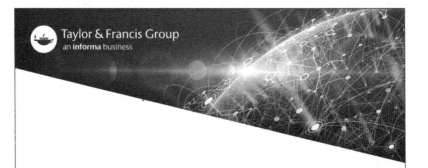